INVESTMENT AND EMPLOYMENT OPPORTUNITIES IN CHINA

Systems Evaluation, Prediction, and Decision-Making Series

Series Editor

Yi Lin, PhD

Professor of Systems Science and Economics
School of Economics and Management
Nanjing University of Aeronautics and Astronautics

Grey Game Theory and Its Applications in Economic Decision-Making
Zhigeng Fang, Sifeng Liu, Hongxing Shi, and Yi Lin
ISBN 978-1-4200-8739-0

Hybrid Rough Sets and Applications in Uncertain Decision-Making
Lirong Jian, Sifeng Liu, and Yi Lin
ISBN 978-1-4200-8748-2

Investment and Employment Opportunities in China
Yi Lin and Tao Lixin
ISBN 978-1-4822-5207-1

Irregularities and Prediction of Major Disasters
Yi Lin
ISBN: 978-1-4200-8745-1

Measurement Data Modeling and Parameter Estimation
Zhengming Wang, Dongyun Yi, Xiaojun Duan, Jing Yao, and Defeng Gu
ISBN 978-1-4398-5378-8

Optimization of Regional Industrial Structures and Applications
Yaoguo Dang, Sifeng Liu, and Yuhong Wang
ISBN 978-1-4200-8747-5

Systems Evaluation: Methods, Models, and Applications
Sifeng Liu, Naiming Xie, Chaoqing Yuan, and Zhigeng Fang
ISBN 978-1-4200-8846-5

Systemic Yoyos: Some Impacts of the Second Dimension
Yi Lin
ISBN 978-1-4200-8820-5

Theory and Approaches of Unascertained Group Decision-Making
Jianjun Zhu
ISBN 978-1-4200-8750-5

Theory of Science and Technology Transfer and Applications
Sifeng Liu, Zhigeng Fang, Hongxing Shi, and Benhai Guo
ISBN 978-1-4200-8741-3

INVESTMENT AND EMPLOYMENT OPPORTUNITIES IN CHINA

Jeffrey Yi-Lin Forrest • Tao Lixin

CRC Press
Taylor & Francis Group
Boca Raton London New York

CRC Press is an imprint of the
Taylor & Francis Group, an **informa** business

CRC Press
Taylor & Francis Group
6000 Broken Sound Parkway NW, Suite 300
Boca Raton, FL 33487-2742

First issued in paperback 2019

ISBN-13: 978-1-4822-5207-1 (hbk)
ISBN-13: 978-1-138-37494-2 (pbk)

Visit the Taylor & Francis Web site at
http://www.taylorandfrancis.com

and the CRC Press Web site at
http://www.crcpress.com

Contents

Synopsis

From the angle of pure economics and based on the most recent statistics from relevant bureaus of mainland China, this volume is a practical guide for investments and employments in China. After introducing the theoretical models, optimization theorems and methods previously published by the authors, the book establishes scientifically sound conclusions and advice through analyzing various investment returns and employment opportunities for three greater regions—East China, Central China, and West China, over 30 provinces, cities, and autonomous regions, six main administrative districts, three special zones, and three major economic sectors.

This volume is intended for such readers as:

1. Scholars, graduate students, and upper-level undergraduate students who are in the areas of economics and finance.
2. All those people who work on evaluations of regional economics and economic sectors. Such capabilities are very important for making forecasts for economic and political reasons.
3. Investors, such as banks, pension funds, hedge funds, and all other kinds of investment firms, which look at the potentials of making investments in China.
4. Since the method used in the book can be used to evaluate other geo- and political regions, policy makers, such as politicians, and business firms will be potential readers who look at political opportunities in the existing imbalances of a particular region to garner economic and/or political gains.

In other words, this volume constitutes a rational reference for all parties from around the world who attempt to make investments or find employment opportunities in mainland China.

Preface

According to modern systems science (Lin, 1999, 2008; Lin et al., 2012), all matters and events exist in the form of systems, and the internal structure of each system determines the output and the functionality of the system. So, the quality of the internal structure should be measured by the effects of the system's output. In terms of the microeconomic systems of corporations, this end seems to be quite obvious. When the profit picture of a company is not good, one can always attribute the problem to many factors of the environment external to the company. Even so, the truly underlying problem is still deeply rooted within the company itself, and within some of the internal structures of the company, such as the product structure, the capital structure, the debt structure, the distribution structure, the organizational human resource structure, the cultural technology structure, the gender structure, the employees' age structure, property rights structure, and so on.

This conclusion also holds true for macroeconomic systems. Although the output efficiency of the macroeconomic system of a society has to be constrained by various factors external to the system, such as the surrounding environment, the state of the international economy, and the economic environment of the world, the conditions and quality of the internal structure, such as regional economic structures, the economic sectorial structure, the industrial structure, the investment structure, the employment structure, the distribution structure, and so on, directly influences the output efficiency of the system. Therefore, such problems as how to adjust and how to optimize macroeconomic system structures have attracted more and more attention from the world of learning.

Recently, the authors of this book have investigated economic systems from the angle of fluid dynamics (Lin and Forrest, 2011) and introduced the concept of comparative gains and the relevant model, developed a set of criteria for testing imbalances existing in macroeconomic systems, and established a theorem on structural optimization (Tao et al., 2011). These works represent some important contributions to the relevant literature on how to adjust and how to optimize the structures of macroeconomic systems.

After having established the theoretical foundation, the rest of the book utilizes the model and the methodology to analyze the investment and employment structures of China by making use of the most recent statistical data. Then, the book explores the directions of the relevant structural adjustments, orders, and intensities so that a practical guide for investment and employment opportunities in China is provided.

It can be reasonably expected that all the conclusions derived in this book constitute a scientifically sound reference for making decisions regarding investment and employment in China.

References

Lin, Y. 1999. *General Systems Theory: A Mathematical Approach*. Kluwer and Plenum Publishers, New York.

Lin, Y. 2008. *Systemic Yoyos: Impacts of the Second Dimension*. Auerbach Publications, an imprint of Taylor & Francis, New York.

Lin, Y., Duan, X. J., Zhao, C. L., and Xu, L. D. 2012. *Systems Science: Methodological Approaches*. CRC Press, an imprint of Taylor & Francs, New York.

Lin, Y. and Forrest, B. C. 2011. *Systemic Structure behind Human Organizations: From Civilizations to Individuals*. Springer, New York.

Tao, L. X., Guo, Y. F., and Du, W. W. 2011. Criteria for omnibearing imbalance of macroeconomic system structures and their strategic optimization. *Kybernetes: The International Journal of Cybernetics, Systems and Management Sciences*, 40(5), 848–853.

Introduction

Economics studies optimal allocation of limited resources. Its essence is about how to materialize a balance between supply and demand. If the supply is greater than the demand, it means a waste of resources so that the supply needs to be reduced. On the other hand, if the supply cannot meet the demand, it means there is a shortage of supply so that the supply of resources needs to be increased.

The scope of limited resources covers a very wide range, including land, capital, technology, means of production, patents, brands, business cultures, labors, talents of business entrepreneurs, and so on. Even with such a wide-ranging scope, limited resources can be classified into two major classes: living labor and materialized labor. The former is reflected on the bodies of workers, while the latter can be attributed to funds (or capital). They are exactly the two major factors of production that appear in the Cobb–Douglas production function.

Each macroeconomic system is composed of the structure of regional economies, the structure of industries, the structure of economic sectors, the structure of income distribution, the structure of investment, the structure of employment, and so on. Any structural imbalance of a macroeconomic system represents a structural imbalance of the system's supply and demand (or a structural imbalance of the daily input and output). The root cause is that there are problems in the structure of investment and in the structure of employment. They are known as imbalances, which need to be adjusted.

In fact, an imbalance in a regional economic structure is reflected in two aspects. One is that the regional investment structure is more or less unstable. In other words, areas that demand more capital inputs receive insufficient investment (insufficient supply), while areas that do not need much capital inputs receive more than sufficient investment (the supply is more than the demand). The second aspect is that the regional employment structure is unstable. In other words, some areas suffer from labor shortages (insufficient supply), while some other areas enjoy abundant supply of available labors (the supply is greater than the demand). Similarly, industrial (economic sectorial) structural imbalances are also reflected in these two aspects.

Structural imbalance and optimization of macroeconomic systems stand for a fundamental problem investigated in macroeconomics. The second author of this book introduced in 2011 the concept of comparative profits, established relevant models, developed criteria for the appearance of comprehensive imbalances, and proved theorems regarding structural optimization. For details, see: Criteria for omnibearing imbalance of macroeconomic system structures and their strategic optimization (L.X. Tao), in: *Kyberentes: The International Journal of Cybernetics, Systems and Management Sciences*, vol. 40, nos. 5–6, pp. 848–853. The first section of this book systematically presents this theory and relevant method.

By simply inserting the most recent statistics available from bureaus of the national government or various provincial governments into the computational models, one can quantitatively describe the imbalances and the severities of imbalances for the corresponding region's economic structure and industrial (economic sectorial) structure. Additionally, he can quantitatively determine the direction, order, time duration, and amount of effort needed to strategically optimize the imbalanced structure based on the established theorems of optimization. Sections 2 and 3 of this book contain a total of six chapters. These chapters make use of the relevant numerical computations of the available statistics to analyze the imbalances in China's regional economic structures and industrial (economic sectorial) structures. Correspondingly, within these chapters, what are outlined are proposals on how to specifically optimize these structures strategically.

Since the structure of any macroeconomic system changes constantly, the investment and employment structures vary accordingly. Therefore, the investment and employment conclusions and suggestions presented in this volume evolve correspondingly. That inevitably indicates that this volume will need to be updated annually along with the availability of new statistics. This volume presents the basic conditions of imbalances in China's regional economic structures and industrial (economic sectorial) structures based on the newly available data of the previous year. That is, the guidelines for investment and employment opportunities in China, as presented in this volume, are most timely and relevant for appropriate decision and policy makers.

Based on the computational results, Section 2 describes the imbalances existing in the three greater areas of East, Central, and West China, and three major industrial structures (Chapter 2). Following that presentation, Chapter 3 describes the East region of China, the optimal area; Chapter 4 paints the structural picture for the Central region of China, the general area; and Chapter 5 discusses the West region of China, the most difficult area. In these chapters, the industrial-specific investment and employment of the relevant cities, provinces, and autonomous regions are looked at in detail. At the end, Section 3 investigates the regional investment and employment opportunities within China's traditional administrative areas according to the tradition and current reform of China (Chapter 6) and the three particular economic areas of the modern time (Chapter 7): the greater Pearl River Delta region, Yangtze River Delta economic zone, and Bohai Economic Rim.

Authors

Jeffrey Yi-Lin Forrest, also known as Yi Lin, holds all his educational degrees (BS, MS, and PhD) in pure mathematics, respectively, from Northwestern University (China) and Auburn University (USA) and had one year postdoctoral experience in statistics at Carnegie Mellon University (USA). Currently, he is a guest or specially appointed professor in economics, finance, systems science, and mathematics at several major universities in China, including Huazhong University of Science and Technology, National University of Defense Technology, Nanjing University of Aeronautics and Astronautics, and a tenured professor of mathematics at the Pennsylvania State System of Higher Education (Slippery Rock campus). Since 1993, he has been serving as the president of the International Institute for General Systems Studies, Inc. Along with various professional endeavors he organized, Dr. Forrest has had the honor to mobilize scholars from over 80 countries representing more than 50 different scientific disciplines.

Over the years, Professor Forrest has and had served on the editorial boards of 11 professional journals, including *Kybernetes: The International Journal of Systems, Cybernetics and Management Science, Journal of Systems Science and Complexity, International Journal of General Systems*, and *Advances in Systems Science and Applications*. Also, he is the editor-in-chief of two book series, one of which is titled *Systems Evaluation, Prediction, and Decision-Making*, published by CRC Press (New York, USA), an imprint of Taylor & Francis since 2008, and the other one titled *Communications in Cybernetics, Systems Science and Engineering*, published by CRC Press (Balkema, the Netherlands), an imprint of Taylor & Francis since 2011.

Some of Dr. Forrest's research was funded by the United Nations, the State of Pennsylvania, the National Science Foundation of China, and the German National Research Center for Information Architecture and Software Technology.

Professor Forrest's professional career started in 1984 when his first paper was published. His research interests are mainly in the area of systems research and applications in a wide-ranging number of disciplines of the traditional science, such as mathematical modeling, foundations of mathematics, data analysis, theory and methods of predictions of disastrous natural events, economics and finance, management science, philosophy of science, and so on. By the end of 2012, he had published over 300 research papers and over 40 monographs and edited special topic volumes by such prestigious publishers as Springer, Wiley, World Scientific, Kluwer Academic (now part of Springer), Academic Press (now part of Springer), and others. Throughout his career, Dr. Jeffrey Forrest's scientific achievements have been recognized by various professional organizations and academic publishers. In 2001, he was inducted into the honorary fellowship of the World Organization of Systems and Cybernetics.

Tao Lixin, a retired professor from Jiangxi University of Finance, was born on November 10, 1944. He received his undergraduate education as a mathematics major from Jiangxi University. Currently, he is on the economics faculty in Huashang School of Guangdong Business College. During his 37 years of university teaching, he has taught 26 different courses in diverse areas such as mathematics, systems science, economics, international trade, management, business studies, and statistics. Currently, Professor Tao has completed over 80 research projects, published over 5,000,000 words of papers in major scholarly journals and monographs covering many scientific disciplines, including econometrics, economics of technology, optimized management, statistical measurement, international trade, law, and so on.

1

THEORETICAL FOUNDATION

Chapter 1

Criteria for Discovering Imbalances in Macroeconomic System Structures

The purpose of this chapter is to propose and demonstrate a set of necessary criteria for the existence of omni-bearing imbalances in macroeconomic system structures, such as regional structure, industrial structures, proportions of urban and rural areas, distribution structures, the structure of property ownerships, and so on, and the relevant method for achieving strategic optimizations, for structural optimizations of various regional macroeconomic systems. At the same time, a universally applicable formula is established for calculating the severities of the omni-bearing imbalances existing in macroeconomic system structures.

Methodologically, this chapter uses statistical data to compare the quantitative outputs of the gains by comparison (GBC) model in order to quantitatively determine the existing structural imbalance, the formation of structural optimization, and the direction and intensity of the necessary intrusive strategic program. It is shown in this chapter that although the output efficiency or the output benefits of the macroeconomic system of a society are constrained by the system's external environment, such as the surrounding environment, the state of the international economy, the development of the world trades, and so on, the internal structure of the system directly determines its output efficiency and benefits. Inversely, the output efficiency and the benefits of the macroeconomic system can be employed to evaluate the quality of the internal structure of that system. This conclusion is a reconfirmation of the results of systems science (Lin, 1999, 2008; Lin et al., 2012) that all matters and events take the form of systems, where the output efficiency of a system is determined essentially by the internal structure of the system. In other words, the system's internal structure must be evaluated and judged by using the output efficiency and benefit.

The quantitative method established in this chapter can be universally applied to investigate structural imbalances that exist potentially in any macroeconomic system. That shows the great significance of this work in terms of theory, methodology, and practical application in the area of structural optimization of global and regional macroeconomic systems.

The first part of this chapter is based on Lin (1999, 2008) and Lin et al. (2012). The second half of this chapter is based on the work by Tao et al. (2011).

1.1 Systems Science and the Systemic Yoyo Model

The basic concepts of modern systems science were first introduced by Ludwig von Bertalanffy (1901–1972). In the 1920s, von Bertalanffy (1924) started to realize that there are similar properties existing in different disciplines and scientific areas so that he began to explore the possibility of describing systems of various objects by using a unified language, which are the thoughts of modern systems. In 1932, von Bertalanffy (1934) developed the theory of open systems, constituting one pillar of the general systems theory. In 1937, he established the key concept of general systems. At the end of the year of 1940, the International Association of General Systems was established. von Bertalanffy's *General System Theory: Foundations, Development, Applications*, published in 1968 by George Braziller, has been treated as the first monograph of systems science. His theory of general systems has constituted the theoretical foundation of systems science and is a qualitative research of systemic thoughts.

1.1.1 What Is Systems Science?

By systems science, it means the science that studies classes, characteristics, mechanisms, and laws of motion of systems, as the matter of subjects. It appeared initially at the end of the 1940s and contains fives main areas of materials:

1. The concepts of systems. That is the general thoughts and theory of systems.
2. The theory of general systems. That is the pure mathematical theory of systems structures and behaviors using the formal language of mathematics.
3. Theoretical analyses of systems. They are particular theories developed to investigate specific systems' structures and behaviors, such as graph theory, game theory, queuing theory, cybernetics, information theory, and so on.
4. Systems methodology. It consists of methods and procedures developed to analyze, plan, and design specific systems by using the theory and technology of systems. It mainly means systems analysis and systems engineering.
5. Applications of systems methods. That is to apply the thinking logic and methods of systems science to various scientific disciplines.

Since the 1960s, the appearance and development of the theory of non-equilibrium self-organizing systems have helped to enrich the contents of general systems theory. The epistemological significance of systems science is that it has promoted and accelerated the development tendency of modern science from a spectrum of relatively isolated disciplines into an organic whole.

Systems science, according to Qian Xuesen (2007), stands for such an area of science and technology that investigates systems and various applications. Like natural science, social science, mathematical science, and so on, it is a new branch of the system of modern science and technology.

Qian Xuesen, a well-known Chinese scientist, has played an important role in the initiation and development of systems science. He explored the general laws of evolution of various matters and social events by using systems logic of thinking and systems methods. On the basis of the existing achievements of systems research, he proposed at the end of 1970s the leveled structures of systems science. He recognized that systems science is composed of three layers, and multiple branches and different technologies, and the following is paraphrased from *Automatic Control and Systems Engineering, Encyclopedia of China*, 1991, published by Press of Encyclopedia of China, Beijing:

1. The layer of engineering technology: Systems engineering, techniques of automation, and technology of communication are the knowledge that directly help to reshape the objective world, where systems engineering contains techniques useful for organizing and managing systems. Corresponding to different types of systems, there are different systems engineering, such as agricultural systems engineering, economic systems engineering, industrial systems engineering, social systems engineering, and so on.
2. The layer of technological science: Operations research, information theory, control theory and cybernetics, and so on, are theories that help guide engineering techniques.
3. The layer of basic science: Systems research stands for such a science that investigates the fundamental attributes and general laws of systems. It is the theoretical basis of all systems works. Systems science is still in the process of being established. The bridge connecting systems science and philosophy is the so-called systems theory or known as systems points of view. It belongs to the category of philosophy. The establishment and development of systems science will definitely widen the coverage of human knowledge, strengthen the human capability to fight against nature for its very own survival, and enhance the development of science, technology, and economies.

1.1.2 Development History of Systems Science

As an area of knowledge, systems science has experienced its very own process of appearance, growth, and formation. And because it is a cross-disciplinary endeavor, it has touched upon a wide range of traditional disciplines, such as mathematics, physics, chemistry, and other branches of natural science, many areas of engineering technology, and quite a few disciplines of social science. Hence, the history of systems science is closely related to the entire development history of the human race. Its ideas can be traced back to the primitive societies of antiquity. The ancient people started their knowing of nature from its wholeness. In other words, the logic thinking of systems science is the very first theory that guided the man in his exploration of nature. As for the establishment of systems science, one has to mention the modern achievement of science and technology. Almost every theory of modern science has been absorbed by systems science and modified to become a part of systems theory. When looking at the history comprehensively, the development of systems science can be roughly divided into three stages: appearance and formation of systems thinking, establishment of quantitative systems scientific methods, and construction of the comprehensive framework of systems science.

1.1.3 Formation of Systemic Thoughts—The Elementary Stage

In ancient times, the level of productivity was low. Man had great difficulties fighting against nature. So, human understanding of nature was staggering at the level of “systems thinking.” People intuitively employed the concept of wholeness to look at the physical world. They related the phenomena of birth, age, sickness, and death with other relevant natural occurrences, forming the world view of unity of man and heaven. Within such a world view, one can find the logic thinking of systems. Both Lao Tzu and Zhuang Tzu philosophies have reflected such a way of reasoning. For instance, both the *Book of Changes* (Wilhalm and Baynes, 1967) and *Tao De Ching* (English and Feng, 1972; Lao Tzu, unknown) have described the integration and transformation of things and events by pointing out that "all things under heaven are born out of 'have,' and 'have' is born out of 'don't have'," "having no name is the start of heaven and Earth, and having name is the mother of all things," "Tao breeds one, one breeds two, two breeds three, and three begets all things of the world." Later, Wang Anshi (1021–1086) studied the evolutionary order of the world as "tian nian sheng shui," which in Chinese is

天年生水

and pointed out in his work "Wu Xing (five elements): Heaven Names Everything" that the heaven first gives birth to water, the Earth then gives birth to fire, the Heaven afterword gives birth to wood, the Earth follows with its birth to gold, and then the Heaven produces Earth. For more about Wang Anshi, refer to Mote (1999). Wang believed that the Heaven and the Earth first give birth to the five elements: water, fire, wood, gold, and Earth, which then led to the formation of everything else in the world. So, ancient Chinese employed such concepts as yin and yang, five elements, and eight trigrams to unify the explanation of various natural phenomena and to form the unity of man and nature. That we can see as realizations of the perspective of wholeness, a viewpoint of movement and change, holistic point of view of the world, and other concepts of systems thinking. Ancient Greek philosopher Democritus (Guthrie, 1979) saw the universe as a unified whole, and studied it holistically. At the same time, he considered the world being composed of individual atoms and that it was the movements and interactions of these atoms that constituted the motion and change of the entire universe. He published a volume on the large system of universe, which can be seen as the earliest work that used the word "system."

No matter whether it was ancient Chinese thinkers or Western philosophers, they had addressed worldly problems from the angle of wholeness. These scholars achieved high in several disciplines and were specialists in different areas of learning. For example, Aristotle and Archimedes of the ancient Greece, and Lao Tzu and Mozi of China, were such scholars. One characteristic of the science of that time is that studies of different disciplines are closely intertwined together. Studies of science and philosophy are written together. The holistic development of science is a product of the then low levels of productivity. However, during this period of time, in terms of science itself the systemic thinking was greatly developed. The systemic points of view of wholeness, movement and change, interconnectedness, and interactions dominated the mainstream of the then scientific research. What needs to be specifically pointed out is that the systemic thinking of the time was forced upon the scholars, because the causes of complex natural phenomena could not be fathomed so that they had no choice but to look at the phenomena from the angle of wholeness, macrocosm, and dialectical thinking. The appearance and development of these early systemic thoughts are closely related to the low levels of productivity and the primitive stages of scientific knowledge and technology.

Although the early man was forced to understand nature by using systemic thoughts and reasoning, these thoughts and logics of reasoning have indeed helped mankind to achieve magnificent successes in science, technology, and commercial goods production. For instance, in the area of civil engineering, Li Bing and his son constructed the well-known Dujiangyan Water Project between 256–251 BC (Zhang and Hu, 2006), that is an irrigation system located in Dujiangyan City, Sichuan. It has been not only a magnificent achievement in the world history of water conservancy construction, but also a successful application of systemic thoughts. The entire irrigation system consists of three main constructions:

1. The fish mouse levee—that divides the water of Min River into inner and outer streams, where the inner stream carries approximately 40%, rising to 60% during flood, of the river's flow into the irrigation system whilst that outer stream drains away the rest, flushing out much of the silt and sediment.
2. The flying sand weir—that has a 200 m-wide opening that connects the inner and outer streams. This opening ensures against flooding by allowing the natural swirling flow of the water to drain out the excess water from the inner to the outer stream.
3. The bottle-neck channel, which Li Bing and his son gouged through the mountain, is the final part of the system. This channel distributes the water to the farmlands to the west, whilst the narrow entrance works as a check gate, creating the whirlpool flow that carries away the excess water over the flying sand weir, to ensure against flooding.

That is, Dujiangyan is a large engineering construction that has the capabilities of flood control, sand drainage, water diversion, and others. Along with over 120 ancillary constructions, the system has also played the role of irrigation. It can be recognized that without the fish mouse levee a large amount of sand and gravels would not be drained into the outer stream; without the bottle-neck channel, no swirling circular flows would be formed so that the sediment could not be drained by being rushed over the flying sand weir. And without the flying sand weir to take away the sediment, the bottle-neck channel would be filled with the sediment so that it could not play the role of irrigation and the water could not be transported into Chengdu plain. Since this project was designed and constructed holistically, it has played the expected multiple roles. It is also because of its holistic design that since its initial construction of more than 2000 years ago, Dujiangyan has been in effective use in the agricultural production and producing economic benefits for Sichuan plain.

In the area of medicine, the theory of Chinese medicine has sufficiently embodied the logic thinking of systems science. The Yellow Emperor's Inner Canon of the ancient Chinese medical theory (2696–2598 BCE) emphasizes on the connection of various human organs, relation between psychological and physiological symptoms, and interaction between human bodily states and the natural environment (Liu, 1994). Each human body structure is seen as a component of nature with all the organs working together as an organic whole. The concepts of yin and yang and the theory of five elements are employed to illustrate the interactions and mutual constraints of the five internal organs. The cause of a disease is jointly analyzed from bodily symptoms, physiological phenomena, and spiritual states of the patient. When treating a disease, the therapeutic principle of correspondence between man and universe is developed by jointly considering the laws of health and nutrition and those of nature. This end in fact represents an emphasis on the interactive effects of subsystems within the human body system and of the system and its environment.

Feeling the pulse, that is treating the human body as a whole, is one of the methods used in Chinese medicine when a patient is diagnosed. Based on the fact that when the body is ill, its blood circulation is somehow affected, the doctor employs the speed, strength, and other characteristics of the wrist arterial pulse to determine where the pathological change is and its severity. The method of acupuncture is developed on the basis that all organs work together closely as a whole; and the problems with different organs can be treated through needling some of the acupuncture points on the ears. That is, no matter whether it is diagnosing or treating a sickness, Chinese medicine has seen the human body as a whole, recognizing that there are close connections between different body parts. What is interesting is that the material supports of these connections include not only the nerves, blood vessels, and various joints between organs, as observed in human anatomy, but also the so-called meridian channels. According to Chinese medicine, the meridian system connects all parts of human body into a unified whole and constitutes the important channels of communication between the parts; and only when the body is metabolically alive and forms a whole, the meridian system appears and functions. Otherwise, the meridian system would disappear. To this end, it is a known fact that as long as a body is opened for the purpose of observing the organization of different parts, the meridian system becomes invisible. The traditional Chinese medicine, which is developed on the viewpoint of wholeness of systems science, has been widely used in China and forms an important part of the world medicine.

Along with the development of productivity, human understanding of nature has been deepened further and further. However, at the same time there has been a tendency of ignoring the research of the holistic characteristics of systems. For instance, let us look at the situation of human body again. Through anatomy not only have various body parts been well recognized along with a deep understanding of their individual structures, but also blood circulations, nerve systems, and lymphoid tissues that connects the parts are found. While our comprehension of each body part is deepened and each specific connection of the parts is discussed, the world of learning seems to have forgotten the need to analyze the whole body and ignored the roles of various parts in the operation of the whole system. Along with the constantly refining knowledge of anatomy, human understanding of muscles has reached a delicate level. At the same time when various physical and chemical experiments on muscles are designed and conducted and when various characteristics of muscle cells have been noted, some special attributes of the muscles that are shown along with the existence of the whole body are rarely considered. The situation has been so extreme that some extraordinary capabilities of the whole body, which cannot be observed in laboratory experiments of local muscle studies, are considered not trustworthy. When some professional researchers saw the performance of a qigong master lying down his naked upper body on a bed of standing nails with extra weights placed on top of his body, they could not believe the fact that the master did not get hurt from the sharp nails. Because no matter whether it is in their theoretical computations or laboratory experiments, they could not make a small piece of muscle stand extra pressure without being pierced through, they deny what they saw in one way or another. However, when this situation is seen from the angle of systems science, it is not difficult to see that experiments on regional muscles cannot reflect the characteristics of the whole system; it is quite normal to observe different attributes of the whole body from those of a piece of local muscle. Ignoring essential differences between the whole and parts has been a negative effect of deepened and refined scientific researches on the development of systems science.

With the deepening understanding of nature, scientific divisions have been finer and finer over time. Scholars in one narrow area no longer know much or anything from another refined area. They can only be known as specialists in their small field. Leonardo da Vinci, an Italian scientist of the middle ages, was a physicist, architect, and an expert in many other scientific areas. After that historical period of time, it became impossible for anybody to be proficient, like da Vinci, in so many different scientific fields as astronomy, physics, and others. Even within the field of physics, modern experts of electromagnetics no longer know much about atomic physics; and theoretical researchers in general are not good at experiments, while laboratory technicians and experts do not have strong background in theoretical analysis. Speaking macroscopically, the reason for the occurrence of this situation is that the more in-depth knowledge is established, the more contents there are. So, even if a person does nothing in life but learns different areas of knowledge, it is still impossible for him to become proficient in many different disciplines. At the same time, because scientific disciplines are so finely defined, it is no longer necessary for any scientific worker to acquire knowledge beyond what he is doing to make a living within his narrow field of specialized knowledge. The abundant existence of narrowly trained scholars is one of the reasons why the development of systems science has been slow.

When seeing from the angle of economic development, no matter whether it is a long lasting slavery system, a feudal society, or the beginning stage of capitalism, the scale of production is generally small and the equipment of manufacturing is simple. So, it is relatively easy to establish harmonic development in economic productions. As long as the intensity of labor is increased, the level of production can be consequently raised. It seems that natural resources are plentiful and could last forever. Any environmental pollution caused by manufacturing commercial goods can be corrected through the effects of nature. So, during this period of time, people focus more on improving individual machines, while ignoring the overall benefits, more on increasing the levels of production, while not considering integrated utilization and resource allocation. That explains why no objective demand exists for the research of systems science that investigates wholes of systems. As a matter of fact, after the magnificent development during the initial period of time, other than some isolated progresses in some regional, industrial, and particular aspects, the overall research of systems science has been in a state of stagnation. As a form of thinking and reasoning, it has been gradually replaced by metaphysics. The situation is like what Friedrich Engels pointed out (1878) to the Greeks, that it is exactly because they were not advanced enough to dissect nature and analyze the resultant parts, nature was still treated as a whole and observed as a whole; holistic connections of natural phenomena were still not confirmed with comprehensive details so that these connections to the Greeks were only consequences of intuition; that is where there were deficiencies in Greek philosophy; and because of these deficiencies, it later had to submit to another point of view.

1.1.4 Quantitative Systems Science Developed for Applications

Just like any scientific theory, the appearance and development of systems science also originated in practical applications and are motivated by calls of solving real-life problems. During the time periods both before and after World War II (WWII), systems science developed quickly at the height of applications. In particular, cybernetics, operations research, and information theory, branches of systems science at the height of applications, gradually appeared after WWII. Before the war, the research on possible applications of systems science had started with some major progress. However, these works were isolated, scattered, and regional. Two particular cases stood out the most. One is the telephone traffic model (Erlang, 1909) proposed by Agner Krarup Erlang (1878–1929). In the twentieth century, the enterprise of telephones developed and grew quickly. When telephone lines were installed, one needed to consider the service efficiency of telephones. If there were too few lines, many telephones would have to share one line, creating jams of conversation; if there were too many lines installed, many of the lines would be idling so that the limited resources would be wasted. By using the method of comparison, Erlang resolved this problem by establishing a clever model. He compared a telephone communication system with a vapor–liquid equilibrium system of water by identifying each telephone that is being picked up to a water molecule that changes from a vapor state to a liquid state. When the telephone is returned to its idling state, it is compared to the molecule that returns back to the vapor state. Assume that the probability for a telephone conversation to finish within a unit time interval is λ. That is, the length of conversation is $1/\lambda$. And further assume that the probability for a telephone to be used within the unit time interval is μ. Then, according to the fact that at the vapor–liquid equilibrium of water, the number of molecules that move from the liquid state to the vapor state is the same as that of molecules that migrate from the vapor state to the liquid state within a unit time interval, Erlang obtained that

$$P_{i-1}\mu = p_i i\lambda$$

where p_i stands for the probability that the system contains i molecules in the liquid state. By applying this general equilibrium relation repeatedly, one obtains

$$P_n = \frac{\mu}{\lambda n} P_{n-1}$$

$$= \left(\frac{\mu}{\lambda n}\right)\frac{\mu}{\lambda(n-1)} P_{n-2} = \cdots = \left(\frac{\mu}{\lambda}\right)^n \frac{1}{n!} P_0$$

Since probability distributions satisfy the condition of summing up to 1, that is, $\Sigma_{i=0}^{n} P_i = 1$, one has

$$P_0\left(1 + \frac{1}{1}\left(\frac{\mu}{\lambda}\right) + \frac{1}{2!}\left(\frac{\mu}{\lambda}\right)^2 + \cdots + \frac{1}{n!}\left(\frac{\mu}{\lambda}\right)^n\right) = 1$$

and

$$P_0 = 1\Big/\left(1 + \frac{1}{1}\left(\frac{\mu}{\lambda}\right) + \frac{1}{2!}\left(\frac{\mu}{\lambda}\right)^2 + \cdots + \frac{1}{n!}\left(\frac{\mu}{\lambda}\right)^n\right)$$

Letting $\rho = \mu/\lambda$ leads to the Erlang formula:

$$P_i = \left(\frac{\rho^i}{i^i}\right)\Big/\left(1 + \rho + \frac{\rho^2}{2!} + \cdots + \frac{\rho^n}{n!}\right)$$

This Erlang formula, which is derived by using statistical computations, indicates that if the probability for a telephone to be used is μ and the average length of telephone conversation is 1/λ, then when the probability for a line is open when a telephone is picked up is given, such as $\rho = 0.9$, i telephones can be served by n telephone lines. Generally, $\mu > \lambda$ holds true. If the total number of telephones is fixed, increasing the number n of telephone lines in the system will make the probability of telephone connections greater; if the number of telephones is increased while all other parameters stay invariant, the probability of telephone connection will drop.

This example not only serves as a representative case of applications of systems science, but also possesses practical significance. The previous Erlang formula is still employed today as a basic reference on which telecommunication networks are designed. The later studies of such problems become the main content of queuing theory. In terms of the methodology of resolving practical problems, the process of model building also provides a useful reference for establishing systems models when systems science is applied in real-life situations.

Another example is the input–output model (Raa, 2009) proposed by Wassily Leontief (1905–1999). The relationship between various departments of a national economy is very complicated. Their individual products serve mutually as raw materials, energies, and so on. The available amount of one product more or less affects the production of many other products; at the same time it is constrained by the availability of many other products. However, for most products, the amount of inputs corresponding to the unit production output, that is the amount of consumption of the various supplies for the production of one unit of products, is relatively stable within a short period of time. That is because the consumption of various raw materials, supplements, energies, and so on, are determined by the level of production and technology, quality of management, relevant natural conditions, and other factors. Within a short period of time, these factors cannot be substantially changed. Even if the consumption coefficients of some products change relatively significantly, the changes tend to follow a pattern that can be determined quite easily.

Leontief's input–output model is established by using mathematics and employs computers to investigate the quantitative relationship between the input and output of various economic activities. In particular, this model plays an important role in studying and analyzing the quantitative dependency relationship between the products and consumptions of various departments (or various products) of a national economy. It has greatly helped to scientifically arrange, predict, and analyze economic activities. This method was initially published in Leontief's paper of 1936, entitled "Quantitative Input and Output Relations in the Economic System of the United States," in *Review of Economics and Statistics*, vol. 18, pp. 105–125. Later in 1941, he published *The Structure of American Economy 1919–1929*, Oxford University Press, New York; and in 1953, he published *Studies in the Structure of the American Economy*, Oxford University Press, New York. In these and relevant papers and books, Leontief not only developed the input–output method, but also constructed the input–output tables for American economy for the years of 1919, 1929, and 1939 based on the published economic statistics. Considering the important roles the input–output method played in economic analysis, Leontief was awarded the 1973 Nobel Prize of Economics. And his method has also been widely applied by various nations.

During WWII, various nations demanded the resolution of how to bring about the optimal overall effects with reasonable local results. Studies of these relevant problems greatly enhanced the development of operations research, various control methods, and game theory. Due to the needs of the war, various real-life problems were posed, such as: How can a defense system be arranged in order to better deter and destroy the enemy's airstrikes? How can targets be searched so that submarines can be found? How much advanced time is needed to fire artilleries in order to deal with high-speed flying objects, such as planes? These and related problems not only represent extreme practicality, but also possess high levels of theoretical values. To meet the needs of the war, a large number of scientists shifted their focus of work to serve the national defense and military operations. Because of the surge in scientific manpower, some of the problems that were related to military operations were resolved quickly along with many new scientific concepts being introduced.

When the war ended, these scholars continued their works at the theoretical heights. They established various applied branches of systems science, such as operations research, management science, cybernetics, information theory, and so on. In particular, Norbert Wiener, who during the war investigated radar and air defense fire control systems, established cybernetics by publishing *Cybernetics: Or Control and Communication in the Animal and the Machine*, (Paris: Hermann & Cie; Cambridge, Mass.: MIT Press) in 1948. Claude Elwood Shannon laid down the foundation for information theory by publishing the paper "A Mathematical Theory of Communication," *Bell System Technical Journal*, vol. 27, pp. 379–423, 623–656, 1948. In the 1950s, various papers and monographs on the methods of operations research appeared, among which Harry H. Goode and Robert Machol published the first monograph in systems engineering, *System Engineering: An Introduction to the Design of Large-scale Systems*, McGraw-Hill, 1957.

With the efforts that followed the initial publications, these and related works have grown into technical disciplines with their respective mathematical foundations. And the establishment of these disciplines made systems science evolve at the height of epistemology into such a quantitative field of knowledge that is constructed on the basis of mathematical science.

The development of a science must find its motivation in economies and technology. After WWII, the economic and scientific developments once again placed systems science in the front of the world of learning from two different angles. One is the expansion in the scale of production, which could no longer be improved by merely increasing the intensity of labor. Due to the increasing ratio of automated productions lines and more diversified manpower involved in various projects, organizational modernization was needed. In particular, the completion of a specific production task might need the collaboration of many different departments. American Apollo space program involved over one million people of different specialties and more than 200 universities. It involved over 100 commercial firms and cost over US$24,000,000,000. In this huge project, the number of personal achievements accomplished by the creative works of individual scientists was small. What accounted the most were the comprehensive goal-oriented researches of large groups of specialists from many disciplines. For the success of these researches, the key was the organization and coordination of people and machines, the smooth connection of various production procedures, and how to optimize the whole. That is, the key was about how to employ the thinking logic and methods of systems science to coordinate research and productions.

On the other hand, due to the limitations of natural resources and working spaces, any improvement of production corresponding to the economic development is not merely for the purpose of economies. Problems related to the production also need to consider the optimization of the whole, allocation of resources, sustainable development of the society, and so on. These problems need to be addressed from multiple angles, such as production and consumption, capital and labor, resource utilization, and environmental pollution by using combined approaches of natural and social sciences. In terms of the economic development itself, the free competitions of the early capitalism have gradually evolved into strengthened macroscopic national controls. Redistributions of production wealth, especially the international trades, have arranged the economic developments of different nations into an organic whole.

All these scenarios indicate that the relationships between the whole and parts, global optimizations, and other related concepts, as advocated by systems science, have become the significant theoretical support for the development of scientific production. During this period of time, the development of systems science can be characterized as follows:

1. Large amounts of mathematical tools are employed so that systems science has been changed from an epistemological discipline into a rigorous science that is practically applicable.
2. The fundamental and applied researches of technology have developed quickly. Aiming at resolving various kinds of particular problems, different methods are established. Although these methods are purposefully designed to resolve practical problems with their individual advantages and constitute a general methodology, no generally applicable theory is formulated and no theoretical system is constructed.

3. In terms of the guiding principle, scholars have either consciously or unconsciously done their thinking and reasoning with the whole in mind. They start with an initial understanding of development of things and events, and follow by applying effective controls of the system until reaching the eventual optimization of the whole. The process can be summarized as from understanding to application, from that the whole is greater than the sum of parts to that imperfect parts can be arranged to form an optimal whole.

1.1.5 Construction of a Comprehensive Systems Science

Dissipative structures and synergetics of physics offer a theory and a method to investigate natural complex systems and a possible way to unify natural and social systems. They provide materials for systems science. Physics is used to discuss reversible, degenerative systems. Both Newton's second law and the second law of thermodynamics describe the evolutionary direction and characteristics of systems. However, the evolutionary directions of these natural systems contradict the phenomena of developments and evolutions widely existing in the biological world and studied in social science. That is, no unified methodology existed for scholars to investigate natural and social systems jointly. To this end, Eugene Paul Wigner (1902–1995), a physicist and mathematician, a co-winner of 1963 Nobel Prize in physics, expressed his opinion (1969):

> What is the most important gap in modern science? It is obviously the separation between physical science and human science. As a matter of fact, there is nothing in common between physicists and psychologists. Perhaps, an exception is some of the tools physicists provided for the superficial psychological attentions, while psychologists warn the physicist about how not get his thinking and discovery affected by his hidden desires.

If one places such a comment from the perspective of the scientific history, he could see that there used to be a barrier that separated the Heavenly bodies from Earth. Then with the coming of modern science, the celestial bodies and Earth are now seen together in a unified whole, known as the universe, without any separation in between. However, the mainstream literature indicates that modern science is largely a science of quantities; it describes a wonderful geometric world in which no man lives. For more details, refer to Lin and OuYang (2010). That is, there is a barrier between the natural scientific world and the physical world with people that is studied in social science.

Ilya Prigogine proposed the theory of dissipative structures. When an open system is far away from an equilibrium state, due to exchanges of matter, energy, information, and so on, with the environment, it can possess some kind of order structure. In naturally existing physical, chemical systems, the phenomena of evolution, similar to those seen in biological systems, can also be observed. These phenomena of evolution can be investigated using the unified language of dissipative structures.

Hermann Haken proposed the theory of synergetics, believing that phase transitions of complex systems originated in the interactions and coordinated effects of subsystems. The concept of order-parameters and the enslaving principle of synergetics are proved to be an effective method to address evolutionary directions of systems.

When discussing the problem of why natural systems evolve in the direction of strengthened orderliness, catastrophe theory, stability theory of differential equations, hypercycle theory, and so on, are employed. When investigating the evolution of nonlinear systems, concepts such as chaos, bifurcation, fractal, and so on, are introduced with relevant theories developed. One characteristic common to all these new concepts, new methods, and new theories of natural science is the new phenomena of evolution emerging out of the nonlinear interactions between various subsystems of complex natural systems. Relevant studies show that these new concepts, methods, and theories are equally applicable to the investigation of natural phenomena, from which of course they are initially proposed, and some social events. So to a degree, they can serve as a unified theory for the study of nature and society, two totally different objective phenomena. Although these contents are not themselves parts of systems science, they can be seen as some of the main materials to be included in the foundations of systems science.

As a matter of fact, the development of systems science in the past century has indicated that only when the study of natural science has deepened to a certain level, and only after some of the theories as mentioned above are successfully established, a corresponding theory of systems science can be developed. At the same time, when such new theories are proposed, the world of learning also likes to enrich them so that they can become systems theories that are able to deal with the widely existing general complex systems. In the rest of this book, we attempt to enrich and improve these new materials of natural science in order to make them parts of systems theory. Of course, our works along this line represent only the start that needs to be followed by many colleagues in the years to come.

Additionally, along with the appearance of complicated production tasks and the new problems arising from the studies of large systems, giant systems, and distribution parameter systems, and so on, the development of systems science at the level of application has been accelerated. The classical control theory, which focuses mainly on single machine automation by using the method of transfer functions, has been evolved into such a modern theory of control, which consists of the concept of state spaces, dynamic programming, Kalman filter, maximum principle, and so on, useful for addressing the problem of automation of multiple machines and of the whole factory.

Nonlinear programming, integer programming, game theory, stochastic processes, and other many new methods of nonlinear operations research, along with modern information theory and technology, make the branches of systems science at the level of techniques, such that control theory, operations research, and information theory, more mature. That of course helps in providing a large amount of materials for the foundation of systems science. In the area of applications, major attention has been given to the study of social systems, economic systems, and other similar systems that involve people. For such systems, theoretical analysis has been combined with computer works so that the method of meta-synthesis can be developed by joining both quantitative and qualitative methodologies. By combining men and machines through the use of scientific discussion halls, the problems of evolution of giant social systems can be possibly resolved both quantitatively and qualitatively.

From the discussions above, it can be seen that in the three levels—fundamental theory, technical basis, and application technology, systems science has grown tremendously in recent years. And in the 1980s, Qian Xuesen (1981) put forward a framework for the system of systems science, analyzed the contents of the science at different levels and their interactions. This work has surely brought systems science into a new era of integrated development. The theoretical system of systems science is currently growing in a rapid growth spurt. The characteristics of this period of development are

1. Various theories that are closely related to systems science are developing tremendously in their respective fields. Nonlinear studies, such as chaos, bifurcation, fractals, and so on, have become part of the scientific frontier. Stochastic control and hierarchical control of complex systems of multiple levels and multiple targets have also become hot research topics in the area of automation.

2. The work of constructing the theoretical framework for systems science has been started by synthesizing advanced theories of different disciplines. Differences between these theories are analyzed, and commonalities are sought. On the basis of the thinking logic of evolution of complex systems, theories and methods from different fields are summarized into a unified general theory, leading to the creation of new concepts and new methods.
3. Due to the need of production development, the research of various real-life systems has been deepened under the guidance of the thinking logic of systems science. Although systems such as artificial intelligence, economic operation, the human brain, and so on, have their respective characteristics, they can surely be analyzed using new logic of reasoning. When one addresses problems from the angle of systems science, he can stand high and go deeper under the surface. When studying complex systems, other than practically applying systems theories, all the new methods developed and new conclusions obtained in the study can also help to enrich the systems theories themselves, making systems science truly able to address the evolution of complex systems.

1.1.6 The Systemic Yoyo Model

By pondering over how modern science has been successfully developed on top of a Cartesian coordinate system, which is a special kind of Euclidean space and plays the role of playground and intuition, and what method is available for us to effectively investigate systems, one can readily see the extreme difficulty here facing the research of systems science. In particular, first, as many variables involved in the study alone would make any of such studies hard to manage. Second, the large-scale complexity involved in the study of systems can make the already tedious and difficult problems even more drastically impossible.

Conventionally, each number is identified with a point on the real number line or a point in a high-dimensional Euclidean space. Each (quantitative) variable is seen as a moving point in the space. So, when as many such moving points and variously combined masses of these points are jointly considered, the problem is likely to be beyond the limited capability of human imagination. Among others, this end can be seen as the core reason for the extreme difficulty faced in dealing with systems. So, the following is a natural question to ask at this particular junction: Can we introduce a systemic intuitive background different of that of Euclidean spaces so that such massive amount of moving particles and masses of various combinations of these particles can be more easily managed? The intuition underneath this question is that because systems science, as the second dimension of science, studies systemhood while the classical science studies thinghood, for details, see (Klir, 1985), the difficulty one faces when dealing with systems is really a difficulty one experiences in the first dimension of science. So, there should be a relatively more manageable means in the second dimension for one to carry out such large-scale tasks. To geometrically comprehend this reasoning and why it might work in dealing with systems, one can imagine a two-dimensional city surrounded by a circular steady wall. If there is no gap on the wall, then it would be difficult for any army to break into the city from within the two-dimensional space. Now, if one smart engineer designs an air-strike by making use of a third dimension, which was not known before, then the forces of his side can be easily parachuted into the city along the third dimension. In other words, the difficulty one faces when dealing with systems is partially and mainly due to the reason that he maintains himself in the first dimension of science without truly taking advantage of systems science, the newly found second dimension of science.

When we study the nature and treat everything we see as a system (Klir, 1985), we find easily that many systems in nature evolve in concert. When one thing changes, many other seemingly unrelated matters alter their states of existence accordingly. This means that changes in nature should be seen as a whole, and the whole evolution of the system of concern should be considered in order to understand how systems evolve as wholes and how systems are related to each other. In the whole evolutions, the most important is discontinuity with which transitional changes (or blow-ups) occur. These blow-ups represent the changes of old structures being replaced by new structures. If the system is described truthfully by a mathematical model, then the model is generally nonlinear; and the blow-ups of the model reflect the destruction of old structures and establishment of new ones. By borrowing the form of calculus, we can write the concept of blow-ups as follows: For a given (mathematical or symbolic) model, that truthfully describes the system of our concern, if its solution $u = u(t;t_0,u_0)$, where t stands for time and u_0 the initial state of the system, satisfies

$$\lim_{t \to t_0} |u| = +\infty,$$

and at the same time moment when $t \to t_0$, the underlying physical system also goes through a transitional change, then the solution $u = u(t;t_0,u_0)$ is called a blow-up solution and the relevant physical movement expresses a blow-up. Thousands of real-life evolutionary systems (Lin, 2008) indicate that disastrous events appear at the moments of blow-ups in the whole evolutions of systems.

For nonlinear models in independent variables of time (t) and space (x, x and y, or x, y, and z), the concept of blow-ups are defined similarly, where blow-ups in the model and the underlying physical system can appear in time or in space or in both.

One of the key features of blown-ups is the quantitative ∞, a mathematical indeterminacy. This mathematical symbol in applications has caused instabilities and calculation spills interrupting each and every working computer. To understand this symbol systemically, let us look at the Riemann ball, Figure 1.1, which is a curvature space. The projection of the point x_i of the ball onto x_i' of the plane connects $-\infty$ and $+\infty$ through a blow-up. In particular, when the dynamic point x_i travels through the North Pole N on the sphere, the corresponding image x_i' on the plane of the point x_i shows up as a reversal change from $-\infty$ to $+\infty$ through a blown-up. So, the planar points $\pm\infty$ stand implicitly for direction changes of one dynamic point on the sphere at the polar point N instead of indeterminacies. Speaking differently, the phenomenon of directionless, as shown by blown-ups of a lower dimensional space, represents a direction change of movement in a higher dimensional curvature space. Therefore, blow-ups can specifically represent implicit

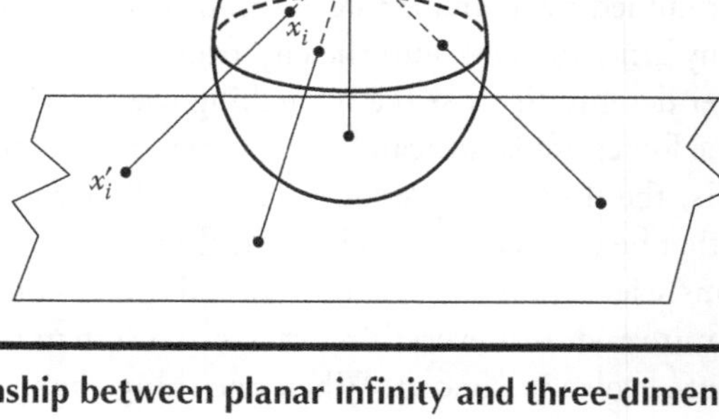

Figure 1.1 The relationship between planar infinity and three-dimensional North Pole.

transformations of spatial dynamics. Through blow-ups, the problems of indeterminacy of a distorted space are transformed into determinant situations of a more general system in a curvature space.

Corresponding to the previous implicit transformation between the plane and the Riemann ball, one can also relate quantitative $\pm\infty$ in one-dimensional space to a dynamic movement on a circle, a curvature space of a higher dimension through the modeling of differential equations.

This discussion indicates that nonlinearity mathematically stands (mostly) for singularities in Euclidean spaces, and physically eddy motions, the movements on curvature spaces. Such motions are about structural evolutions, a natural consequence of uneven evolutions of materials.

Next, we look at the general dynamic system and how it is related to eddy motions. Newton's second law of motion is

$$m\frac{d\vec{v}}{dt} = \vec{F}.$$

Based on Einstein's concept of uneven time and space of materials' evolutions, we can assume $\vec{F} = -\nabla S(t,x,y,z)$, where $S = S(t, x, y, z)$ stands for the time–space distribution of the external acting object. Let $\rho = \rho(t, x, y, z)$ be the density of the object being acted upon. Then, the previous equation for a unit mass of the object being acted upon can be written as

$$\frac{d\vec{u}}{dt} = -\frac{1}{\rho(t,x,y,z)}\nabla S(t,x,y,z),$$

where $\vec{u}$ is used to replace the original $\vec{v}$ in order to represent the meaning that each movement of some materials is a consequence of mutual reactions of materials' structures. Now, if ρ is not a constant, then the previous equation becomes

$$\frac{d(\nabla x \times \vec{u})}{dt} = -\nabla x \times \left[\frac{1}{\rho}\nabla S\right] \neq 0,$$

which stands for an eddy motion because of the cross products. In other words, a nonlinear mutual reaction between materials' uneven structures and the unevenness of the external forcing object will definitely produce eddy motions.

At this junction, it is important to note that eddy motions are confirmed not only by daily observations of surrounding natural phenomena, but also by laboratory studies from as small as atomic structures to as huge as nebular structures of the universe.

What is shown above implies that eddies come from the unevenness of the internal structures of materials. So, if the world is seen at the height of structural evolutions, then the world is simple with only two forms of motions: clockwise rotation and counterclockwise rotation. And the vertical vectority in structures has been very practically implemented in the common form of motion of the universe.

Now, a fundamental problem is why all structures of the universe are in rotational movements. According to the concept of uneven space and time of Einstein's theory of relativity, we can assume that all materials have uneven structures. Out of these uneven structures, there naturally exist gradients. With gradients, there appear forces. Combined with uneven arms of forces, the carrying materials will have to rotate in the form of moments of forces.

Based on what is discussed above, we can now imagine that the universe be entirely composed of eddy currents, where eddies exist in different sizes and scales and interact with each other. That is, the universe is a huge ocean of eddies, which change and evolve constantly. One of the most important characteristics of spinning fluids, including spinning solids, is the difference between the structural properties of inwardly (converging) and outwardly (diverging) spinning pools and the discontinuity between these pools. Due to the stirs in the form of moments of forces, in the discontinuous zones, there exist sub-eddies and sub-sub-eddies (Figure 1.2, where sub-eddies are created naturally by the large eddies *M* and *N*). Their twist-ups (the sub-eddies) contain highly condensed amounts of materials and energies. In other words, the traditional frontal lines and surfaces (in meteorology) are not simply expansions of particles without any structure. Instead, they represent twist-up zones concentrated with irregularly structured materials and energies (this is where the so-called small probability events appear, and the small-probability information is observed and collected so that such information (event) should also be called irregular information (and event)).

Based on what is discussed above, the concepts of black holes, big bangs, and converging and diverging eddy motions are coined together (Wu and Lin, 2002) in the model shown in Figure 1.3 for each object and every system imaginable. In particular, each system or object considered in a study is a multi-dimensional entity that spins about its invisible axis. If we fathom such a spinning entity in our three-dimensional space, we will have a structure as shown in Figure 1.3a. The side of black hole sucks in all things, such as materials, information, energy, and so on. After funneling through the short narrow neck, all things are spit out in the form of a big bang. Some of the materials spit out from the end of the big bang will never return to the other side, and others will (Figure 1.3b). Such a structure, as shown in Figure 1.3a, is called a yoyo due to its general shape. More specifically, what this model says is that each physical entity in the universe, be it a tangible or intangible, a living being, an organization, a culture, a civilization, and so on, can all be seen as a kind of realization of a certain multi-dimensional spinning yoyo with an invisible spin field around it. It stays in a constant spinning motion as depicted in Figure 1.3a. If it does stop its spinning, it will no longer exist as an identifiable system. What Figure 1.3c shows is that due to the interactions between the eddy field, which spins perpendicularly to the axis of spin of the model, and the meridian field, which rotates parallel to axis of spin (Figure 1.3b), all the materials returning to the black-hole side travel along a spiral trajectory.

As for why materials in the universe and matters of the world rotate in the first place and continue to do so, according to the concept of uneven space and time of Einstein's theory of relativity (Einstein, 1987), we know that all materials have uneven structures. Out of these uneven structures, there naturally exist gradients. With gradients, there will appear forces. Combined

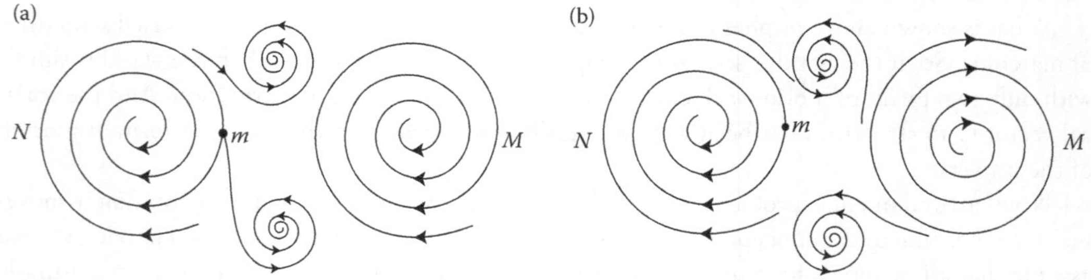

Figure 1.2 Appearance of sub-eddies.

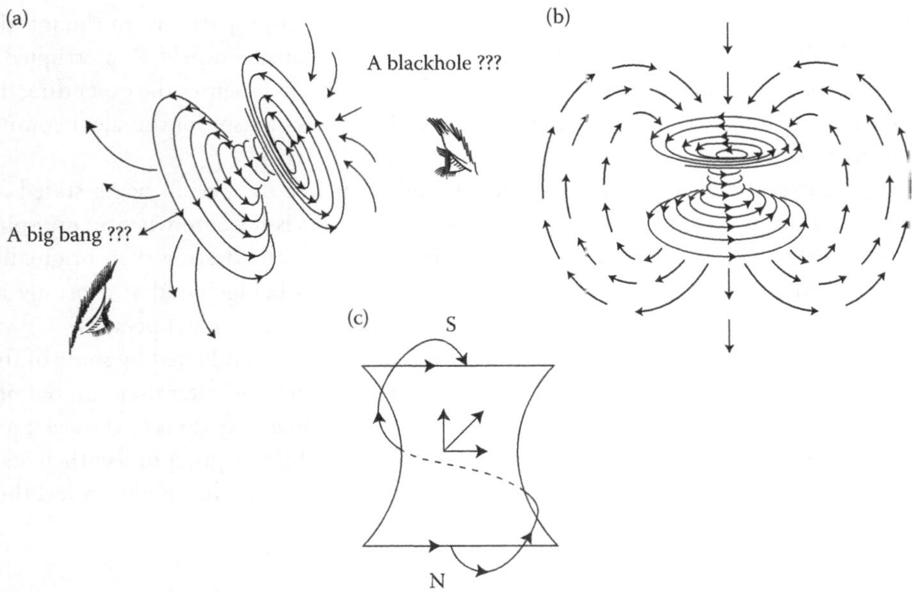

Figure 1.3 The eddy motion model of the general system.

with uneven arms of forces, the carrying materials will have to rotate in the form of moments of forces (Lin, 2008, p. 31).

The multi-dimensional yoyo model in Figure 1.3 is manifested in different areas of life. For example, each human being, as we now see it, is a three-dimensional realization of such a spinning yoyo structure of a higher dimension. To illustrate this end, let us consider two simple and easy-to-repeat experiments.

1.1.6.1 Experiment #1: Feel the Vibe

Let us imagine we go to a sporting event, say a swim meet. Assume that the area of competition contains a pool of the Olympic size and along one long side of the pool there are about 200 seats available for spectators to watch the swim meet. The pool area is enclosed with a roof and walls all around the space.

Now, let us physically enter the pool area. What we find is that as soon as we enter the enclosed area of competition, we immediately fall into a boiling pot of screaming and jumping spectators, cheering for their favorite swimmers competing in the pool. Now, let us pick a seat a distance away from the pool deck anywhere in the seating area. After we settle down in our seat, let us purposelessly pick a voluntary helper standing or walking on the pool deck for whatever reason, either for her beauty or for his strange look or body posture, and stare at him intensively. Here is what will happen next: Magically enough, before long, our stare will be felt by the person from quite a good distance; she/he will turn around and locate us in no time out of the reasonably sized and boiling audience.

By using the systemic yoyo model, we can provide one explanation for why this happens and how the silent communication takes place. In particular, each side, the person being stared at and us, is a high-dimensional spinning yoyo. Even though we are separated by space and possibly by informational noise, the stare of one party on the other has directed that party's spin field of the yoyo structure into the spin field of the yoyo structure of the other party. Even though the later party initially did not know the forthcoming stare, when her/his spin field is interrupted by the sudden intrusion of another unexpected spin field, the person surely senses the exact direction and location where the intrusion is from. That is the underlying mechanism for the silent communication to be established.

When this experiment is done in a large auditorium where the person being stared at is on the stage, the afore-described phenomenon does not occur. It is because when many spin fields interferes with the field of a person, these interfering fields actually destroy their originally organized flows of materials and energy so that the person who is being stared at can only feel the overwhelming pressure from the entire audience instead of from individual persons.

This easily repeatable experiment in fact has been numerously conducted by some of the local high school students. When these students eat out in a restaurant and after they run out of topics to gossip about, they play the game they call "feel the vibe." What they do is to stare as a group at a randomly chosen guest of the restaurant to see how long it takes the guest to feel their stares. As described in the situation of swim meet earlier, the chosen guest can almost always feel the stares immediately and can locate the intruders in no time.

1.1.6.2 Experiment #2: She Does Not Like Me!

In this case, let us look at the situation of human relationship. When an individual A has a good impression about another individual B, magically, individual B also has a similar and almost identical impression about A. When A does not like B and describes B as a dishonest person with various undesirable traits, it has been clinically proven in psychology that what A describes about B is exactly who A is himself (Hendrix, 2001).

Once again, the underlying mechanism for such a quiet and unspoken evaluation of each other is based on the fact that each human being stands for a high-dimensional spinning yoyo and its rotational field. Our feelings toward each other are formed through the interactions of our invisible yoyo structures and their spin fields. So, when person A feels good about another person B, it generally means that their underlying yoyo structures possess the same or very similar attributes, such as spinning in the same direction, both being either divergent or convergent at the same time, both having similar intensities of spin, and so on. When person A does not like B and lists many undesirable traits B possesses, it fundamentally stands for the situation that the underlying yoyo fields of A and B are fighting against each other in terms of either opposite spinning directions, or different degrees of convergence, or in terms of other attributes.

Such quiet and unspoken evaluations of one another can be seen in any working environment. For instance, let us consider a work situation where quality is not and cannot be quantitatively measured, such as a teaching institution in the USA. When one teacher does not perform well in his line of work, he generally uses the concept of quality loudly in day-to-day settings in order to cover up his own deficiency in quality. When one does not have honesty, he tends to use the term honesty all the time. It is exactly as what Lao Tzu (exact time unknown, Chapter 1) said over 2000 years ago: "The one who speaks of integrity all the time does not have integrity."

When we tried to repeat this experiment with local high school students, the following is what we found: When two students A and B, who used to be very good friends, turned away from each other, we ask A why she does not like B anymore. The answer is exactly what we expect: "Because she does not like me anymore!"

With this systemic yoyo model in place, we are now ready to move onto the main model that will be used throughout the rest of this book.

1.2 A New Understanding of "Poverty Emerges from Surplus"

In his monumental work "Socialism from Utopia to Science," Engels (Marx and Engels, 1979, p. 492) quotes St Simon as saying "poverty emerges from surplus." In most textbooks, this statement is interpreted as that under the then prevalent, specific circumstances, the excessive surpluses led to reduction in production output, which in turn caused factories to close, workers to lose their jobs, levels of incomes to be lowered, so that demand became further insufficient. That of course led to more excessive surpluses, more factories to close, and so on. So, over time, the society became further impoverished.

Now, from the angle of the systems yoyo model, as introduced in the first half of this chapter and (Lin, 2008), it is natural to realize that this statement that "poverty emerges from surplus" contains its significance more than the implied particular historical background, the specific time moment, and the individual event. It should in fact be a maxim that encompasses a rich philosophical content and a wide range of practical applicability.

The word "surplus" of course indicates an excess of supply in the economy. And economic surplus in essence means an excessive supply of labor or an excessive supply of capital or both. It is because any factor of production, no matter how complex it is, can be brought down to the levels of labor (including both intellectual and physical) and capital. Therefore, the excessive supply of any product, no matter if it is of a form or not, can also be attributed to excessive supplies of labor and/or capital. By excessive supply of labor, it implies that a large amount of labor is not or cannot be applied to any means of production. In some other cases, although all available labor engages meaningfully in different means of production without any visible excess, the quantity of the produced products is more than the demand of the whole society. Then in terms of the effective demand of the whole society, the supply of labor is no doubt in excess. By excessive supply of capital, it means a shortage of investment demand, and either the rate of return on investment is too low or the value of capital is so low that it is provided without much expense. When a shortage in investment demand appears, it implies that investors do not see a profitable future in the economy, leading to production declines and economy slumps, where workers lose their employment, and poverty becomes a natural consequence. For the situations where either the rate of return on investment is too low or the value of capital is so low that it is provided without much expense, they are phenomena of planned economies. Because capitals are cheap and can be obtained without incurring much loss, it leads to the phenomenon that all levels of production units blindly fight to acquire the nearly free capitals and squander what is attained on unnecessary places and projects. Here, the word "unnecessary" in essence means that the capitals spent on the particular places and projects are excessive or superfluous. In the recent history of China, several decades of planned economy had led to huge waste of the limited amount of capital. That explained why during that period of time, China had been financially poor.

One basic proposition of economics says that resources are limited, while need (desire) is unlimited. That is why the research of economics has to investigate how to optimize the distribution of the limited resources. Any excess means a waste of resources; and the limited resources should be allocated rationally to create additional wealth and to satisfy additional needs. The consequence of wasting resources (or producing excessive supplies) is of course economic poverty.

If one understands the statement that "poverty emerges from surplus," then many social- and economic phenomena can be explained. For example, let us look at the problem of why the planned economy in China cannot be sustained. During the time of several decades when China was a planned economy, the life quality of the ordinary people was not improved much, and the nation continued to be among the most poor nations and under-developed. By employing the aforedescribed new comprehension of the statement that "poverty emerges from surplus," it can be seen that what happened during that particular time period was that the limited human resources and capital resources were supplied at extreme low costs or at no cost so that huge wastes were consequently created. These wastes were caused by the so-called excessive supply of various resources on many places and projects. As another example, let us look at the problem of why the tides of migrant workers have existed powerfully in China for several decades without any sign of weakening. The obvious reason is of course that the economic conditions in the countries are too deprived, while the cause of the destitute conditions is the excessive supply of labor. In other words, the supply of capital is severely insufficient.

Here, we mentioned that the supply of capital is insufficient. Because the destitute conditions are caused by an excessive supply of labor, then what would insufficient supply cause? In fact, insufficiency will limit development. That is, development is constrained by insufficiency. Here, the so-called insufficiency means insufficient economic supply, that is, the insufficient supply of goods as caused by insufficient supplies of labor and capital. The sparse population and lack of capital have been two major bottlenecks of the economic development of Russia, including the former Soviet Union. So, importing a large amount of human labor, including human talents, and capital will be fundamental for the economic growth of Russia. For many developing countries, although they have abundant supplies of labor, they are short of talents and capital, which severely limit these nations' economic and social developments.

Excesses and insufficiencies of the supply are relative in terms of the effective demands. When the supply is more than the effective demand, then an excess will be produced. That in turn causes poverty. When the effective demand is more than the supply, then an insufficiency in the supply is created, which will restrain the development. Therefore, for any nation, especially for a developing nation that is still suffering from poverty and backwardness, it is essential for it to seek a dynamic balance between economic demand and supply.

Related to what is considered in this chapter, a different model is introduced in Lin (2008).

1.2.1 Criterion for Finding Structural Imbalances in Economic Systems

To find such a criterion, let us suppose that the total economic system is W, the total market demand in term t is $D_W(t)$, and the total market supply is $S_W(t)$. If $D_W(t) = S_W(t)$, then it indicates that the total market supply and the total market demand in term t are in balance. To discuss an imbalance of the system's structure, let us suppose that the overall system consists of n subsystems W_i ($i = 1, 2, \ldots, n$). Their demands to the overall system and supplies from the overall system in term t are $D_i(t)$ and $S_i(t)$, respectively. Then the structure of demands and supplies of the economic system W in term t can find its expression in the following three vectors:

Demand structure vector:

$$d(t) = (d_i(t), d_2(t), \ldots, d_n(t)), \quad d_i(t)\frac{D_i(t)}{D_W(t)} \quad i = 1,2,\ldots,n$$

Employment structure vector:

$$S^1(t) = (s_1^1(t), s_2^1(t), \ldots, s_n^1(t)); \quad s_i^1(t) = \frac{S_i^1(t)}{S_W^1(t)} \quad i = 1,2,\ldots,n$$

and Capital structure vector:

$$S^2(t) = (s_1^2(t), s_2^2(t), \ldots, s_n^2(t)); \quad s_i^2(t) = \frac{S_i^2(t)}{S_W^2(t)} \quad i = 1,2,\ldots,n$$

where $d_i(t)$, $s_i^1(t)$, and $s_i^2(t)$ symbolize the proportion of subsystem W_i's contribution of demand to the overall system in terms of the total demand in term t, and the proportions of supply of labor and capital it obtains from the overall system in term t, respectively. An even more important concept is the following proportion:

$$G_i^j(t) = \frac{d_i(t)}{s_i^j(t)}, \quad j = 1,2 \tag{1.1}$$

which is called the gains by comparison (GBC) or the comparative gains.

When $j = 1$, $G_i^1(t)$ is referred to as the comparative gain of the subsystem W_i's labor. When $j = 2$, $G_i^2(t)$ is referred to as the comparative gain of the subsystem W_i's capital.

If $G_i^j(t) = 1$, it indicates that the proportion of subsystem W_i's contribution of demand to the overall system in term t is exactly equal to that of the supply of labor and capital it obtains from the overall system. This means that the very subsystem has realized equilibrium of supply and demand without any gain or loss under its level of productivity in term t, its "need and enjoyment" meet the average scale of the society exactly. It also indicates that a unit labor time of creating value is exactly equal to a unit of socially necessary labor time. It reflects the average unit of productivity of the society in term t. In other words, it stands for the fact that the gain of capital has reached the average level of the society

If $G_i^j(t) \geq 1$, it indicates that the effective demand of subsystem W_i in term t is excessive, the supply of labor or capital is relatively insufficient, its "need and enjoyment" exceed the social average, little input produces abundant output, and a unit of labor time is shorter than a unit of the socially necessary labor time. That represents the fact that the subsystem's development direction of social productivity in term t is along that of optimal adjustment and improvement of the economic structure. In other words, it indicates that the gain of capital exceeds the social average.

If $G_i^j(t) < 1$, it indicates that the effective demand of subsystem W_i in term t is insufficient, while the supply of labor or capital is relatively excessive, the subsystem's "need and enjoyment" are lower than the social average, a lot of input produces a little output, a unit labor time is longer than a unit of the socially necessary labor time, and the excessive supply is not recognized by the society. In other words, it indicates that the gain of capital is smaller than the social average, causing waste of the social capital.

By combining what is discussed above, it can be seen that the comparative gains $G_i^j(t)$ given in Equation 1.1 can be used as a criterion for telling structural imbalances existing in a macroeconomic system. By substituting actual statistical data into Equation 1.1, the comparative gains of the subsystems can be calculated. Obviously, the subsystems with their comparative gains $\neq 1$ are structurally out of balance. The further away from 1 a subsystem's comparative gain is, the worse its structural imbalance is.

Additionally, the concept of comparative gains and formula in Equation 1.1 are not confined to any particular macroeconomic system. In fact, the macroeconomic system of any chosen society is an exceptionally complex system; that is, a pansystem as first studied by the famous Chinese scholar Professor Xuemou Wu (1996), who defines pansystems as general systems of a very wide significance, relations, and their various combinations.

Besides industrial structures, there can also be economic sectorial structures, regional structures, ownership structures, and income structures. From whatever angle that is considered, a society constitutes a macroeconomic system, which is subject to influences of structural imbalance for one reason or another, where the imbalances need to be measured and can be made of use. The concept of comparative gains and the relevant formula are, therefore, universally useable in the study of macroeconomic systems. In other words, the concept of comparative gains can serve as a criterion for omni-bearing imbalances existing in the structure, such as the industrial structure, economic sectorial structure, regional structure, property ownership structure, income structure, and so on, of macroeconomic systems.

1.2.2 Criteria for Optimizing Structures of Macroeconomic Systems

Contemporary systems science holds that all matters and events are of the form of systems, and that the efficiency and functionality of a system depend on the internal structure of the system (Lin, 1999, 2008; Lin et al., 2012). Therefore, the structure of a system must be judged and evaluated by using the system's output, efficiency, and gains. Although the output gains of the macroeconomic system of a society are constrained by the external environment of the system, such as the surrounding environment, international economy, or even the global economy, the state of the system's internal structure directly determines its output gain. Inversely, the output gain of the macroeconomic system can also be used as a reference to the evaluation of its internal structure.

Economic growth rate, unemployment rate, inflation rate, Gini coefficient, and the degree of economic stability are usually used as measures of the output efficiency of a macroeconomic system. Obviously, an ideal macroeconomic system is one with low unemployment rate, low inflation rate, low Gini coefficient, which enjoys a high rate of continuous economic growth. Such an ideal economic system is not only the objective after which human society has long been pursuing, but also the theoretical result from optimizing the external environment and the internal structure of the system.

According to the point of view of the operations research, each macroeconomic system has multiple objectives. As constrained by its external environment, the system with all its objectives reaching their respective extrema, for example, the unemployment rate, the inflation rate, and Gini coefficient are all equal to zero, the rate of economic growth hits the highest level possible, and so on, or with optimal output gain, can be referred to as an optimal economic system. In this case, the internal structure of the system is referred to as the optimal structure.

By taking the economy as a whole, high rate of unemployment and low rate of growth signify an excess in production, depression of the economy, and an insufficiency in the effective demand. In other words, on the whole, the supply of labor and capital is relatively excessive, which represents a gross imbalance. However, the gross imbalance with excessive supply over demand is caused by a structural imbalance. A gross imbalance with either gross excess or gross insufficiency is caused by an imbalance of its internal structure. The premise of optimizing a macroeconomic system is to solve the imbalance problem of its internal structure. To sum up what we just discussed, we have:

Theorem:

A necessary and sufficient condition for achieving the optimal structure for a macroeconomic system is the following: For any $i \in I = \{1,2,\ldots,n\}, j = 1,2,$

$$G_i^j(t) = 1; \quad \text{that is, } d_i(t) = s_i^j(t) \tag{1.2}$$

1.2.3 Strategic Optimization of Structures of Macroeconomic Systems

From the previous theorem, it follows that the direction of a strategic optimization of the structure of the macroeconomic system of concern is as follows: For any $i \in I$ and $j = 1,2$, let

$$\text{Either } G_i^j(t) \to 1 \quad \text{or} \quad s_i^j(t+1) \to d_i(t) \tag{1.3}$$

Equation 1.3 signifies that the resource allocation of the forthcoming term should be determined by the demand of the current term; in other words, the supply should be determined by the predicted demand. In a much broader sense, this conclusion coincides with the planning-programming budgeting system that clearly stipulates that determine the scale of investment by using the output in order to realize the optimization of the industrial structure.

In addition, the concept of comparative gains, as defined in Equation 1.1, in fact stands for an irresistible force, which possesses the social functionality of directing people involved in the process to spontaneously moving toward structural evolution and structural optimization. As for the owners of resources, these individuals or organizations naturally hope to invest as little as possible and gain as much return as possible by making their respective $G_i^j(t) > 1$. In such a process of making their respective $G_i^j(t) > 1$, the social and economic system evolves and optimizes spontaneously through the mechanism of competition. That is Adam Smith's system without macroeconomic control. In such a system, the equilibrium of supply and demand can be realized spontaneously by the market mechanism of competition that is made up of various enterprises and markets. For various reasons, however, this equilibrium is not real to the economy, leading to a disequilibrium economy (Li, 1991).

Practically, there are two approaches that can be employed to reach the strategic optimization for the structure of the macroeconomic system, as defined in Equation 1.3: self-adjustment and reallocation through the market mechanism, and directional adjustment and reallocation through the planning mechanism. The former allows people to accept the guidance of the forces of comparative gains so that resources are able to flow spontaneously and freely without any administrative interference. The latter shows the part the government can play in readjusting the structure of the macroeconomic system in order to achieve structural equilibrium (optimization), to establish and to perfect the market system of free competition, to institute the necessary rules for market competition, to orderly maintain the market environment of competition, and to consciously transfer excessive resources existing in underdeveloped areas, industries, business lines, and sectors with low productivities to developed areas, industries, business lines, and sectors with high productivities. So, a structural equilibrium of the macroeconomic system could be potentially secured.

Specifically for China, this implies that the conventional ideas and the traditional logic of thinking must be thoroughly changed in terms of making choices of macroeconomic institutions and policies. First, major adjustments in favor of free movement of the population and talents must be made with respect to personnel, labor, employment systems, and the policies regarding employment and domicile. Second, the policies of finance, banking, and investment should favor the areas, business lines, industries, and property ownerships with the comparative gains ≥1. Third, in respect to the policy of income distribution, there should be no more extreme equalitarianism.

Based on the analysis presented in this chapter, if follows that if China could be free of all forms of government interferences in its future reform, measure the structure of its macroeconomic system timely and earnestly according to the requirements of Equation 1.1, and undeviatingly encourage the aforementioned macroeconomic control, then what is expected includes the achievement of a long-term gross equilibrium of the macroeconomic system and an omni-bearing structural optimization.

2

REGIONS AND INDUSTRIES IN EAST, CENTRAL, AND WEST CHINA

In Section 2, we look at the investment and employment problems in the three greater geographic areas of China: the east, the central plain, and the west by analyzing the most recent statistical data. On the basis of the theory, model, and methodology presented in Section 1, we consider the optimal area, the general area, and the most difficult area in terms of investment and employment for the coming years.

This section constitutes a practical guide for investment and employment decision making in these respective areas.

Chapter 2

Investment and Employment in East, Central, and West China

In this chapter, we look at the investment and employment opportunities in three greater areas of China, known as East China, Central China, and West China. Their respective and relative geographic locations are shown in Figure 2.1.

2.1 Regions of East, Central, and West China

2.1.1 Introduction

From Chapter 1, we have the following concept of comparative gains and the relevant model:

$$G_i^j(t) = \frac{d_i(t)}{s_i^j(t)}, \quad \text{for } j = 1, 2 \tag{2.1}$$

From this model, it follows that

When $j = 1$, $G_i^1(t)$ is referred to as the comparative gains of labor of the ith subsystem.
When $j = 2$, $G_i^2(t)$ is referred to as the comparative gains of capital of the ith subsystem.

where $d_i(t), s_i^1(t), s_i^2(t), i = 1, 2, \ldots, n$, are known as the supply–demand structure, employment structure, and the investment structure of the macroeconomic system, respectively.

It is shown that this model can be employed as a criterion for checking whether or not a macroeconomic system suffers from an imbalance. That is,

1. If for any i, $G_i^j(t) = 1$ holds true, then the supply–demand structure of the macroeconomic system is seen as in a total equilibrium. That is, the supply–demand structure is optimal.

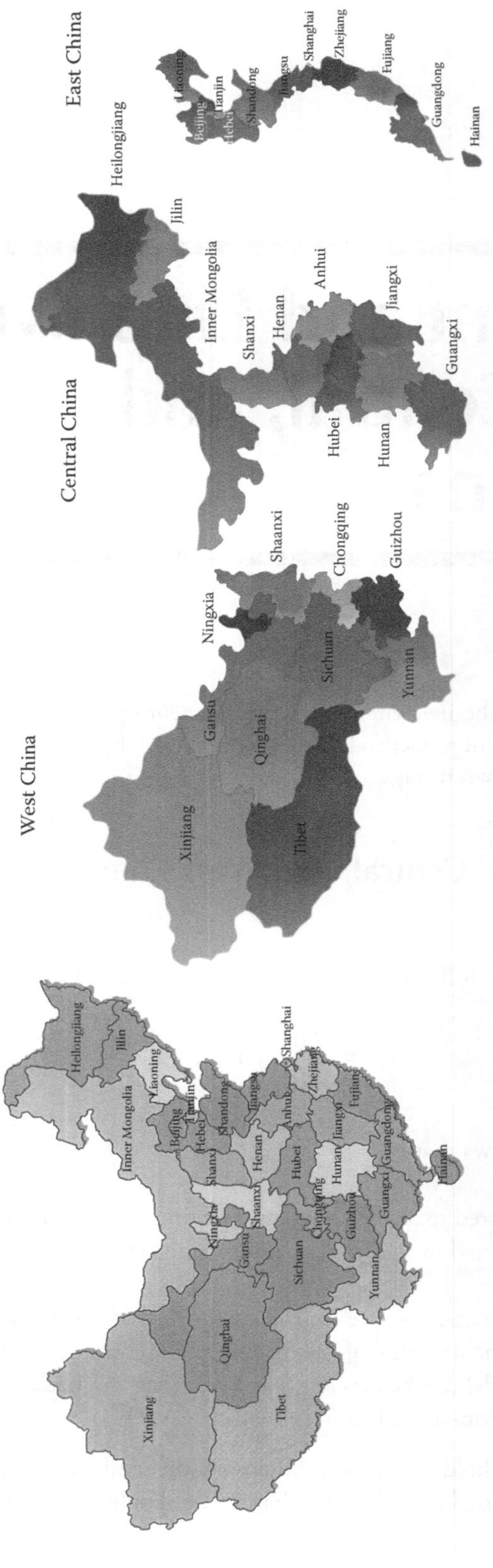

Figure 2.1 Respective and relative locations of East, Central, and West China.

2. If for any i one always has $G_i^j(t) \approx 1$, or more specifically, $\left|G_i^j(t) - 1\right| \leq 0.05$, then the supply–demand structure of the macroeconomic system is seen as in basically equilibrium state. That is, the supply–demand structure is in a suboptimal state.
3. Also, if there is a certain i such that $\left|G_i^j(t) - 1\right| > 0.05$, then an imbalance (or nonoptimal) appears in the supply–demand structure of the ith subsystem of the macroeconomic system.
   a. If $G_i^j(t) > 1.05$, then the ith subsystem suffers from such an imbalance that the supply cannot meet the demand.
   b. If $G_i^j(t) < 0.95$, then the ith subsystem suffers from such an imbalance that the supply is more than the demand.

Here, the degree of severity of the supply–demand imbalance can be measured and ordered by using the absolute values of $G_i^j(t)$, which can in turn be employed as the theoretical basis for direction and strength of making adjustment to the investment and employment structures. In the following, we will apply this theory and methodology to analyze the imbalances that exist in the economic structures and sectorial structures of the three greater areas—East, Central, and West China, and establish the investment and employment guidelines based on our results on structural optimization.

2.1.2 Current State of the Regional Economies of East, Central, and West China

As is well known, the regional GDP is computed by using either the expenditure approach or the final product approach. That is, it is obtained by summing up the values of all products and labors consumed by the society. This value, of course, represents the true demand (or the effective demand) generated by all labor and capital of the region. Mainland China contains a total of 31 provinces, municipalities, and autonomous regions, which excludes Hong Kong and Macao. The GDP of each region is exactly equal to the demand that the region provides to the entire society, constituting the regional demand structure $d_i(t)$ of China, see the second column in Table 2.2.

In order to analyze the current states of the regional economic structures of East, Central, and West China, we compute respectively the demand structure vector $d(t)$, the employment structure vector $s^1(t)$, and the investment structure vector $s^2(t)$ for each of the provinces, municipalities, and autonomous regions based on the relevant statistics of the recent year. For the detailed statistics of the GDPs, total values of fixed properties, and employment situations of the provinces, municipalities, and autonomous regions, see Table 2.1. Then, we substitute the computed vectors $d(t)$, $s^1(t)$, and $s^2(t)$ into Equation 2.1 to calculate the comparative gains $G_i^1(t) = \left(d_i(t)/s_i^1(t)\right)$ of labor and the comparative gains $G_i^2(t) = \left(d_i(t)/s_i^2(t)\right)$ of capital, respectively. The results are shown in Table 2.2.

From Table 2.2, it follows that no matter whether it is the supply–demand of labor or the supply–demand of capital, there are severe imbalances in the regional economic structure of East, Central, and West China.

In fact, from the fifth column (the comparative gains of labor) and the seventh column (the comparative gains of capital) of the table, it follows that the comparative gains of all of the 31 provinces, municipalities, and autonomous regions from around China except three (Shandong, Hainan, and Heilongjiang) are not equal to 1. That is, almost all of the 31 provinces, municipalities, and autonomous regions of China are in an imbalanced state. If a region whose comparative gains $G_i^j(t)$ are between 0.9500 and 1.0500 is seen as basically balanced and the rest are seen

Table 2.1 GDPs, Values of Fixed Properties, and Employments of Various Provinces, Municipalities, and Autonomous Regions (2011)

Area	*Province/Municipality/ Autonomous Region (i)*	*Regional GDP (0.1 Billion Yuan)*	*Value of Fixed Properties (0.1 Billion Yuan)*	*Employments (10,000)*
East China	1. Beijing	16251.93	5578.9	2143.9
	2. Tianjin	11307.28	7067.7	688
	3. Hebei	24515.76	16389.3	2157.4
	4. Liaoning	22226.70	17726.3	2675.8
	5. Shanghai	19195.69	4962.1	2186
	6. Jiangsu	49110.27	26692.6	6451
	7. Zhejiang	32318.85	14185.3	4810.8
	8. Fujian	17560.18	9910.9	2208.9
	9. Shandong	45361.85	26749.7	4315.9
	10. Guangdong	53120.28	17069.2	5943.6
	11. Hainan	2522.66	1657.2	383.2
	Total	293491.45	147989.2	33964.5*
Central China	12. Shanxi	11237.55	7073.1	1225.2
	13. Inner Mongolia	14359.88	10366.2	1120.1
	14. Jilin	10568.83	7441.7	1283.7
	15. Heilongjiang	12582	7475.4	1672.1
	16. Anhui	15300.65	12455.7	2028.8
	17. Jiangxi	11702.82	9087.6	1942.4
	18. Henan	26931.03	17769.0	2892.4
	19. Hubei	19632.26	12557.3	2701.7
	20. Hunan	19669.56	11880.9	2413.3
	21. Guangxi	11720.87	7990.7	1489.4
	Total	153705.45	104097.6	18769.1*
West China	22. Chongqing	10011.37	7473.4	1580.5
	23. Sichuan	21026.68	14222.2	2647.7
	24. Guizhou	5701.84	4235.9	789.9
	25. Yunnan	8893.12	6191.0	1753.6
	26. Tibet	605.83	516.3	153.4
	27. Shaanxi	12512.30	9431.1	1427.2
	28. Gansu	5020.37	3965.8	747
	29. Qinghai	1670.44	1435.6	233.1
	30. Ningxia	2102.21	1644.7	321.3
	31. Xinjiang	6610.05	4632.1	771.5
	Total	74154.21	53747.8	10425.2*
Entire Nation	Total	521351.11	305834.6	63158.8

Source: Data are from China Statistical Abstract (2012) (XLS, DOC).

Note: *: These values are not the sums of the relevant data. Instead, they are the means of comparative gains of labor of relevant regions.

Table 2.2 The Regional Economic Structures and Measures of Comparative Gains of the 31 Provinces, Municipalities, and Autonomous Regions of East, Central, and West China (2011)

Area	*Province/ Municipality/ Autonomous Region (i)*	*Demand Structure $d_i(t)$ (%)*	*Employment Structure $s_i^1(t)$(%)*	*Comparative Gain of Labor $G_i^1(t)$*	*Investment Structure $s_i^2(t)$ (%)*	*Comparative Gain of Capital $G_i^2(t)$*
East China	1. Beijing	0.031	0.0339	0.9145	0.018	1.722
	2. Tianjin	0.022	0.0109	2.0183	0.023	0.957
	3. Hebei	0.047	0.0342	1.3743	0.054	0.870
	4. Liaoning	0.043	0.0424	1.0142	0.025	1.720
	5. Shanghai	0.037	0.0346	1.0694	0.016	2.313
	6. Jiangsu	0.094	0.1021	0.9207	0.087	1.080
	7. Zhejiang	0.062	0.0762	0.8136	0.046	1.348
	8. Fujian	0.037	0.0350	1.0571	0.032	1.156
	9. Shandong	0.087	0.0683	1.2738	0.087	1.000
	10. Guangdong	0.102	0.0941	1.0840	0.056	1.821
	11. Hainan	0.005	0.0061	0.8197	0.005	1.000
	Total	0.5629	0.5378	1.0467*	0.4839	1.1633**
Central China	12. Shanxi	0.022	0.0194	1.1340	0.023	0.957
	13. Inner Mongolia	0.028	0.0177	1.5819	0.034	0.824
	14. Jilin	0.020	0.0203	0.9852	0.024	0.833
	15. Heilongjiang	0.024	0.0265	0.9057	0.024	1.000
	16. Anhui	0.029	0.0321	0.8411	0.041	0.707
	17. Jiangxi	0.022	0.0308	0.7143	0.030	0.733
	18. Henan	0.052	0.0458	1.1354	0.058	0.897
	19. Hubei	0.038	0.0428	0.8879	0.041	0.927
	20. Hunan	0.038	0.0382	0.9948	0.039	0.974
	21. Guangxi	0.022	0.0236	0.9322	0.026	0.846
	Total	0.2948	0.2972	0.9919*	0.3404	0.8660**
West China	22. Chongqing	0.019	0.0250	0.76	0.024	0.792
	23. Sichuan	0.040	0.0419	0.9547	0.047	0.851
	24. Guizhou	0.011	0.0125	0.88	0.014	0.786
	25. Yunnan	0.017	0.0278	0.6115	0.020	0.850
	26. Tibet	0.001	0.0024	0.4167	0.002	0.500
	27. Shaanxi	0.024	0.0226	1.0619	0.031	0.774
	28. Gansu	0.010	0.0118	0.8475	0.013	0.769
	29. Qinghai	0.003	0.0037	0.8108	0.005	0.600
	30. Ningxia	0.004	0.0051	0.7843	0.005	0.800
	31. Xinjiang	0.013	0.0122	1.0656	0.015	0.867
	Total	0.142	0.1651	0.8601*	0.1757	0.8082**

Source: Data are from Shanghai Statistical Abstract (2012) (XLS, DOC).

Note: * and **: These values are not the sums of the relevant data. Instead, they are the means of relevant regions, where * the means of the comparative gains of labor and ** the means of comparative gains of capital of relevant regions.

as imbalanced, then other than a small number (10 to be exact) of provinces, municipalities, and autonomous regions that are situated in balanced or basically balanced states, all the rest, over 52/62 = 84%, are classified as regions with supply–demand imbalances. In particular, Tibet, Shanghai, Tianjin, Beijing, Guangdong, Zhejiang, Fujian, Qinghai, Yunnan, Jiangxi, Anhui, Chongqing, Gansu, Guizhou, Ningxia, Shaanxi, Henan, Guangxi, Sichuan, and other provinces, municipalities, and autonomous regions suffer from severe supply–demand imbalances with Tibet having the lowest comparative gains of labor (0.4167) and Tianjin the highest comparative gains of labor (2.0183). That implies that both Tianjin and Tibet represent the two extreme levels of productivity in Chinese economy. The comparative gain of capital of Tibet is the lowest (0.500), while the value of Shanghai is the highest (2.313). That indicates that Shanghai and Tibet stand for the two extremes of the most and least efficient results of investment.

In terms of the three greater regions of East, Central, and West China, the data in the tables show that although the supplies of labor and capital have increased tremendously when compared to the past, say, those in 2008 or those in particular in 1999, (for details, see the fourth part of this section: Achievements and Lessons of the Regional Economic Development), severe imbalances still exist. The mean of the comparative gains of labor of the 11 provinces and municipalities of East China is 1.0467. That means that the regional labor structures are basically balanced with quite good effects, while the comparative gain of capital is 1.1633 ($\gg$1). That together means that East China is the region with the best values of labor and capital, and the development direction of the social productivity of the current time. At the same time, it also indicates that the supplies of labor and capital to the east area are relatively insufficient; additional labor needs to be introduced into this area. That constitutes the focus of the next round of investments.

The mean comparative gain of labor of the 10 provinces and regions of the central area is equal to 0.9919 ($\approx$1) and the mean comparative gain of capital is 0.8660 ($\ll$1). These values indicate that the supply of labor in this area is basically balanced. However, the supply of capital is severely excessive. This excess of capital should be transferred to the east area as quickly as possible. For the 10 provinces, municipalities, and regions of the west area, other than Xinjiang and Shaanxi, where the supply of labor is slightly insufficient, both the supply of labor ($0.8601 \ll 1$) or the supply of capital ($0.8082 \ll 1$) are in abundant excess. It represents the area with the lowest social productivity, the worst return on capital, and the poorest and most backward region of China.

Since comparative gain is defined to be the ratio of two percentages, the gain = 1 means that the supply is equal to the demand, and that the supply is balanced with the demand. When the comparative gain of labor of a certain region is greater than 1, it means that this region has produced relatively more social demand with relatively less amount of labor. So, the region's productivity of labor is relatively high, and the comparative gain of labor is relatively greater. For example, Tianjin occupies 1.09% of the national labor, while it creates 2.2% of the social demand of the nation. Hence, the comparative gain of labor for Tianjin is equal to 2.0183 (= 0.0220/0.0109). That explains why the labor productivity of Tianjin is far beyond that of any other 30 provinces, municipalities, and autonomous regions of China. The labor productivity of Tibet is the lowest, where the comparative gain of labor is only equal to 0.4167 (= 0.001/0.0024). That explains that with 0.24% occupation of the national labor Tibet only produces 0.1% social demand of the nation.

So, based on the magnitudes of comparative gains of labor, the 31 provinces, municipalities, and autonomous regions of China can be ordered as follows:

1	2	3	4	5	6	7	8
Tianjin	Inner Mongolia	Hebei	Shandong	Henan	Shanxi	Guangdong	Shanghai
9	10	11	12	13	14	15	16
Xinjiang	Shaanxi	Fujian	Liaoning	Hunan	Jilin	Sichuan	Guangxi
17	18	19	20	21	22	23	24
Jiangsu	Beijing	Heilongjiang	Hubei	Guizhou	Gansu	Anhui	Hainan
25	26	27	28	29	30	31	
Zhejiang	Qinghai	Ningxia	Chongqing	Jiangxi	Yunnan	Tibet	

Similarly, if a region's comparative gain of capital is greater than 1, it means that this region produces relatively more social demand with relatively less input of capital. So, the return on capital for this region is relatively good. For example, with an occupation of 1.6% of the national investment, Shanghai creates 3.7% of the national social demand. Hence, the comparative gain of capital for Shanghai is 2.313 (= 0.037/0.016). That implies that the effect of investment in Shanghai is far beyond that of any other 30 provinces, municipalities, and autonomous regions of China. On the other hand, the return on capital in Tibet is the worst, where the total amount invested in Tibet in 2011 is equal to 0.2% of the total national investment, however, it only creates 0.1% of the national social demand.

So, based on comparative gains of capital, the 31 provinces, municipalities, and autonomous regions of China can be ordered as follows:

1	2	3	4	5	6	7	8
Shanghai	Guangdong	Beijing	Liaoning	Zhejiang	Fujian	Jiangsu	Shandong
9	10	11	12	13	14	15	16
Hainan	Heilongjiang	Hunan	Tianjin	Shanxi	Hubei	Henan	Hebei
17	18	19	20	21	22	23	24
Xinjiang	Sichuan	Yunnan	Guangxi	Jilin	Inner Mongolia	Ningxia	Chongqing
25	26	27	28	29	30	31	
Guizhou	Shaanxi	Gansu	Jiangxi	Anhui	Qinghai	Tibet	

2.1.3 The Main Results of this Section

Based on the previous discussion of the three greater geographic areas of China, Table 2.2 implies that no matter how we make the comparison, either using comparative gains of labor or the comparative gains of capital, the three greater areas can be ordered as follows:

East China, Central China, and West China.

Therefore, in terms of these three greater areas, the optimal area for investment and employment purposes is the east, the general area is the central plain, and most difficult area is the west.

This is the current state of the orderings of the greater east, central and west areas of China, and their 31 provinces, municipalities, and autonomous regions, based on their regional economic structures and comparative gains of labor and comparative gains of capital. It also represents a good reference for the development direction of employment and investment in mainland China.

2.1.4 Achievements and Lessons of the Regional Economic Development

Similar to what are shown previously, we have also, respectively, evaluated the regional economic structures and calculated the measures of comparative gains for the year 1999 and for the year 2008 by using the relevant data of the greater areas of East, Central, and West China, and the relevant 31 provinces, municipalities, and autonomous regions. For details, see Tables 2.3 and 2.4. A preliminary report can be found in Tao (2001a), where some policy suggestions regarding how to optimize the then current economic structures were given.

By comparing Table 2.2 with Tables 2.3 and 2.4, we can see what experiences and lessons can be and should be learned from the evolution of China's regional economic structures in the past ten plus years so that the future adjustment and optimization of these regional economic structures can be carried out with more precision.

First, let us look at the structural evolution of China's East, Central, and West areas.

In terms of the employment structure, the employment rate of the east area had risen to 40.72% in the year 2008 from 37.74% in the year 1999, and reached 53.78% in 2012. This indicates that during all these years the east area has indeed received a good number of labors from the central and west areas. Specifically, the attraction to the labor of central and west areas has been particularly strong.

In terms of the comparative gains of labor, although the changes were quite minor, the east area experienced some slight drop (from 1.5003 to 1.4337), the central area enjoyed some slight increase (from 0.7620 to 0.7767), and the west area suffered from a continued fall (from 0.6274 to 0.5822). If these changes are reflected in the angle of income distribution, it can be seen that before 2008 the distribution discrepancies were quite large. In particular, the income difference between the east area and other two areas was once enlarged.

However, in recent years, the phenomenon of comparative gains of labor has been improved and the income difference between the areas has been disappearing quickly.

Although the comparative gains of labor of the east area have experienced some drop (from 1.5003 to 1.4337 then to 1.0467, the comparative gains of the central area experienced a large magnitude of increase (from 0.7620 to 0.7767 and then to 0.9919), which is nearly equal to the state of optimization. In comparison, not only did the west area end its fall in its comparative gains of labor, but also did it enjoy a much larger magnitude of increase (falling from 0.6274 to 0.5822 and then rising to 0.8601). This fact means that after over a decade of economic structural adjustment, China's employment structure has been evolving successfully along the direction of optimization with noticeable achievements.

In terms of investment structures, the proportion of investment given to the central and west areas within the entire nation's grand total has been increasing in the past ten plus years, which changed from the original 0.2467 and 0.1551 to 0.3183 and 0.1555, respectively. However, the comparative gains of capital for these areas have been disappointing to say the least. They fell respectively from the original 1.1994 and 0.8923 to 0.8916 and 0.8520. On the other hand, for the east area, the opposite held true: The comparative gains of capital increased from the original 0.9465 to 1.1093. That implies that although in the past ten plus years the investment in the east

Table 2.3 The Regional Economic Structures and Measures of Comparative Gains of the 31 Provinces, Municipalities, and Autonomous Regions of East, Central, and West China (2009)

Area	*Province/ Municipality/ Autonomous Region (i)*	*Demand Structure $d_i(t)$ (%)*	*Employment Structure $s_i^1(t)$ (%)*	*Comparative Gain of Labor $G_i^1(t)$*	*Investment Structure $s_i^2(t)$ (%)*	*Comparative Gain of Capital $G_i^2(t)$*
East China	1. Beijing	0.0251	0.0100	2.5100	0.0404	0.6213
	2. Tianjin	0.0165	0.0067	2.4627	0.0199	0.8291
	3. Hebei	0.0521	0.0544	0.9577	0.0611	0.8527
	4. Liaoning	0.0476	0.0287	1.6585	0.0386	0.2332
	5. Shanghai	0.0460	0.0108	4.2593	0.0641	0.7176
	6. Jiangsu	0.0878	0.0575	1.5270	0.0843	0.0415
	7. Zhejiang	0.0612	0.0426	1.5376	0.0676	0.9053
	8. Fujian	0.0405	0.0261	1.6590	0.0374	0.0829
	9. Shandong	0.0874	0.0752	1.1622	0.0767	1.1395
	10. Guangdong	0.0966	0.0602	1.6047	0.1014	0.9527
	11. Hainan	0.0054	0.0052	1.1154	0.0067	0.8050
	Total	0.5662	0. 3774	1.5003*	0. 5982	0.9465**
Central China	12. Shanxi	0.0172	0.0230	0.7478	0.0165	0.0424
	13. Inner Mongolia	0.0145	0.0163	0.8896	0.0120	1. 2083
	14. Jilin	0.0190	0.0176	1.0795	0.0173	1. 0983
	15. Heilongjiang	0.0331	0.0269	1.2305	0.0259	1. 2780
	16. Anhui	0.0332	0.0530	0.6264	0.0243	1. 3663
	17. Jiangxi	0.0224	0.0314	0.7643	0.0157	1. 4268
	18. Henan	0.0522	0.0833	0.6267	0.0417	1. 2518
	19. Hubei	0.0440	0.0412	1.0680	0.0428	1. 0280
	20. Hunan	0.0380	0.0559	0.6798	0.0305	1. 2459
	21. Guangxi	0.0223	0.0397	0.5617	0.0200	1. 1151
	Total	0.2959	0. 3883	0.7620*	0. 2467	1.1994**
West China	22. Chongqing	0.0169	0.0262	0.6450	0.0181	0.9337
	23. Sichuan	0.0423	0.0717	0.6318	0.0423	1.000
	24. Guizhou	0.0104	0.0316	0.3639	0.0108	0.9630
	25. Yunnan	0.0212	0.0364	0.6236	0.0229	0.9258
	26. Tibet	0.0012	0.0020	0.6000	0.0018	0.6667
	27. Shaanxi	0.0170	0.0285	0.6386	0.0203	0.8374
	28. Gansu	0.0106	0.0190	0.6000	0.0123	0.8618
	29. Qinghai	0.0027	0.0039	0.6423	0.0040	0.6750
	30. Ningxia	0.0028	0.0043	0.6512	0.0044	0.6364
	31. Xinjiang	0.0133	0.0107	1.2430	0.0182	0.7316
	Total	0.1384	0. 2343	0.6274*	0. 1551	0.8923**

Source: Data are from Shanghai Statistical Abstract (2012) (XLS, DOC).

Note: * and **: These values are not the sums of the relevant data. Instead, they are the means of relevant regions, where * the means of the comparative gains of labor and ** the means of the comparative gains of capital of relevant regions.

Table 2.4 The Regional Economic Structures and Measures of Comparative Gains of the 31 Provinces, Municipalities, and Autonomous Regions of East, Central, and West China (2008)

Area	*Province/ Municipality/ Autonomous Region (i)*	*Demand Structure* $d_i(t)$ *(%)*	*Employment Structure* $s_i^1(t)$ *(%)*	*Comparative Gain of Labor* $G_i^1(t)$	*Investment Structure* $s_i^2(t)$ *(%)*	*Comparative Gain of Capital* $G_i^2(t)$
East China	1. Beijing	0.0321	0.0161	1.9938	0.0235	1.3660
	2. Tianjin	0.0194	0.0069	2.8116	0.0235	0.8255
	3. Hebei	0.0495	0.0501	0.9880	0.0526	0.9411
	4. Liaoning	0.0411	0.0288	1.4271	0.0626	0.6565
	5. Shanghai	0.0419	0.0123	3.4065	0.0340	1.2324
	6. Jiangsu	0.0926	0.0601	1.5408	0.0904	1.0243
	7. Zhejiang	0.0657	0.0506	1.2984	0.0587	1.1193
	8. Fujian	0.0331	0.0285	1.1614	0.0351	0.9430
	9. Shandong	0.0950	0.0734	1.2943	0.0886	1.0722
	10. Guangdong	0.1091	0.0751	1.4547	0.0715	1.5259
	11. Hainan	0.0045	0.0057	0.7895	0.0046	0.9783
	Total	0.5838	0.4072	1.4337*	0.5263	1.1093**
Central China	12. Shanxi	0.0212	0.0217	0.9770	0.0225	0.9422
	13. Inner Mongolia	0.0237	0.0151	1.5695	0.0351	0.6695
	14. Jilin	0.0196	0.0157	1.2484	0.0351	0.5584
	15. Heilongjiang	0.0254	0.0229	1.1092	0.0229	1.1092
	16. Anhui	0.0271	0.0493	0.5497	0.0269	1.0074
	17. Jiangxi	0.0198	0.0305	0.6492	0.0201	0.9851
	18. Henan	0.0563	0.0800	0.7038	0.0654	0.8609
	19. Hubei	0.0346	0.0394	0.8782	0.0336	1.0298
	20. Hunan	0.0341	0.0523	0.6520	0.0343	0.9942
	21. Guangxi	0.0219	0.0385	0.5688	0.0223	0.9821
	Total	0.2838	0.3654	0.7767*	0.3183	0.8916**
West China	22. Chongqing	0.0156	0.0252	0.6190	0.0227	0.6872
	23. Sichuan	0.0382	0.0668	0.5719	0.0412	0.9272
	24. Guizhou	0.0102	0.0316	0.3228	0.0106	0.9623
	25. Yunnan	0.0174	0.0367	0.4741	0.0165	1.0545
	26. Tibet	0.0012	0.0022	0.5455	0.0020	0.6000
	27. Shaanxi	0.0209	0.0267	0.7828	0.0286	0.7308
	28. Gansu	0.0053	0.0190	0.2789	0.0109	0.4862
	29. Qinghai	0.0029	0.0038	0.7632	0.0037	0.7838
	30. Ningxia	0.0034	0.0042	0.8095	0.0054	0.6296
	31. Xinjiang	0.0128	0.0112	1.1429	0.0140	0.9143
	Total	0.1324	0.2274	0.5822*	0.1554	0.8520**

Source: Data are from Shanghai Statistical Abstract (2012) (XLS, DOC).

Note: * and **: These values are not the sums of the relevant data. Instead, they are the means of relevant regions, where * the means of the comparative gains of labor and ** the means of the comparative gains of capital of relevant regions.

area has dropped when compared to the grand total investment nationwide, the effect of the investment in the east area has risen clearly. Specifically, the return on investment in Shanghai has been ranked number 1 that is a quite distant above number 2 in the nation with the comparative gain of capital = 2.313.

Second, let us look at the structural evolution of the 31 provinces, municipalities, and autonomous regions.

In terms of the comparative gains of labor, the most recent ordering has been given above. However, in the year 1999, the ordering from the highest to the lowest was the following:

1	2	3	4	5	6	7	8
Shanghai	Beijing	Tianjin	Fujian	Liaoning	Guangdong	Zhejiang	Jiangsu
9	10	11	12	13	14	15	16
Xinjiang	Heilongjiang	Shandong	Hainan	Jilin	Hubei	Hebei	Inner Mongolia
17	18	19	20	21	22	23	24
Jiangxi	Shanxi	Hunan	Ningxia	Chongqing	Qinghai	Shaanxi	Sichuan
25	26	27	28	29	30	31	
Henan	Anhui	Yunnan	Tibet	Gansu	Guangxi	Guizhou	

In this list, what changed most in the past ten plus years are Inner Mongolia, Beijing, Fujian, and Shanghai. Inner Mongolia moved from the 16th position in 1999 to the 4th position in 2008 and the current 2nd position, while Beijing, Shanghai, and Fujian, respectively, moved from the 1st, 2nd, and the 4th positions in 1999 to the most recent 8th, the 18th, and the 11th positions. As for all other provinces, municipalities, and autonomous regions, their respective rakings basically stayed the same except some minor changes.

In terms of the comparative gains of capital, the ordering from the highest to the lowest in 1999 was:

1	2	3	4	5	6	7	8
Jiangxi	Anhui	Heilongjiang	Henan	Hunan	Liaoning	Inner Mongolia	Shandong
9	10	11	12	13	14	15	16
Guangxi	Jilin	Fujian	Shanxi	Jiangsu	Hubei	Sichuan	Guizhou
17	18	19	20	21	22	23	24
Guangdong	Chongqing	Yunnan	Zhejiang	Gansu	Hebei	Shaanxi	Tianjin
25	26	27	28	29	30	31	
Hainan	Xinjiang	Shanghai	Qinghai	Tibet	Ningxia	Beijing	

Changes in this respect have three major characteristics. First, other than isolated cases, such as Tianjin and Hebei, the returns on investment for the east area all increased from the original states of not good, where the comparative gains of capital were once far less than 1, to either good or supply–demand balance. The most change occurred to Beijing, which jumped from the original last place in 1999 to the second place in 2008 and the most recent third place. On the other hand, Guangdong, Shanghai, Zhejiang, Jiangsu, Hainan, and Shandong moved upward, respectively, from the original 17th, 27th, 20th, 13th, 25th, and 8th places in 1999 to the 1st, 3rd, 4th, 9th, 14th, and 6th places in 2008, and the most recent 2nd, 1st, 5th, 7th, 9th, and 8th places. The second characteristic is that in the east area, other than Heilongjiang, Hubei, and Anhui, all fell from the ranking of high efficiency (with their respective comparative gains of capital much greater than) 1 to low efficiency (with their respective comparative gains far less than 1). Here the most change happened to Jiangxi, which fell from the original first place in 1999 to the 12th place in 2008 and the most recent 28th, which is the fourth to the end of the list. As for Henan, Hunan, Inner Mongolia, Jilin, and Shanxi, they fell respectively from their 4th, 5th, 7th, 10th, and 12th places in 1999 to 21st, 11th, 26th, 30th, and 17th places in 2008, and to the most recent 15th, 11th, 22nd, 21st, and 13th places. The third characteristic is that in the west area, other than Xinjiang with a major improvement in its comparative gains (it rose from the 26th place in 1999 to the 20th place in 2008 and to the most recent 17th place), not only did the effects of investment in all other provinces, municipalities, and autonomous regions not improve, but also they were ranked toward the end of the ordering when compared to other regions from around China. Also, the comparative gains of capital of the entire west area dropped from the original 0.8923 in 1999 to 0.8520 in 2008 and to the most recent 0.8082.

Third, let us look at the problems that should be addressed and lessons that should be learned from the evolution of China's economic structure

Lesson # 1: The main reason for the existence of imbalances in China's economic structure is the past way of allocating resources when planned economy was applied.

Lesson #1.1: There are difficult hurdles for human resources to flow.

Since there has been a shortage of labor resource in the east area, human talents should have been allowed to relocate freely without any constraint, such as the household registration system, employment system, and others. The labor resource naturally flows to geographic places, industries, economic sectors, and companies with higher comparative gains of labor. So, the supply–demand balance of the labor resource can basically be reached by replying on the market rules of the economy. What the government and the society can do is to relax the household registration system, and the constraints of the employment system, while increasing the investment in education and job training so that more workers can choose to relocate freely, while having the capabilities and skills to adapt to structural transformations and upgrades of industries. In fact, after having experienced the policy adjustments in recent years, the relevant obstacles in this regard have been correspondingly relaxed, leading to the significant progresses and achievements in the recent years. Even so, further relaxation is necessary in order to allow labors to move more freely.

Lesson #1.2: The blindness and freeness of planned investments.

Over the past 60 plus years since the time modern China was in existence, China experienced over 30 years of the blindness and freeness of investment in its planned economic system. Despite the recent changes in this tradition, but little has actually improved. That is the fundamental reason why imbalances in the effect of investment returns have appeared. There have not been sufficient reasoning and justification done before making an investment on such questions as: what the return on the investment is? Where the investment should be made? How much should be committed? What projects should be funded and what should not? For the worthy projects, how much should be invested? How soon the investment can be recovered? What the risk of losing the investment? and so on. Otherwise, knowing from the fact that the west area has poor investment returns and that the area is the most difficult place to invest, why should anyone continue to increase the investment in West China? And, why could the increased investment to the central plain of China have caused wide ranging decline in the investment returns? According to the data of 1999, it should be admitted that the choices of investment regions are not wrong, but the problem might very well be about the selected investment projects. Otherwise, how could an excess in investment have appeared? And, how could the phenomenon of large-scale, wide-ranging decline in the investment efficiency have appeared?

Lesson #1.3: Strong political colors exist in economic decision makings.

Strong political color in economic decision making has been one of the important drawbacks of China's planned economy. That is why such decision as increasing investment to central plain and west area of China was made. The important reason might be rooted in the two strategic decisions on the development of the west and the rise of central plain made by the central government at the beginning of this century. In fact, in the next 5 years, 10 years, or even longer period of time, if the central government wants to play its role in the adjustment of optimization of China's regional structures, then its focus of attention should not be on the strategy of the so-called development of the west. And, it should not do anything that is in the opposite direction of the strategic optimization of China's regional structures as described in Equation 2.3. Instead, it should concentrate its main attention on completing and perfecting afore-mentioned markets, policies, and the relevant system of regulations so that the market forces can play the role of naturally distributing resources. In the area of macroeconomic regulation and control, the government can and should provide guidance in terms of regional resource allocation and even can get directly involved with human resources and investments. However, no matter whether the government is involved with the distribution of resources directly or indirectly, it should follow the guidance of the concept of comparative gains of labor and capital in order to overcome the problem of making blind decisions (Tao, 2001a).

As a matter of fact, from Table 2.4, it follows that even more than 4 years ago, the comparative gains of capital of the central plain and the west area had been far smaller than 1, which were 0.8916 and 0.8520, respectively. This indicated the fact that the investments in these areas had been in abundant excess. Even so, the investments in these two areas continued to increase proportionally when compared to the entire nation. They increased, respectively, from 31.83% and 15.54% in 2008 to the most recent 34.04% and 17.57%.

Lesson # 2: Choose the path of expanding the domestic demand by lifting the levels of income and consumption of the Chinese people.

The current economic states of China's regional structures, as shown in Tables 2.2 through 2.4, indicate that all the provinces, municipalities, and autonomous regions that are located along the national border, such as those neighboring a foreign nation or next to the ocean, tend to have better gains from investment and labor. These regions include Liaoning, Hebei, Beijing, Tianjin, Shandong, Jiangsu, Shanghai, Zhejiang, Fujian, Guangdong, and Guangxi of the coastal provinces and municipalities of the southeast, Yunnan province of the southwest, Xinjiang of the northwest, Heilongjiang and Jilin of the Northeast, and Inner Mongolia and Shanxi of the north. At the same time, the provinces, municipalities, and autonomous regions that are not adjacent to national borders, such as Jiangxi, Anhui, Henan, Shaanxi, Qinghai, Hunan, Sichuan, Chongqing, Guizhou, and others, tend to suffer from poor gains of labor and capital. For both Ningxia and Tibet, although they are also physically located on the border with India, Mongolia, and other foreign countries, because of their difficult conditions of transportation with these foreign economies they do not have much trade exchanges with the foreign lands so that their gains of labor and capital have been quite poor.

This explains that since the reform was implemented, especially during the decade-plus years since entering the new century, the economic growth of China has been mainly dependent on the income from foreign trades and exports without much influence from the domestic demand. Along with the development and changes of the world economy, in order to maintain its viable, long-term, stable, and fast growth, China can no longer continue to rely on its economic development model that is dominated mainly by exports. Instead, it should make relevant adjustment, where the key is to expand the domestic demand. China has the largest human population in the world, and an extremely huge consumption space and scale. However, the current problem is that the levels of income and consumption of Chinese people are too low, causing the effective demand to be very low. This fact tells us that there is an urgent need to raise the levels of incomes and consumption for those low and middle income residents, especially those people who live in the countryside, so that the domestic demand can be expanded. That is not only a long-term strategy for China to maintain a long-term, sustaining, stable, healthy, and fast growth, it also represents the strategic selection for China to adjust and to optimize its regional economic structures.

2.1.5 Directions, Focuses, and Measures for Strategic Optimization

First, let us look at the strategic direction and path.

Based on the theory presented in Chapter 1, the following result describes the optimal structure for a macroeconomic system: The macroeconomic system is of its optimal structure if, and only if, for any $i \in I = \{1,2,\ldots,n\}$ and $j = 1, 2$, the following equation holds true

$$G_i^j(t) = 1 \quad \text{or} \quad d_i(t) = s_i^j(t) \tag{2.2}$$

From this theorem, it follows readily that the strategic direction of optimization for the structure of a macroeconomic system, including any regional economic structure, is that if for any $i \in I = \{1,2,\ldots,n\}$ and $j = 1, 2$, within their corresponding time periods, the structure needs to make

$$G_i^j(t) \to 1 \quad \text{or} \quad s_i^j(t+1) \to d_i(t) \tag{2.3}$$

which means that the allocation of resources, such as those of labor and capital, of the next time period should be determined by the demand of the current time period. That is, the supply is determined by the signaled demand. This conclusion coincides perfectly with how structural optimization is materialized from employing the planning and programming budgeting system (PPBS) that has been widely applied in the western world, which specifically stipulates that investment is determined by the output.

There are two ways to walk through the strategic optimization of regional economic structures as described by the previous equation.

The first way is to allow the market mechanism to automatically make the necessary adjustment and resource allocation. The second way is to make the oriented adjustment and resource allocation through a planned mechanism. The former way allows people to be propelled along the direction of the forces of the comparative gains so that resources flow spontaneously and freely without being interfered administratively. By employing this approach, the excessive resources of underdeveloped regions (provinces, municipalities, and autonomous regions) should naturally flow to more developed areas. The second way means that the government plays a role in the process of structural adjustment in order to realize structural balance (optimization). That is to establish a completely competitive market system with good markets, factors markets, and so on, where rules of market competition are well formulated, the environment of market competition is well maintained, macroeconomic policies and strategies for regional economic development are introduced so that the excessive resources of the areas of underdeveloped productivities are purposefully and consciously transported to areas with relatively more developed productivities. By doing so, structural balances in regional economies will be facilitated.

Second, let us look at the strategic focus and difficulty.

The so-called optimization of an economic structure stands for such a process for the originally unbalanced economic structure to become basically balanced through various methods or measures. Based on the current states of China's regional economic structure, as described above, and according to an analysis on the previously presented data, the emphases and difficulties of China's regional economic structures should include:

1. The movement of labor, especially the movement of labor in the countryside, represents both an emphasis and a difficulty. It is because the movement involves not only an adjustment of all interest groups, but also a lift in the skill set of the country labor.
2. In terms of the distribution of labor, the excessive labor of the central plain and the west area should be moved eastward as much as possible in order to achieve a balance in the supply and demand relationship. The emphasis order of the labor should be as follows:

   Tianjin, Inner Mongolia, Hebei, Shandong, Henan, Shanxi, Guangdong, Shanghai, Xinjiang, Shaanxi, Fujian, and Liaoning.
3. In the area of capital allocation, no matter whether it is the introduction of foreign capital or the investment of domestic funds, the main investment direction in the coming years will be in the regions of the east area and provinces and regions along the national border, especially in the provinces, municipalities, and autonomous regions with high returns on investments. According to their relative emphases, these regions can be ordered as follows:

   Shanghai, Guangdong, Beijing, Liaoning, Zhejiang, Fujian, Jiangsu, Shandong, Hainan, and Heilongjiang.

The principle behind economic structural adjustments is that the further away the comparative gain of a province, or a municipality, or an autonomous region is from 1, the more adjustment emphasis the province, the municipality, or the autonomous region should be given, and the more adjustment strength the province, the municipality, or the autonomous region should acquire.

It is our belief that if in the process of future reforms the regional economic structures can be measured and calculated realistically, promptly, and accurately by using the model in Equation 2.1, and if macroeconomic adjustments can be carried out unswervingly in the direction of $G_i^j(t) \rightarrow 1$, then along with the strategic adjustments and optimization of the relevant sectorial structures, urban–rural structures, ownership structures, and other structures, strategic optimizations of the regional economic structures of China's east, central, and west areas can be materially realized in the near future.

2.2 Economic Sectors of China

2.2.1 Current State of Investment and Employment and Strategic Optimization

The afore-described theory, model, and method can also be employed to explore the investment and employment opportunities in various economic sectors. To this end, all needed is to compute and associate the current states (see Table 2.5) and relevant comparative gains (see Table 2.6) of China's economic sectorial structures (i.e., the supply–demand structures of economic sectors), sectorial investment, and employment structures. All relevant methods of computation are the same as used above. The results are summarized in Tables 2.6 and 2.7.

Both Tables 2.6 and 2.7 indicate that the overall picture of the supply and demand of China's top three economic sectors for both 2008 and 2011 have shown improvements. In fact, among the six comparative gains of labor and capital, four were evolving in the right directions where the comparative gains of the primary and secondary economic sectors and the comparative gains of capital of the primary and the tertiary sectors were approaching 1, and other than the comparative gain of capital of the primary sector all other three gains did not improve much. This end implies

Table 2.5 The Total Output Values, Investments on Fixed Assets, and State of Employments in China's Top Three Economic Sectors (for the Year of 2011)

Sectors (i)	*Total Output of Sector (0.1 Billion Yuan)*	*Total Investment on Fixed Properties (0.1 Billion Yuan)*	*Employment (10,000 People)*
Primary sector	47712.0	5064.5	30654
Secondary sector	220591.6	76961.3	21109
Tertiary sector	203260.1	90802.7	25717
Total	471563.7	172828.4	77480

Source: Data from *China Statistics Abstract* 2012 (XLS, DOC).

Table 2.6 China's Economic Sectorial Structure and Relevant Measures of Comparative Gains (t = Year of 2011)

Sectors (i)	*Demand Structure $d_i(t)$ (%)*	*Employment Structure $s_i^1(t)$ (%)*	*Comparative Gain of Labor $G_i^1(t)$*	*Investment Structure $s_i^2(t)$ (%)*	*Comparative Gain of Capital $G_i^2(t)$*
Primary sector	10.1	34.8	0.29	6.1	1.66
Secondary sector	46.8	29.5	1.78	40.1	1.17
Tertiary sector	43.1	35.7	1.21	53.8	0.80
Total	100	100	—	100	—

Source: Data from *China Statistics Abstract* 2012 (XLS, DOC).

Table 2.7 China's Economic Sectorial Structure and Relevant Measures of Comparative Gains (t = Year of 2008)

Sectors (i)	*Demand Structure* $d_i(t)$ *(%)*	*Employment Structure* $s_i^1(t)$ *(%)*	*Comparative Gain of Labor* $G_i^1(t)$	*Investment Structure* $s_i^2(t)$ *(%)*	*Comparative Gain of Capital* $G_i^2(t)$
Primary sector	11.31	39.56	0.2859	2.93	3.86
Secondary sector	48.62	27.24	1.7849	44.53	1.0918
Tertiary sector	40.07	33.20	1.2037	52.54	0.7627
Total	100	100	—	100	—

Source: Data from China Statistics Yearbook 2009, Edited by D. Guo and Z. L. Xiong, 2009, published by Press of China Statistics, Beijing.

that the adjustment on China's economic sectorial structure during the three years from 2008 and 2011 had seen positive effects. However, all of the six gains still suffered from severe imbalances, because $|G_i^j - 1| \gg 0.05$, for $i = 1, 2, 3, j = 1, 2$. In particular, the comparative gains of labor and capital of the primary sector were still far away from 1. This means that it is still a challenging task to adjust and to optimize China's economic sectorial structure, where additional intensity is needed along the right direction.

In terms of the comparative gains of labor, although the labor supply of the primary sector in 2011 was slightly better than that of 2008, there were still severe excesses ($G_i^j(t) < 0.2860 < 0.29 \ll 1$). The comparative gains were not respectively as good as those (0.29/1.59<)1/5 and (0.29/1.21<)1/4 of the secondary and tertiary sectors. However, the labor supplies of the secondary and tertiary sectors were far from sufficient ($G_2^1(t) \gg 1$). This implies not only that the levels of productivity of the primary sector, which is the agricultural production, are still too low, the labor resources in the rural areas are too much in excess, while the relevant effective demands are severely insufficient, but also that the primary sector (agriculture) is a relatively poor subsystem, where the interests of a lot of rural residents were adversely affected. So, the overall national economy and the stability of the society had to be affected. That also implies that the employment structures have to be adjusted as quickly as possible so that the large amount of excessive labors of the primary economic sector (agriculture) can be transferred to the secondary and the tertiary economic sectors.

In terms of the comparative gains of capital, the data in the tables show that although the investment situation in the tertiary economic sector in 2011 had been improved slightly from that in 2008, there was still excessive investment ($G_3^2(t) = 0.80 \ll 1$). As for the primary economic sector, although the investment situation in 2011 was improved quite majorly when compared to that in 2008, the magnitude of the investment was still severely insufficient ($G_1^2(t) = 1.66 \gg 1$), which was worse than 1/2 of the tertiary economic sector (0.80/1.66 < 0.49) and 71% of the secondary economic sector (1.17/1.66 < 0.705). That implies that additional investment of capital is still due to the primary economic sector (agriculture); the shortage has not only severely limited the growth of the primary economic sector, but also brutally damaged the economic interests of the majority of the rural population. Of course, this has been a common phenomenon in China. That is why the central government has repeatedly emphasized on the urgency of resolving the problems of agriculture. In particular, this implies that the investment structure needs to be adjusted as quickly as possible so that the large amount of excessive and ineffective investments in the tertiary economic sector can be transferred to the primary and the secondary economic sectors, especially the primary sector.

2.2.2 Shifts of Capital and Rural Labors: Keys of Optimization of Sectorial Structure

From the previous analysis of Table 2.6, it follows that the key to strategically optimize China's economic sectors is the transfer of capital and rural labor. On the one hand, the large amount of excessive labor in the primary economic sector (agriculture) needs to be transferred to the secondary and the tertiary economic sectors as quickly as possible. On the other hand, the scale of investment in the tertiary economic sector needs to be controlled, while more intensified and more increased investments should be going into the secondary and particularly the primary economic sector. To this end, the specific problem is how to materialize the desired transfer of capital and rural labor from one economic sector to another.

As is well known, it takes long time to adjust and to optimize the sectorial structure. Similar to other structural optimization problems considered in macroeconomic systems, it belongs to the category of such problems as long-term strategic planning that has to consider the overall system and its operations, where the direction, specific path, and emphasis of optimization need to be selected along with the objectives, difficulties, and intensity of the optimization. Since the relevant optimization policies, laws, and measures need to be established, it is a task to be successfully accomplished only over time, step after step toward the predetermined goals.

First, let us look at the strategic direction and path.

According to the previously stated theorem on the structural optimization of macroeconomic system, we have that the structure of a macroeconomic system is optimal if and only if for any $i \in I = \{1,2,\ldots,n\}, j = 1,2$, the following equation holds true:

$$G_i^j(t) = 1, \quad \text{that is, } d_i(t) = s_i^j(t) \tag{2.4}$$

and the corresponding direction of structural optimization of a macroeconomic system, including sectorial structures: For any $i \in I = \{1,2,\ldots,n\}, j = 1,2$, we let Equation 2.3 to be

$$G_i^j(t) \to 1, \quad \text{or} \quad s_i^j(t) \to d_i(t) \tag{2.5}$$

As a matter of fact, we have analyzed the state of evolution of China's employment structure by using the relevant data for 1985 and 1999 (Tao, 2001b). Our results clearly indicate that China, as a nation and as a culture, is moving forward nonstoppably to become a modern civilization, where the proportional weight of the primary economic sector is falling, while those of the secondary and tertiary economic sectors are rising. The force behind the changes is the comparative gains of labor and capital. The comparative gains (>1) of the secondary and tertiary economic sectors represent a significant attraction to the comparative gains (<1) of the primary economic sector. As stated earlier, in the recent past 30 plus years, the comparative gains of the primary economic sector have been falling further and further, forcing large amounts of labor of the primary economic sector (i.e., countryside) to rush unconsciously to relatively more developed coastal cities and the secondary and tertiary economic sectors, forming the noticeable waves of labor migrations. That direction of labor movement is exactly what the comparative gains of labor and capital are telling. In essence, this labor movement represents a tide of development seeking for societal righteousness, directed by an "invisible hand" to actively look for social balance. It is exactly such forces as comparative gains that stimulate the evolution and development of the society and the mankind.

There are two ways to materialize the strategic optimization of structure: One is to make use of the automatic adjustment and regulation of the market mechanism, and the other to employ the oriented adjustment and distribution of the mechanism of the planned economy.

As we have discussed earlier, the first way is to accept and follow the forces of the comparative gains so that various resources move and flow spontaneously and freely without any interference of the government. In such a case, in terms of the adjustment and optimization of sectorial structures, the excessive supplies of resources from backward economic productions, where the comparative gains are less than 1, will surely and naturally flow to more advanced economic productions.

The second way of materializing the strategic optimization of sectorial structures means that the government is fully responsible for the structural adjustment and plays an active role in the process of optimizing the structural balance (optimization). That is, the government develops and establishes a complete market system of free competition, consisting of commodity markets, various factors markets, and so on, with rules of competition well introduced so that the market order can be maintained. In particular, the government formulates the detailed plans, and introduces macroeconomic policies, including such policies that regulate the development of industry, and the development strategies for regional economies so that excessive resources of the industries with relatively backward production levels and qualities (where the comparative gains are less than 1) can be purposefully and consciously reallocated to industries with relatively more advanced levels and qualities of production, which will naturally encourage the formation of balanced sectorial structures.

Second, let us look at the strategic aims, emphases, and difficulties.

China has determined to reach the living quality of moderately developed countries for its dense population by the middle of this century. As for the top three economic sectors of the moderately developed countries, no matter whether it is their proportional weights in their respective domestic GDPs (the demand structures) or their proportional weights of employments in their corresponding national total employments (the employment structures), the basic situation is that the primary economic sector takes less than 10%, the secondary economic sector less than 40%, and the tertiary economic sector higher than 50%. Correspondingly, Table 2.7 shows that in terms of these two structures, China is still far from that of the moderately developed countries, especially in the structure of employment. Because of this reason, we believe that for China, its strategic objective for industrial adjustment and optimization should consist of three kinds: the near term, the mid-term, and long term. As for the near-term objective, by the year 2020, China's demand structure and employment structure should, respectively, reach 9%, 45%, 46%, and 28%, 33%, 39%. As for the mid-term objective, by the year 2030, China's demand structure and employment structure should, respectively, reach 8%, 42%, 50%, and 20%, 35%, 45%. As for the long-term objective, by the year 2045, China's demand structure and employment structure should, respectively, reach 6%, 38%, 56%, and 10%, 34%, 56%. This implies that on the basis of 2011, in the following 8 years, nearly 8.8% of the national labor force needs to gradually move out of the primary economic sector (from the countryside). So, the first adjustment focus of the employment structure is to transfer the excessive labor out of the primary economic sector, while the difficulty is how the secondary and the tertiary economic sectors would absorb the large magnitude of rural labor migrating from the primary economic sector.

Similarly, the objectives of the investment structures of China's primary, secondary, and tertiary economic sectors for 2020 should be set, respectively, at 8%, 46%, and 46%. This implies that on the basis of 2011, in the 8 years that are forthcoming, nearly 7.8% of the total investment in the entire society needs to be gradually moved out of the tertiary economic sector, where 1.9% should be transferred into the primary economic sector and 5.9% into the secondary economic sector. Hence, both the focus and difficulty expected for the adjustment of the investment structure are about how to move the excessive capital out of the tertiary economic sector. Evidently, in the coming years, the focus of China's investment will be directed to the primary and the secondary economic sectors. Especially, the reorientation of investment will aim at high-tech industries in these two economic sectors in the form of venture capitals.

Third, let us look at the specific strategic steps and measures.

From the criterion for optimal structure as stated in Equation 2.1, it follows that the strategic objective for the adjustment of China's sectorial structure of the next 8 years is still far away from the optimal structure (1,1,1). In particular, there are still large differences in the employment structures of the three economic sectors, and the investment structures of the primary and tertiary economic sectors. Even so, this strategic objective is set along the direction of reaching the structural optimization as outlined in Equation 2.3. Hence, what is established here can be used as a reference for accomplishing the initial adjustment objective of China's sectorial structure. Because of the severe industrial imbalances of the historical reason, it is not possible to optimize the structure with only one step. Instead, it will take quite a long period of time for China to gradually materialize its structural optimization for its industries.

In order for China to materialize the strategic optimization of its sectorial structure, according to our analysis above, the selection of macroeconomic systems and implementation of macroeconomic policies have to be done by completely altering the traditional beliefs and methods. The first is about the system of how human resource is allocated and how talents are employed. Correspondingly, in terms of the policies regarding employment, household registration, and others, major adjustment needs to be done in order to make the population and talents move freely. The second is to make sure that all the fiscal, monetary, and investment policies give preference to economic sectors with their respective comparative gains of labor and capital greater than 1 so that the limited resources can be effectively employed to produce additional GDP instead of being distributed without any focus producing wastes and inefficient output. The third is about the policies of distribution. These policies have to consider costs, efficiencies, and contributions instead of being purely fair. Of course, the most important and most key is what will be discussed in the following: Increase the investment, especially that of capital, in education, which represents the key for successfully transporting rural labors.

In order to speed up the strategic optimization process of China's sectorial structure, China needs to pay specific attention to three important factors that are influential to the process of adjustment and optimization of its sectorial structure. The first is its ability to migrate the rural labor; the second is its shortage of high-quality talents; and the third is where its limited capital is going to come from and where this limited resource is going to be used.

In terms of the severely imbalanced current state of China's sectorial structure, as shown in Table 2.6, we believe that

1. The process of optimization of China's sectorial structure is to a large degree determined by the speed and magnitude of movement of rural labors. Our work indicates that there are five factors that could potentially affect this movement: the quantity of supply (origins), the motivation of movement, capability of movement, space scale of movement, and the system's

resistance to the movement. Here, the "capability of movement" factor is the key, located at the central position of all the five factors. It is because this factor affects and dominates all other four factors. In particular, an increasing capability of movement can enlarge the space of movement (employment space) for the rural labors, enhance the motivation for rural labors to move, raise the supply of excessive rural labors, overcome to a certain degree some of the system's resistance to the movement of rural labors, and strengthen the stability of the movement of rural labors. Therefore, the movement capability of rural labors represents the dominant factor that affects the speed of China's recent urbanization. It is also the key factor that influences the optimization of China's sectorial structure.

This end means that the movement capability of rural labors needs to be raised. In other words, the deposit of rural human capitals needs to be increased. This deposit represents the "qualitative capability" of rural labors' movement. Along with the accelerated development of modernized productions, the renewing speed of knowledge and technology has been increasing, which imposes more and higher demands on the quality of labor. So, in order to smoothly move rural labors, these labors have to meet the qualitative requirements of the modernizing sectorial development; otherwise, there would be a gap between the supply of low quality rural labors and the societal demand for highly skilled workers, constituting a chasm between the quality threshold for employments and the human capitals of the rural and urban areas. The appearance of such a chasm will not be beneficial for the movement of the rural labors. It could further hinder the optimization progress of the sectorial structure in China. In other words, increasing the quality of rural labors and the level of rural human capital represents the basic guarantee for speeding up the optimization process of China's sectorial structure.

To this end, there is a need to accelerate the investment system toward education of the entire society, especially toward the education system in rural China with increased capital input so that the compulsory rural education systems will be financially funded by the relevant provinces instead of the relevant counties. At the same time, fiscal inputs toward the rural compulsory education and toward the job training of rural laborers should be increased. Additionally, the strength of fiscal supports to the education systems of the economically poor and backward regions of the central and west China needs to be raised so that all necessary supports for rural education and the salaries of the teachers of the rural education systems can be dispensed promptly. Also, there is a need to make an overall plan for the job training of rural laborers, to establish an agency to coordinate the training and movement of rural laborers, and, based on the market demand and societal demand, organize job training bases to complete the job training system in rural China. Through providing localized job training that is especially designed to meet the regional market demands, the quality of the rural laborers will be practically improved, and the capability of movement of these laborers will be inevitably lifted. That of course will help to satisfy the demand for labor in the optimization process of China's sectorial structure.

2. There is a need to create a good environment and a system, consisting of such components as labor, human resource, employment, household registration, and so on, that encourages free flow of human talents and creates employment opportunities of fair competition in any economic sector. This environment and system will be different from the one that currently exist and that hinders the workers of the primary economic sector (the countryside) to move into the secondary and tertiary economic sectors (the cities).
3. Capital represents a scarce resource in China's future development. For a long period of time, the limited capital has been wasted within the tertiary economic sector except education.

The investment focus of the present and future China should not continue to throw its scarce capital in the tertiary economic sector, where the return on investment has been extremely poor. Instead, the focus should be within the primary economic sector that has been ignored for several decades, and the secondary economic sector that has suffered from adequate capital input for a long period of time. Especially, the focus should be the high-tech industries of these economic sectors.

One of the important principles of the World Trade Organization (WTO) is that of nondiscrimination. That is a principle widely accepted by the international world. Hence, in order to make up the shortage of capital, there is a need to expand the sources of capital, where capitals of the people and funds from foreign entities should be actively considered. In this regard, there should not be any discriminant policy toward a specific economic sector. Instead, more effort should be given to create venture capital funds and to allow owners of capital to compete fairly and freely for investment opportunities in any economic sector.

Also, the investment of capital should be oriented toward education. Learning from the international experience, the reason why Japan could rise quickly after its defeat in World War II and become the number 2 economic power in the world is especially its capital investment in education. In November 1962, the Ministry of Education of Japan announced that funding for education is considered a part of economic investment, while making the theory of investment in education an action principle for the Japanese government to conduct its business in education.

The so-called theory of investment in education is to treat expenses on education as the most important investment on human assets, which are expected to produce future returns. That is, people are treated as a resource and education is seen as a means to develop economy. The acceptance and prevalence of this theory produced a long-lasting effect in Japanese society, causing all major Japanese corporations and government to continuously invest in education. The relevant statistics indicate that during the time period from 1950 to 1972, the funding for education in Japan increased 25 fold, and in the fiscal budget of Japanese government, the funding for education amounted to over 20%, representing the maximum among all capitalist countries (Wu, 1994). The overall funding for education amounted to about 6% of the GDP and 6.8% at the maximum time. In comparison, China spent less than 4%, generally somewhere 1–3%, of its GDP on the education system.

4. China possesses an extremely abundant supply of human resources. However, due to educational hysteresis and long period of inadequate policies on human talents, especially those regarding intellectuals and those regarding the income distribution of intellectuals, the resource of high-quality talents qualified for large-scale strategic adjustment and upgrading of China's economic sectorial structure is abnormally lacking. Because of this reason, China should, on the one hand, improve its quality of education at all costs comprehensively in the elementary level and in specialty education, vocational education, various degree education, continuing education, and so on, by increasing its capital input (we suggest to increase the annual input of over 6% of its national GDP into education), and by eliminating the monopolies in the area of education so that the market of education can be normalized and the treatment of teachers can be greatly raised. On the other hand, China should materially adjust its policies regarding human talents and how talents can be scientifically evaluated, selected, introduced, cultivated, and used (Tao, 2001c). Specifically, there is a strong need to scientifically evaluate the value of intellectuals (Gu and Zhang, 2000). On such a basis, realistically raise the economic status of intellectuals so that the domestic intellectuals truly

want to stay in China, and intellectuals from abroad desire to move to and live in mainland China (Gu, 2001). Only in this way can the difficult problem for the secondary and tertiary economic sectors to absorb the large amount of excessive labors from the primary economic sector can be well resolved, and a steady support of human resource for the diversification and strategic optimization of China's economic sectorial structure and urbanization will be provided.

Chapter 3

East China

The Optimal Area

From what has been discussed in the previous chapter, it follows that if China is divided into the three greater areas of east, center, and west, then the east area is the optimal place for investment and employment. In this chapter, the fundamental theory, model, and method as presented in Chapter 1 will be employed to analyze the recent statistical data from the year 2011 to calculate, analyze, and look at the investment and employment opportunities in this particular area. In terms of this area's geographic location, see Figure 3.1.

3.1 Regional Industries and Economic Sectors of the East

3.1.1 Current State of the Regional Economic Structure of East China

The 2011 GDP of East China and the relevant investment and employment situations are shown in Table 3.1.

Based on the data presented in the previous table, we can calculate and obtain the data in Table 3.2 by using the relevant formulas discussed in Chapter 1.

Although eastern China represents the most optimal area of investment and employment in Mainland China, the Table 3.2 indicates that the regional economic structure still experiences severe imbalances, no matter whether it is in terms of the supply and demand of labor or the supply and demand of capital.

In fact, from the fifth column (the comparative gains of labor) and the seventh column (the comparative gains of capital) of the table, it follows that among the 22 relevant data of the 11 provinces, municipalities, and regions, only three of them are in relative equilibrium: the comparative gain of capital of Hainan = 1, and the comparative gains of capital of Guangdong and Liaoning are, respectively, 1.034 and 0.965, while all others, amount to 89% (= 19/22), are in various states of imbalance or severe imbalance. Here, the most severe imbalance appears in such areas as Tianjin, Beijing, Hebei, Liaoning, Shanghai, Zhejiang, Siangsu, and others. In fact, Zhejiang's comparative gain of labor is the lowest (= 0.775), Hebei's comparative gain of capital is the lowest

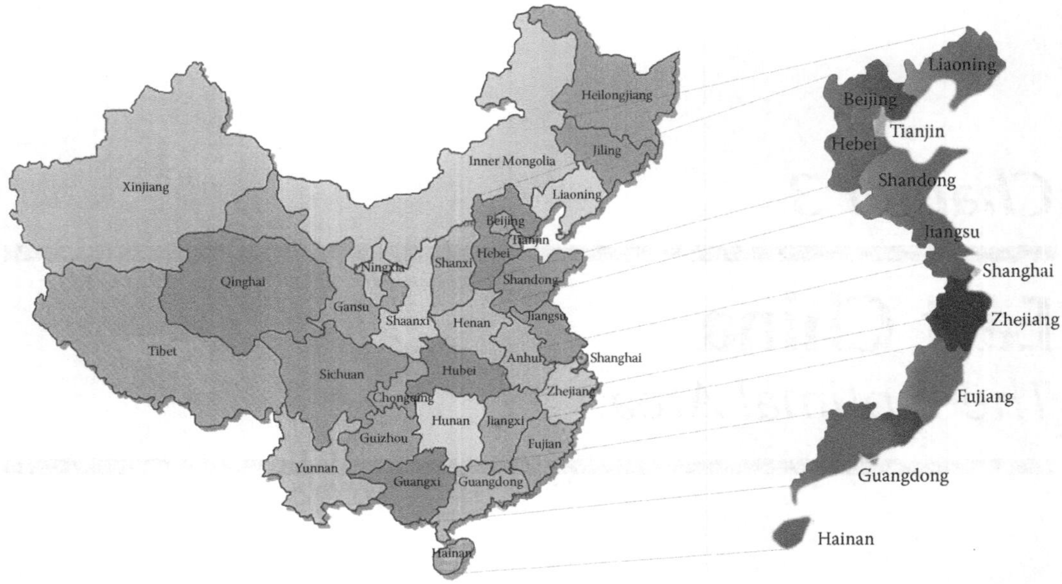

Figure 3.1 Geographic location of east China.

Table 3.1 GDPs, Values of Fixed Properties, and Employments of Various Provinces and Municipalities of East China (2011)

Area	*Province/Municipality/ Autonomous Region (i)*	*Regional GDP (0.1 Billion Yuan)*	*Value of Fixed Properties (0.1 Billion Yuan)*	*Employments (10,000)*
East China	1. Beijing	16251.93	5578.9	2143.9
	2. Tianjin	11307.28	7067.7	688
	3. Hebei	24515.76	16389.3	2157.4
	4. Liaoning	22226.70	17726.3	2675.8
	5. Shanghai	19195.69	4962.1	2186
	6. Jiangsu	49110.27	26692.6	6451
	7. Zhejiang	32318.85	14185.3	4810.8
	8. Fujian	17560.18	9910.9	2208.9
	9. Shandong	45361.85	26749.7	4315.9
	10. Guangdong	53120.28	17069.2	5943.6
	11. Hainan	2522.66	1657.2	383.2
	Total	293491.45	147989.2	33964.5*

Source: Data are from *China Statistical Abstract 2012* (XLS, DOC).

Note: *: these values are not the sums of the relevant data. Instead, they are the means of comparative gains of labor of relevant regions.

Table 3.2 The Regional Economic Structures and Measures of Comparative Gains of the Provinces and Municipalities of East China (2011)

Area	*Province/ Municipality/ Autonomous Region (i)*	*Demand Structure* $d_i(t)$ *(%)*	*Employ-ment Structure* $s_i^1(t)$ *(%)*	*Comparative Gain of Labor* $G_i^1(t)$	*Investment Structure* $s_i^2(t)$ *(%)*	*Comparative Gain of Capital* $G_i^2(t)$
East China	1. Beijing	0.055	0.063	0.879	0.038	1.458
	2. Tianjin	0.039	0.020	1.950	0.048	0.812
	3. Hebei	0.084	0.064	1.313	0.111	0.757
	4. Liaoning	0.076	0.079	0.962	0.120	0.633
	5. Shanghai	0.065	0.064	1.016	0.034	1.912
	6. Jiangsu	0.167	0.190	0.879	0.180	0.928
	7. Zhejiang	0.110	0.142	0.775	0.096	1.146
	8. Fujian	0.060	0.065	0.923	0.067	0.896
	9. Shandong	0.155	0.127	1.220	0.181	0.856
	10. Guangdong	0.181	0.175	1.034	0.115	1.574
	11. Hainan	0.009	0.011	0.818	0.011	1.000
	Total	1.000	1.000	—	1.000	–

(= 0.757); and for Shanghai and Tianjin, their comparative gains of capital are among the highest and are, respectively, equal to 1.950 and 1.912.

In terms of the magnitudes of comparative gains of labor, the 11 provinces, municipalities, and regions of this area can be ordered as follows:

1	2	3	4	5	6
Tianjin	Hebei	Shandong	Guangdong	Shanghai	Liaoning
7	8	9	10	11	
Fujian	Beijing	Jiangsu	Hainan	Zhejiang	

In terms of the magnitudes of the comparative gains of capital, these 11 provinces, municipalities, and regions of this area can be ranked as follows:

1	2	3	4	5	6
Shanghai	Guangdong	Beijing	Zhejiang	Hainan	Jiangsu
7	8	9	10	11	
Fujian	Shandong	Tianjin	Hebei	Liaoning	

3.1.2 Investment and Employment Opportunities in East China

First, in terms of labor, the comparative gains of the first five provinces and municipalities are over 1, representing the most ideal places for employment. The rest six provinces and municipalities experience excessive labor so that limitations should be imposed to control the import of labors. In particular, for the two provinces of Zhejiang and Hainan, they should try different methods to transport their excessive labors to the first five provinces and municipalities.

In other words, no matter whether the job seekers, who look for employment opportunities in the eastern area of China, are from eastern China, or other parts of the Mainland China, or the rest of the world, the order of the provinces and municipalities in terms of their employment opportunities is given as follows:

1	2	3	4	5
Tianjin	Hebei	Shandong	Guangdong	Shanghai

Without special skills, please do not look for employment opportunities in the other six provinces and municipalities, because these provinces and municipalities are experiencing excessive labor supplies.

This is the direction of employment.

Second, in terms of capital, the comparative gains of the first four provinces and municipalities are more than 1, representing ideal places for potential investments. The other seven provinces and municipalities experience excessive investment capital so that the injection of capital should be controlled and reduced. Specifically, the five provinces and municipalities—Liaoning, Hebei, Tianjin, Shandong, Fujian—have suffered from excessive capital which should be relocated to the four previously ranked provinces and municipalities.

In other words, no matter whether the investors, who look for investment opportunities in the eastern area of China, are from eastern China, or other parts of the Mainland China, or the rest of the world, the order of the provinces and municipalities in terms of their investment opportunities is given as follows:

1	2	3	4
Shanghai	Guangdong	Beijing	Zhejiang

Unless there are particular niches, please do not go to the other seven provinces and municipalities.

This is the direction of investment.

Third, by combining the two criteria discussed above, it follows that the most optimal provinces and municipalities in this area in terms of investment and employment opportunities are Shanghai and Guangdong. It is because their comparative gains of labor and capital are all more than 1. The secondary choices are Tianjin, Beijing, and Shandong. And the last choice is Liaoning.

In the following sections, we will analyze computationally the investment and employment problems of these 11 provinces and municipalities of eastern China by making use of the recent statistics (2011) and produce our best conclusions for the reader to make his/her appropriate decisions.

Note: Because we have no access to relevant statistics of regions internal to individual provinces, municipalities, and autonomous regions, we have not been able to produce any computational conclusions on investment and employment regarding these finer divisions of China. Instead, we can only focus on the investigation of the corresponding investment and employment problems in terms of economic sectors of the individual provinces, municipalities, and autonomous regions. The same situation holds true for the rest of this book.

3.2 Shanghai

Shanghai is the second largest city by population in China and one of the four municipalities directly controlled by the Chinese central government with a total population of near 24 million as of 2013. It represents one of the global centers of finance. Its container port is the busiest in the world.

The city is located in the Yangtze River Delta in East China. It sits at the mouth of the Yangtze River in the middle portion of the Chinese coast. It is roughly equidistant from Beijing and Hong Kong. As a whole, the city consists of a peninsula between the Yangtze River and Hangzhou Bay, and a number of smaller islands. The municipality borders Jiangsu province and Zhejiang province in the north, south, and west, and is bounded by East China Sea in the east. For details, see Figure 3.2.

The city proper is bisected by Huangpu River, a branch of the Yangtze. The historic center of the city, known as the Puxi area, is located on the western side of Huangpu River; the newly developed district, known as Pudong, containing the financial district Lujiazui, is located on the eastern bank.

Shanghai's land area is mostly flat except some hills in the southwest corner with an average elevation of 4 m. The city has many rivers, canals, streams, and lakes, and is known for its rich water resources as part of the Taihu drainage area.

Historically, it was during the Song Dynasty (AD 960–1279) in 1074 that Shanghai was upgraded from a small village to a market town. During the time period between Yuan Dynasty in 1292 and 1927, the area was designated merely as a county seat administered by Songjiang prefecture. And in 1927 Shanghai officially became a city.

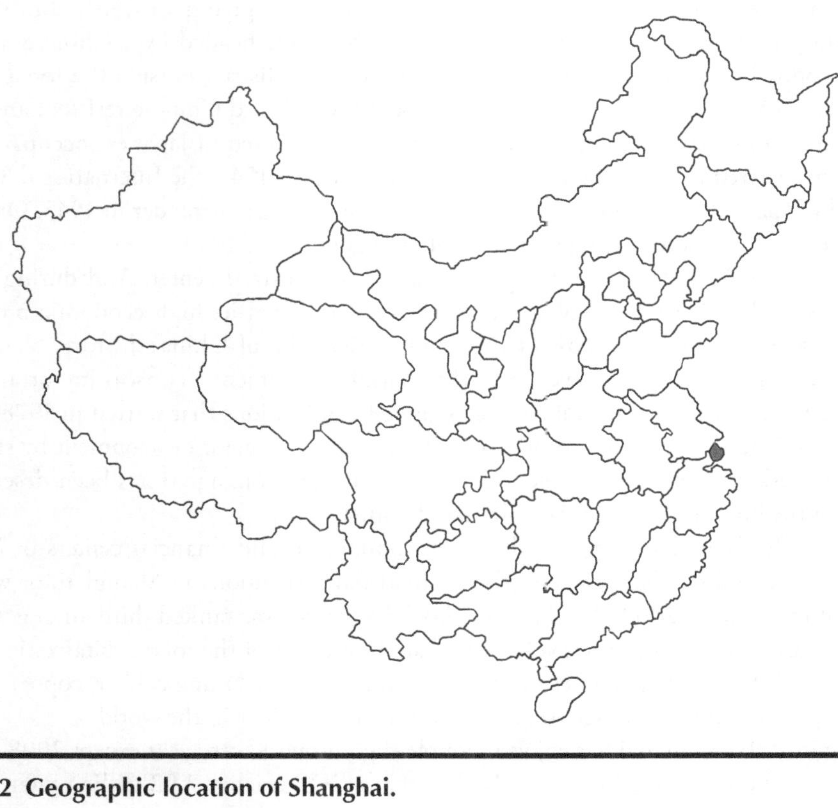

Figure 3.2 Geographic location of Shanghai.

In 1554 a city wall, which measured 10 m high and 5 km in circumference, was built to protect the town from raids by Japanese pirates. And in 1602, which was during Wanli reign (1573–1620), Shanghai received an important psychological boost from the erection of a City God Temple. During the Qing Dynasty, Shanghai became one of the most important sea ports in the Yangtze Delta region. By 1735 Shanghai had become the major trade port in the lower Yangtze River region.

International attention of Shanghai grew in the nineteenth century when Europeans recognized the city's economic and trade potential. The first opium war with British forces ended with the 1842 Treaty of Nanjing. That allowed the British to dictate the treaty ports, including Shanghai, for international trade. The Treaty of the Bogue, which was signed in 1843, and the Sino-American Treaty of Wanghia, which was signed in 1844, forced Chinese concession for Europeans and Americans to visit China and trade on Chinese soil.

In 1854, the Shanghai Municipal Council was created to manage all foreign affairs. In 1863, the British settlement in northern Huangpu District and the American settlement in southern Hongkou District joined together to form the Shanghai International Settlement. The French elected to stay out of the Shanghai Municipal Council and maintained its own concession in the south and southwest. Citizens of many countries and all continents came to Shanghai to live and work during the ensuing decades. Fleeing from the newly established Soviet Union in the 1920s and 1930s, around 20,000 White Russians and Jews took up residence in Shanghai. By 1932, Shanghai had become the world's fifth largest city and home to 70,000 foreigners. In the 1930s, some 30,000 Jewish refugees from Europe arrived in the city. The Sino-Japanese War elevated Japan to become another foreign power in Shanghai with the Treaty of Shimonoseki. Shanghai was then the most important financial center in the Far East.

Under the rule of the Republic of China (1911–1949), Shanghai gained its political status as a municipality in 1927. The new city government, which was headed by a Chinese mayor and municipal council, established a new city-center in Yangpu district outside the foreign concessions. On January 28, 1932, Japanese forces attacked the city and Chinese resisted; and a cease-fire was brokered in May. The Battle of Shanghai in 1937 resulted in Japanese occupation of the Chinese administered parts of Shanghai; and on December 8, 1941, the International Settlement was taken by Japanese force and remained occupied until Japan's surrender in 1945. On May 27, 1949, the People's Liberation Army took control of Shanghai.

During the 1950s and 1960s, Shanghai became an industrial center. And during the most riotous times of the Cultural Revolution, Shanghai maintained its high economic productivity and relative social stability. In most of the People's Republic of China's history, Shanghai has been a major contributor of tax revenue to the central government. Its fiscal importance to the central government also restrained it from economic liberalizations that started in 1978. In 1991, Shanghai experienced its first round of economic reforms and major development by starting its massive city works that are still visible today. In recent years, Shanghai has been described as a showpiece of the booming economy of Mainland China.

Economically, Shanghai is one of the main commercial and financial centers of Mainland China. As of the end of 2009, there were 787 financial institutions in Shanghai, of which 170 were foreign-invested. In 2009, the Shanghai Stock Exchange was ranked third among worldwide stock exchanges in terms of trading volume and sixth in terms of the total capitalization of listed companies. And the trading volumes of six key commodities, including rubber, copper, and zinc, on the Shanghai Futures Exchange were, respectively, ranked first in the world.

Starting in 1992, Shanghai has enjoyed double-digit growth every year except 2008 and 2009 during the global economic crisis. In 2011, Shanghai's GDP reached 1.92 trillion yuan (US$297 billion), which was equivalent to a GDP of 82,560 yuan (US$12,784) per capita. The top three largest service industries in Shanghai are financial services, retail, and real estate with manufacturing and agricultural sectors accounted, respectively, for 39.9% and 0.7% of the total output.

Because of its particular location, Shanghai has the world's busiest container port, which handled 29.05 million twenty-foot equivalent units (TEUs) in 2010.

As one of the main industrial centers of China, Shanghai plays a key role in China's heavy industries. A large number of industrial zones constitute the backbone of Shanghai's secondary industry. Heavy industries accounted for 78% of the GDP of Shanghai in 2009. China's largest steelmaker Baosteel Group and Jiangnan Shipyard, one of China's oldest shipbuilders, are both located in Shanghai. Auto manufacture represents another important economic sector in Shanghai. For example, SAIC Motor is one of the largest automotive corporations in China. It has strategic partnerships with both Volkswagen and General Motors. At the same time, the conference sector is also growing rapidly. For example, in 2012, Shanghai hosted more than 780 international gatherings.

With such brief background information of Shanghai in place (for more reading on the history of Shanghai, please consult with Danielson, 2004; 2010; Elvin, 1977; Horesh, 2009; Johnson, 1995), let us next look at how all economic data tell us about Shanghai in terms of labor and capital expenditure.

3.2.1 Current State of Investment and Employment Structures of Shanghai

The previous theory, model, and method can also be employed to investigate the investment and employment situations in the economic sectors and industries of individual provinces, municipalities, and autonomous regions. To this end, we only need to do the necessary computations for the structures of economic sectors and industries (the supply-and-demand structures of the economic sectors and industries), the current states of the investment and employment structures (such as what is shown in Table 3.3 for Shanghai) for each of the provinces, municipalities, and autonomous regions, and the corresponding comparative gains. And then, what is left is to make the comparisons. The detailed computations are carried out as described before. The results are shown in Table 3.4.

Table 3.4 indicates that first, in terms of the three major sectors, among the six relevant data points of the three sectors, only two are basically in equilibrium, where the comparative gains of labor of the secondary and the tertiary sectors are, respectively, 1.024742 and 1.030677, while all the other four (4/6 = 67%) are in extremely imbalanced states. In particular, the comparative gain of labor of the first sector is way too low, and is equal to only 0.192806, which is not even equal to one-fifth of those of the same indicator of the secondary and tertiary sectors. In terms of the capital investment, the comparative gains of capital of the primary and the secondary sectors are way too low, and are, respectively, 1.58579 and 1.78377, while the capital investment in the tertiary sector is in extreme excess (0.75944).

Second, in terms of the employment situation, the relevant data of the 21 industries of the Shanghai city should contain 42 data points. However, because of the lack of raw data, we only collected 36 of them. So, we can only base our discussion on these available statistics. Among all the industries, only one or two are basically in equilibrium, where the comparative gain of labor in wholesale and retail is 0.963054, and the comparative gain of capital in health, safety, and welfare is 1.03974. All the rest (34/36 ≥ 94%) are situated in extremely imbalanced states. In particular, the comparative gains of the six industries, such as production/supply of electricity, gas, and water; mining; finance; real estate/housing; information transfer, computer, and

Table 3.3 GDPs, Total Investments in Fixed Properties, and Employment Situation of Shanghai's Three Economic Sectors and Industries (2011)

Sectors (i)	*Total Output of Sector (0.1 Billion Yuan)*	*Total Investment on Fixed Properties (0.1 Billion Yuan)*	*Employment (10,000 People)*
Total	19195.69	4517.09	1104.33
Primary sector	124.94	18.54	37.28
Secondary sector	7927.89	1045.86	445.08
Tertiary sector	11142.86	3452.69	621.97
Industries (*i*)			
Agriculture, forest, animal husbandry, fishery	304.85	18.6	37.28
Mining	11.62	0.6	0.05
Manufacturing		1097.5	341.41
Production/supply of electricity, gas, and water	1746.2	159	6.27
Construction	719.3	12.9	97.35
Communication, transportation, storage, postal service	868.31	492.36	58.14
Information transfer, computer, and software areas	784.77	14.39	27.57
# Information Transfer			
Wholesale and retail	3040.99	52.75	181.66
Hotels and restaurants	279.34	35.88	47.23
Finance	2277.4	14.85	28.41
Real estate/housing	1019.68	2204.96	35.59
Leasing and commercial services	912.6	52.52	62.31
Scientific research, technology service, and geological survey	447.02	39.03	29.99
Water conservation, environment, and public facility management	53.4	370.76	11.2
# Public Facility Management			
Residents and other services	205.06	3.34	58.86
Education	437.09	37.85	30.31
Health, safety, and welfare	286.89	64.93	20.65

Table 3.3 (continued) GDPs, Total Investments in Fixed Properties, and Employment Situation of Shanghai's Three Economic Sectors and Industries (2011)

Sectors (i)	*Total Output of Sector (0.1 Billion Yuan)*	*Total Investment on Fixed Properties (0.1 Billion Yuan)*	*Employment (10,000 People)*
Social, sports, and entertainment	115.21	39.31	10.65
Public administration and social organizations	415.1	29.76	19.4

Source: Data are from *Shanghai Statistical Yearbook 2012* (XLS, DOC).

software areas; and public administration and social organizations, are all over 1. Especially, the comparative gain of labor in the production/supply of electricity, gas, and water even reached 16.02218, which is more than 80 times of the lowest value of this indicator that is reached by residents and other services.

At the same time, the imbalance in terms of the comparative gains of capital seems to be way too severe. The comparative gains are all over one in 14 industries such as finance; residents and

Table 3.4 Structures and Comparative Gain Measures of Shanghai's Three Economic Sectors and Industries (*t* = 2011)

Sectors (i)	*Demand Structure $d_i(t)$ (%)*	*Employment Structure $s_i^1(t)$ (%)*	*Comparative Gain of Labor $G_i^1(t)$*	*Investment Structure $s_i^2(t)$ (%)*	*Comparative Gain of Capital $G_i^2(t)$*
Total	100	100	–	100	–
Primary sector	0.650875	3.375803	0.192806	0.410441	1.58579
Secondary sector	41.30036	40.30317	1.024742	23.1534	1.78377
Tertiary sector	58.04876	56.32103	1.030677	76.43616	0.75944
Industries (*i*)					
Agriculture, forest, animal husbandry, fishery	1.588117	3.375803	0.470441	0.41177	3.85681
Mining	0.060534	0.004528	13.37000	0.013283	4.55732
Manufacturing		30.91558	0	24.29662	
Production/supply of electricity, gas, and water	9.096834	0.567765	16.02218	3.519965	2.58435
Construction	3.747195	8.815300	0.425079	0.285582	13.1213

continued

Table 3.4 (continued) Structures and Comparative Gain Measures of Shanghai's Three Economic Sectors and Industries ($t = 2011$)

Sectors (i)	*Demand Structure $d_i(t)$ (%)*	*Employment Structure $s_i^1(t)$ (%)*	*Comparative Gain of Labor $G_i^1(t)$*	*Investment Structure $s_i^2(t)$ (%)*	*Comparative Gain of Capital $G_i^2(t)$*
Communication, transportation, storage, postal service	4.523463	5.264731	0.859201	10.89994	0.41500
Information transfer, computer, and software areas	4.088261	2.496536	1.637573	0.318568	12.8332
# Information Transfer					
Wholesale and retail	15.84205	16.44979	0.963054	1.167787	13.5659
Hotels and restaurants	1.455223	4.276801	0.340260	0.794317	1.83204
Finance	11.86412	2.572601	4.611723	0.328751	36.0884
Real estate/housing	5.312026	3.222769	1.648280	48.81373	0.10882
Leasing and commercial services	4.754192	5.642335	0.842593	1.162695	4.08894
Scientific research, technology service, and geological survey	2.328752	2.715674	0.857523	0.864052	2.69515
Water conservation, environment, and public facility management	0.278187	1.014190	0.274295	8.207939	0.03389
# Public Facility Management					
Residents and other services	1.068261	5.329929	0.200427	0.073941	14.4474
Education	2.277022	2.744651	0.829622	0.837929	2.71744
Health, safety, and welfare	1.494554	1.869912	0.799264	1.43743	1.03974
Social, sports, and entertainment	0.600187	0.964386	0.622351	0.870251	0.68967
Public administration and social organizations	2.162465	1.756721	1.230966	0.658831	3.28227

Source: Data are from *China Statistics Abstract 2012* (XLS, DOC).

other services; wholesale and retail; construction; information transfer, computer, and software areas; mining; leasing and commercial services; agriculture, forest, animal husbandry, fishery; public administration and social organizations; education; scientific research, technology service, and geological survey; production/supply of electricity, gas, and water; hotels and restaurants; health, safety, and welfare. In particular, the comparative gains of capital of the five industries—finance; residents and other services; wholesale and retail; construction; information transfer, computer, and software areas—all reached levels of more than 10.

On the other hand, the values of this indicator of the industries, such as "Water conservation, environment, and public facility management," "residents and other services," "wholesale and retail," "construction," and "information transfer, computer, and software areas," are too low, reaching 0.03389. This value is equal to 0.0009 (≈ 0.03389/36.0884) for "Water conservation, environment, and public facility management," 0.0024 (≈ 0.03389/14.4474) for "residents and other services," 0.0025 (≈ 0.03389/13.5659) for "wholesale and retail," 0.0026 (≈ 0.03389/13.1213) for "construction," and 0.0027 (≈ 0.03389/12.8334) for "information transfer, computer, and software areas" when compared to finance.

So, in terms of the magnitudes of the comparative gains of labor, the 18 industries of Shanghai with available data can be ranked as follows:

1	2	3	4
Production/supply of electricity, gas, and water	Mining	Finance	Real estate/housing
5	6	7	8
Information transfer, computer, and software areas	Public administration and social organizations	Wholesale and retail	Communication, transportation, storage, postal service
9	10	11	12
Scientific research, technology service, and geological survey	Leasing and commercial services	Education	Health, safety, and welfare
13	14	15	16
Social, sports, and entertainment	Agriculture, forest, animal husbandry, fishery	Construction	Hotels and restaurants
17	18		
Water conservation, environment, and public facility management	Residents and other services		

In terms of the magnitudes of the comparative gains of capital, the 18 industries of Shanghai with available data can be ranked as follows:

1	2	3	4
Finance	Residents and other services	Wholesale and retail	Information transfer, computer, and software areas
5	6	7	8
Leasing and commercial services	Public administration and social organizations	Education	Scientific research, technology service, and geological survey
9	10	11	12
Hotels and restaurants	Health, safety, and welfare	Social, sports, and entertainment	Communication, transportation, storage, postal service
13	14	15	16
Real estate/ housing	Water conservation, environment, and public facility management	Production/ supply of electricity, gas, and water	Construction
17	18		
Agriculture, forest, animal husbandry, fishery	Mining		

3.2.2 Investment and Employment in Shanghai's Economic Sectors and Industries

First, in terms of the comparative gains of labor, the values of this indicator of the six industries:

Production/supply of electricity, gas, and water	Mining	Finance
Real estate and housing	Information transfer, computer, and software areas;	Public administration and social organizations

are over 1, representing relatively optimal industries for employment. In all the other 12 industries, there has been an excessive supply of labor so that appropriate control on the increase of labor into these industries should be imposed. Specifically, the excessive labor in the following industries should be transferred into the previous six industries.

Residents and other services	Water conservation, environment, and public facility management	Hotels and restaurants	Construction
Social, sports, and entertainment	Health, safety, and welfare	Education	Leasing and commercial services
Scientific research, technology service, and geological survey	Communication, transportation, storage, postal service	Wholesale and retail	

In other words, no matter whether the job seekers, who look for employment opportunities in Shanghai, are from eastern China, or other parts of the Mainland China, or the rest of the world, they should first consider the tertiary economic sector and the secondary economic sector with the relevant industries ranked as follows:

1	2	3
Production/supply of electricity, gas, and water	Mining	Finance
4	5	6
Real estate/housing	Information transfer, computer, and software areas	Public administration and social organizations

Without special skills, do not look for employment opportunities in the primary economic sector and other industries, because those areas are experiencing excessive labor supplies.

That is the direction of employment.

Second, in terms of capital, the following 14 industries have enjoyed the comparative gains of more than 1, representing relatively more optimal areas of additional investment.

Finance	Residents and other services	Wholesale and retail	Construction
Information transfer, computer, and software areas	Mining	Leasing and commercial services	Agriculture, forest, animal husbandry, fishery
Public administration and social organizations	Education	Scientific research, technology service, and geological survey	Production/supply of electricity, gas, and water
Hotels and restaurants	Health, safety, and welfare		

However, for the following four industries, excessive investment of capital has been given. Further capital investment in these industries should be curbed and redirected to the previous 14 industries.

Social, sports, and entertainment	Communication transportation, storage, postal service	Real estate/ housing	Water conservation, environment, and public facility management

In other words, no matter whether the investors, who look for investment opportunities in Shanghai, are from eastern China, or other parts of the Mainland China, or the rest of the world, they should first consider the following industries, ranked from the most optimal to the least.

Finance	Residents and other services	Wholesale and retail	Construction
Information transfer, computer, and software areas	Mining	Leasing and commercial services	Agriculture, forest, animal husbandry, fishery
Public administration and social organizations	Education	Scientific research, technology service, and geological survey	Production/supply of electricity, gas, and water
Hotels and restaurants	Health, safety, and welfare		

At the same time, try not to invest additional capital in the following four industries, because there has been a severe over supply of capital in these industries.

That is the direction of capital investment.

Social, sports, and entertainment	Communication, transportation, storage, postal service	Real estate/ housing	Water conservation, environment, and public facility management

Third, by combining the two criteria discussed above, it follows that the industries in Shanghai that are most optimal for investment and employment are the following, because the comparative gains of labor and capital of each of these industries are over 1.

Finance	Production/supply of electricity, gas, and water	Mining
Information transfer, computer, and software areas	Public administration and social organizations	

The secondary choice is the industry of wholesale and retail. And the last choice is the industry of water conservation, environment, and public facility management.

3.3 Guangdong Province

Located on the South China sea coast of China, this province is also known as Canton or Kwangtung with a total of 4300 km (2700 mi) of coastline. In January 2005, it became the most populous province in China with 79.1 million registered permanent residents and 31 million migrants who lived in the province for at least six months of the year (Chinadaily.com, 2012). The city of Guangzhou, the provincial capital, and Shenzhen, an economic hub, are among the most populous and important cities in China. The Pearl River Delta is the convergent point of three

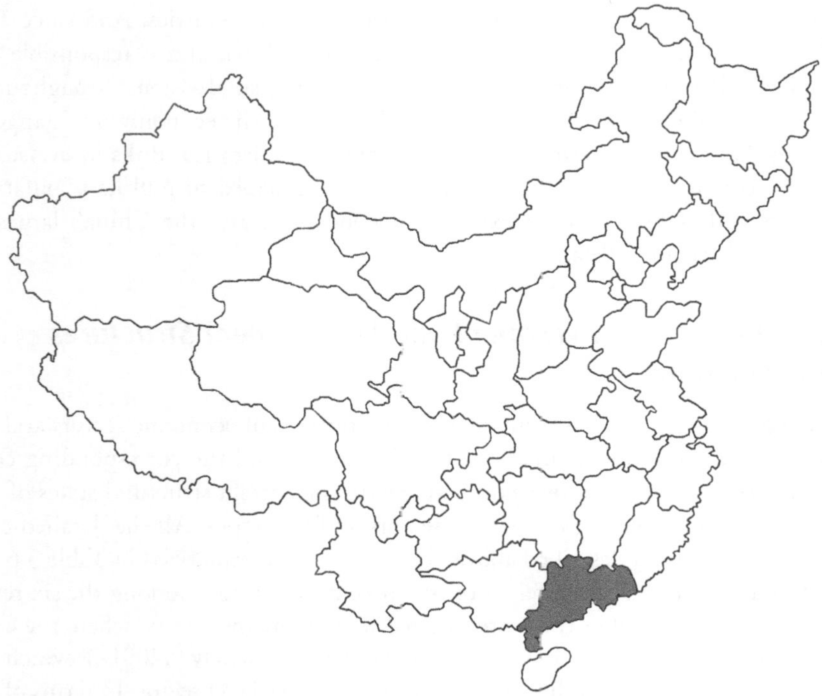

Figure 3.3 Geographic location of Guangdong.

upstream rivers: the East River, North River, and West River, and is filled with hundreds of small islands. The province is geographically cut off from the north by Nan Mountains, which consist of mountain ranges. For details, see Figure 3.3.

Starting in the year 1989 Guangdong has topped the total GDP rankings among all provinces. In 2011, Guangdong's GDP reached CNY 5267 billion (US$815.53 billion), making the province's economy roughly the size of the Netherlands and over half of India. Guangdong has also been the host of the largest import and export fair in China, known as the Canton Fair in the city of Guangzhou.

Historically, Chinese administration in the region began in the Qin Dynasty nearly 2000 years ago. During the Three Kingdoms period, the Wu Kingdom made Guangdong into a province in the year 226. Over time, the demographics of Guangdong slowly shifted with more Han Chinese and some Persian women massively migrating into the region from the north during periods of political turmoil and nomadic incursions. Since the 1500s, Guangdong has had extensive trade connections with the rest of the world. Macau, which is located on the southern coast of Guangdong, was the first European settlement established in 1557. In the 1800s, the First Opium War was triggered by the opium trades through Guangzhou, which opened the era of foreign incursion and intervention in China. At the same time, Guangdong was a major exit port for Chinese laborers to migrate to Southeast Asia and the West in the nineteenth century (Liu, 1995, pp. 364). Based on its direct contact with the West, Guangdong had been the center of anti-Manchu and anti-imperialist activities, where the founder of modern China, Sun Yat-sen, was from. During the early 1900s, Guangdong was the staging area for Kuomintang (KMT) to prepare for its Northern Expedition to bring the then warlords of China back under the control of the central government. In recent decades, Guangdong has experienced an extremely rapid economic growth and enjoyed the highest gross domestic product in China.

Economically, Guangdong is comparable in size to many countries. And since 1989, it has been the largest province in terms of GDP in Mainland China, and is responsible for 11% of China's overall GDP. Its GDP is larger than all but six country subdivisions throughout the world. In 1978, Deng Xiaoping's open door policy quickly changed the economy of Guangdong based on its access to the ocean, bordering with Hong Kong, and historical links to overseas Chinese. The province's economic boom began in early 1990s and helped to pull its population density higher. It is not only China's largest exporter of goods, but also the China's largest importer (Thechinaperspective.com, 2012a).

3.3.1 Current State of Investment and Employment Structures of Guangdong

For Guangdong Province, we first compute the total outputs of economic sectors and industries, the investment and employment situations (see Table 3.5), and the corresponding comparative gains. By comparing the comparative gains, we obtain the current structural states of investment and employment for Guangdong's economic sectors and industries. All the detailed calculations are similar to what has been described above. The results are summarized in Table 3.6.

Table 3.6 indicates that first, in terms of the three major sectors, among the six relevant data points of the three sectors, all of them are situated in extreme instability, where the comparative gain of labor of the primary sector is way too low and equal to merely 0.209174, which is not even reaching one-fifth of that same indicator of the secondary (tertiary) sector. In terms of the capital investment, the comparative gains of the primary and the secondary sectors are much too low, and are respectively equal to 3.95016 and 1.50548, while the capital investment in the tertiary sector is in extreme excess (0.68913).

Second, in terms of the industrial situations, the relevant data of the 21 industries of Guangdong Province should contain 42 data points. However, because of the lack of raw data, we only collected 36 of them. So, we can only base our discussion on these available statistics. Among all the industries, only one (the comparative gain of labor of the industry of education is equal to 1.007340) is basically in equilibrium, and all the other 31 industries (31/32 ≥ 96%) are situated in extremely imbalanced states. In particular, the comparative gains of labor of the following 10

Table 3.5 GDPs, Total Investments in Fixed Properties, and Employment Situation of Guangdong's Three Economic Sectors and Industries (2011)

Sectors (i)	*Total Output of Sector (0.1 Billion Yuan)*	*Total Investment on Fixed Properties (0.1 Billion Yuan)*	*Employment (10,000 People)*
Total	53210.28	16843.83	5960.7434
Primary sector	2665.2	213.58	1427.34
Secondary sector	26447.38	5561.01	2526.48
Tertiary sector	24097.7	11069.23	2006.92
Industries (*i*)			
Agriculture, forest, animal husbandry, fishery	4384.44	213.58	1427.3414

Table 3.5 (continued) GDPs, Total Investments in Fixed Properties, and Employment Situation of Guangdong's Three Economic Sectors and Industries (2011)

Sectors (i)	*Total Output of Sector (0.1 Billion Yuan)*	*Total Investment on Fixed Properties (0.1 Billion Yuan)*	*Employment (10,000 People)*
Mining		74.71	13.1537
Manufacturing		4523.42	2235.8595
Production/supply of electricity, gas, and water		934.31	34.452
Construction	1797.78	28.56	243.0193
Communication, transportation, storage, postal service	2090.36	1750.8	162.384
Information transfer, computer, and software areas	1504.15	389.1	58.0551
# Information Transfer			
Wholesale and retail	5681.17	421.94	792.02
Hotels and restaurants	1192.28	288.2	213.2807
Finance	2916.13	30.59	56.4258
Real estate/housing	3321.31	5532.58	71.3297
Leasing and commercial services	1932.21	202.94	88.6564
Scientific research, technology service, and geological survey	546.1	124.04	31.3549
Water conservation, environment, and public facility management	229.24	1609.34	20.491
# Public Facility Management			
Residents and other services	855.12	15.02	167.1762
Education	1226.17	279.16	136.3576
Health, safety, and welfare	754.75	127.83	60.2708
Social, sports, and entertainment	327.76	189.88	23.82
Public administration and social organizations	1520.95	107.81	125.2953

Source: Data are from *Statistical Yearbook of Guangdong Province 2012* (XLS, DOC).

Table 3.6 Structures and Comparative Gain Measures of Guangdong's Three Economic Sectors and Industries ($t = 2011$)

Sectors (i)	*Demand Structure* $d_i(t)$ (%)	*Employment Structure* $s_i^1(t)$ (%)	*Comparative Gain of Labor* $G_i^1(t)$	*Investment Structure* $s_i^2(t)$ (%)	*Comparative Gain of Capital* $G_i^2(t)$
Total	100	100	–	100	–
Primary sector	5.008807	23.94567	0.209174	1.268001	3.95016
Secondary sector	49.70352	42.38532	1.172659	33.01512	1.50548
Tertiary sector	45.28768	33.66895	1.345087	65.71682	0.68913
Industries (*i*)					
Agriculture, forest, animal husbandry, fishery	8.239836	23.94569	0.344105	1.268001	6.49829
Mining		0.220672		0.443545	
Manufacturing		37.50974		26.85506	
Production/supply of electricity, gas, and water		0.577982		5.546898	
Construction	3.378633	4.076997	0.82870	0.169558	19.9262
Communication, transportation, storage, postal service	3.928489	2.724224	1.442058	10.39431	0.37795
Information transfer, computer, and software areas	2.826803	0.973957	2.902389	2.310045	1.22370
# Information Transfer					
Wholesale and retail	10.67683	13.28727	0.803538	2.505012	4.26219
Hotels and restaurants	2.240695	3.578089	0.626227	1.711012	1.30957
Finance					
Real estate/housing	6.241858	1.196658	5.216076	32.84633	0.19003
Leasing and commercial services	3.631272	1.487338	2.441457	1.204833	3.01392
Scientific research, technology service, and geological survey	1.026305	0.526023	1.951065	0.736412	1.39366
Water conservation, environment, and public facility management	0.430819	0.343766	1.253234	9.554478	0.04509

Table 3.6 (continued) Structures and Comparative Gain Measures of Guangdong's Three Economic Sectors and Industries ($t = 2011$)

Sectors (i)	*Demand Structure* $d_i(t)$ (%)	*Employment Structure* $s_i^1(t)$ (%)	*Comparative Gain of Labor* $G_i^1(t)$	*Investment Structure* $s_i^2(t)$ (%)	*Comparative Gain of Capital* $G_i^2(t)$
# Public Facility Management					
Residents and other services	1.607058	2.804620	0.573004	0.089172	18.0220
Education	2.304385	2.287594	1.007340	1.657343	1.39041
Health, safety, and welfare	1.418429	1.011129	1.402817	0.758913	1.86904
Social, sports, and entertainment	0.615971	0.399615	1.541413	1.127297	0.54641
Public administration and social organizations	2.858376	2.102008	1.359831	0.640056	4.46582

Source: Data are from *China Statistics Abstract 2012* (XLS, DOC).

industries are over 1 with those of finance and real estate and housing even gone beyond 5, equaling, respectively, 5.789406 and 5.216076. These values are respectively (5.789406/0.344105 ≈) 17 times and (5.216076/0.344105 ≈) 15 times more than that of the industry of "agriculture, forest, animal husbandry, fishery."

Finance	Real estate/housing	Information transfer, computer, and software areas
Scientific research, technology service, and geological survey	Social, sports, and entertainment	Communication, transportation, storage, postal service
Health, safety, and welfare	Public administration and social organizations	Water conservation, environment, and public facility management
Education		

As for the comparative gains of capital, the imbalance seems to be intolerable. The values of this indicator of the following 12 industries are greater than 1. In particular, the comparative gains of capital of such industries as "finance," "construction," and "residents and other services" are all greater than 10, while the value of this indicator of the industry "water conservation, environment, and public facility management" is much too low and equal to merely 0.04509, which is only (0.04509/30.1768<) 0.0015 of the value of "finance" (0.04509/19.9262 ≈), 0.002 of the value of "construction," and (0.04509/18.0220 ≈) 0.0022 of the value of "Residents and other services."

Finance	Construction	Residents and other services	Agriculture, forest, animal husbandry, fishery
Public administration and social organizations	Wholesale and retail	Leasing and commercial services	Health, safety, and welfare
Scientific research, technology service, and geological survey	Education	Hotels and restaurants	Information transfer, computer, and software areas

So, in terms of the magnitudes of the comparative gains of labor, the 15 industries of Guangdong Province with available data can be ranked as follows:

1	2	3	4
Finance	Real estate/housing	Information transfer, computer, and software areas	Scientific research, technology service, and geological survey
5	6	7	8
Social, sports, and entertainment	Communication, transportation, storage, postal service	Health, safety, and welfare	Public administration and social organizations
9	10	11	12
Water conservation, environment, and public facility management	Education	Construction	Wholesale and retail
13	14	15	
Hotels and restaurants	Residents and other services	Agriculture, forest, animal husbandry, fishery	

In terms of the magnitudes of the comparative gains of capital, the 16 industries of Guangdong Province with available data can be ranked as follows:

1	2	3	4
Finance	Construction	Residents and other services	Agriculture, forest, animal husbandry, fishery

5	6	7	8
Public administration and social organizations	Wholesale and retail	Leasing and commercial services	Health, safety, and welfare
9	10	11	12
Scientific research, technology service, and geological survey	Education	Hotels and restaurants	Information transfer, computer, and software areas
13	14	15	16
Social, sports, and entertainment	Communication, transportation, storage, postal service	Real estate/housing	Water conservation, environment, and public facility management

3.3.2 Investment and Employment in Guangdong's Economic Sectors and Industries

First, in terms of the comparative gains of labor, the values of this indicator of the following 10 industries are over 1, representing relatively ideal areas of employment. The supplies of labor in all other industries are in excess and should be reduced and redirected to the previously listed industries. Especially, the excessive labor in "agriculture, forest, animal husbandry, fishery," "residents and other services," "hotels and restaurants," "wholesale and retail," and "construction" should be relocated into the previously listed industries.

Finance	Real estate/housing	Information transfer, computer, and software areas	Scientific research, technology service, and geological survey
Social, sports, and entertainment	Communication, transportation, storage, postal service	Health, safety, and welfare	Public administration and social organizations
Water conservation, environment, and public facility management	Education		

In other words, no matter where the job seeker, who is looking for employment opportunities in Guangdong, is from, the eastern part of China, the other parts of China, or the rest of the world, his/her employment opportunity is potentially in the tertiary and the secondary economic sector with the industries ranked as follows:

1	2	3	4	5
Finance	Real estate/ housing	Information transfer, computer, and software areas	Scientific research, technology service, and geological survey	Social, sports, and entertainment
6	7	8	9	10
Communication, transportation, storage, postal service	Health, safety, and welfare	Public administration and social organizations	Water conservation, environment, and public facility management	Education

Without special skills, do not waste time and effort in looking for employment opportunities in the primary economic sector and other five industries, because those areas are experiencing excessive labor supplies.

That is the direction of employment.

Second, in terms of the comparative gains of capital, the following 12 industries have enjoyed the comparative gains of more than 1, representing relatively more optimal areas of additional investment.

Finance	Construction	Residents and other services	Agriculture, forest, animal husbandry, fishery
Public administration and social organizations	Wholesale and retail	Leasing and commercial services	Health, safety, and welfare
Scientific research, technology service, and geological survey	Education	Hotels and restaurants	Information transfer, computer, and software areas

However, for the following three industries, excessive investment of capital has been given. Further capital investment in these industries should be curbed and redirected to the previous 12 industries.

In other words, no matter whether the investors, who look for investment opportunities in Guangdong, are from eastern China, or other parts of the Mainland China, or the rest of the world, they should first consider the following industries, ranked from the most optimal to the least:

1	2	3	4
Finance	Construction	Residents and other services	Agriculture, forest, animal husbandry, fishery
5	6	7	8
Public administration and social organizations	Wholesale and retail	Leasing and commercial services	Health, safety, and welfare
9	10	11	12
Scientific research, technology service, and geological survey	Education	Hotels and restaurants	Information transfer, computer, and software areas

At the same time, try not to invest additional capital in the following industries, because there has been a severe over supply of capital in these industries.

That is the direction of capital investment.

Water conservation, environment, and public facility management	Real estate/ housing	Communication, transportation, storage, postal service

Third, by combining the two criteria discussed above, it follows that the industries in Guangdong that are most optimal for investment and employment are the following, because the comparative gains of labor and capital of each of these industries are over 1.

Finance	Information transfer, computer, and software areas	Leasing and commercial services	Scientific research, technology service, and geological survey
Public administration and social organizations	Health, safety, and welfare	Education	

As for each of other industries, at least one of the comparative gains is far from 1.

3.4 Tianjin

Tianjin is the largest coastal city in northern China that is governed as one of the four municipalities of China under direct administration of the central government. It borders with Hebei Province and Beijing. It is bounded in the east by the Bohai Gulf of the Yellow Sea. The city lies at the northern end of the Grand Canal of China, which connects with the Yellow River and Yangtze River. Within the Tianjin municipality, at the the confluence of the Ziya River, Daqing River, Yongding River, North Grand Canal, and South Grand Canal forms the Hai River, which enters the Pacific Ocean within the municipality. In terms of urban population, Tianjin is the fourth largest in China after Shanghai, Beijing, and Guangzhou. Its main urban area is located along the Hai River that connects to the Yellow and Yangtze Rivers via the Grand Canal, and Binhai on the coast of Bohai Sea. For details, see Figure 3.4.

Historically, where Tianjin is located was created by sedimentation of various rivers entering the sea at Bohai Gulf. Along with the opening of the Grand Canal in the Sui Dynasty, Tianjin became a trading center. In 1858 with the assigning of the Treaties of Tianjin and the ratification by the emperor of China in 1860, the city was formally open to foreign trade. Starting in the year 1900, Qing General Yuan Shikai helped to transform Tianjin into a modern city by establishing the first modern Chinese police force and by supervising China's first modern democratic elections for a county council. Tianjin became a municipality of China in 1927. In 1937, Tianjin fell to the occupation of Japan and was ruled by the North China Executive Committee, a puppet state based in Beijing. The Japanese occupation lasted until August 15, 1945, when Japan surrendered, which marked the end of World War II.

Economically, Tianjin's GDP reached ¥1.12 trillion in 2011, representing China's highest per-capita GDP of $13,393, followed by Shanghai of $12,784 and Beijing of $12,447, levels on par

Figure 3.4 Geographic location of Tianjin.

with some developed countries. Currently, many major projects are underway to extend the metro system and to improve road and rail links with the nearby national capital, Beijing. Even though the financial crisis of 2007 spread rapidly over the world and the current economic growth of the world is extremely slow, Tianjin's economic growth has maintained at a steady speed. For instance, Tianjin total output value in 2008 was RMB 635.438 billion, representing an increase of 16.5% over the previous year. The total output per capita in Tianjin was US$ 7800 that neared the middle level developed countries. So as of this writing, over 21,000 companies in Tianjin have made foreign investments, amounting to over US$47.2 billion. Along with active development of financial reform and innovation, Tianjin has become the pilot city with the most industrial investment funds in China. Currently, Tianjin is developing itself into an eco-friendly city. For more details, please consult Chinagate (2007) and Tianjinplus.com (2012).

3.4.1 Current State of Investment and Employment Structures of Tianjin

For Tianjin City, we first compute the total outputs of economic sectors and industries, the investment and employment situations (see Table 3.7), and the corresponding comparative gains. Then, by comparing the comparative gains, we obtain the current structural states of investment and employment for Tianjin's economic sectors and industries. All the detailed calculations are similar to what has been described above. The results are summarized in Table 3.8.

Table 3.8 indicates that first, in terms of the three major sectors, among the six relevant data points of the three sectors, all of them are situated in extreme imbalances, where the comparative gain of labor of the primary sector is way too low and equal to merely 0.147307. That is not even

Table 3.7 GDPs, Total Investments in Fixed Properties, and Employment Situation of Tianjin's Three Economic Sectors and Industries (2011)

Sectors (i)	*Total Output of Sector (0.1 Billion Yuan)*	*Total Investment on Fixed Properties (0.1 Billion Yuan)*	*Employment (10,000 People)*
Total	11307.28	7510.67	763.16
Primary sector	159.72	150.33	73.18
Secondary sector	5928.32	3229.73	315.99
Tertiary sector	5219.24	4130.61	373.99
Industries (*i*)			
Agriculture, forest, animal husbandry, fishery		150.33	73.18
Mining		228.26	11.12
Manufacturing		2720.53	228.21
Production/supply of electricity, gas, and water		239.76	6.16
Construction	497.48	41.18	70.5
Communication, transportation, storage, postal service	632.1	549.71	27.83
Information transfer, computer, and software areas	172.1	72.29	5.42
# Information Transfer			
Wholesale and retail	1463.89	127.79	109.1
Hotels and restaurants	194.52	58.09	34.89
Finance	756.5	39.62	12.54
Real estate/housing	411.46	1569.83	15.47
Leasing and commercial services	277.57	446.95	26.49
Scientific research, technology service, and geological survey	332.7	23.53	21.5
Water conservation, environment, and public facility management	72.65	850.44	6.76
# Public Facility Management			
Residents and other services	237.05	100.48	41.81

continued

Table 3.7 (continued) GDPs, Total Investments in Fixed Properties, and Employment Situation of Tianjin's Three Economic Sectors and Industries (2011)

Sectors (i)	*Total Output of Sector (0.1 Billion Yuan)*	*Total Investment on Fixed Properties (0.1 Billion Yuan)*	*Employment (10,000 People)*
Education	248.65	70.65	30.5
Health, safety, and welfare	119.61	21.09	15.14
Social, sports, and entertainment	57.2	105.06	4.65
Public administration and social organizations	243.24	95.08	21.89

Source: Data are from *China's Statistics Abstracts 2012* (XLS, DOC).

Table 3.8 Structures and Comparative Gain Measures of Tianjin's Three Economic Sectors and Industries ($t = 2011$)

Sectors (i)	*Demand Structure $d_i(t)$ (%)*	*Employment Structure $s_i^1(t)$ (%)*	*Comparative Gain of Labor $G_i^1(t)$*	*Investment Structure $s_i^2(t)$ (%)*	*Comparative Gain of Capital $G_i^2(t)$*
Total	100	100	–	100	–
Primary sector	1.412541	9.589077	0.147307	2.001552	0.70572
Secondary sector	52.42923	41.40547	1.266239	43.00189	1.21923
Tertiary sector	46.15823	49.00545	0.941900	54.99656	0.83929
Industries (*i*)					
Agriculture, forest, animal husbandry, fishery		9.589077		2.001552	
Mining		1.457099		3.039143	
Manufacturing		29.90330		36.22220	
Production/supply of electricity, gas, and water		0.807170		3.192258	
Construction	4.399643	9.237906	0.476260	0.548287	8.02435
Communication, transportation, storage, postal service	5.590204	3.646680	1.532957	7.319054	0.76379
Information transfer, computer, and software areas	1.522028	0.710205	2.143083	0.962497	1.58133

Table 3.8 (continued) Structures and Comparative Gain Measures of Tianjin's Three Economic Sectors and Industries ($t = 2011$)

Sectors (i)	*Demand Structure* $d_i(t)$ *(%)*	*Employment Structure* $s_i^1(t)$ *(%)*	*Comparative Gain of Labor* $G_i^1(t)$	*Investment Structure* $s_i^2(t)$ *(%)*	*Comparative Gain of Capital* $G_i^2(t)$
# Information Transfer					
Wholesale and retail	12.94644	14.29532	0.905610	1.701446	7.60908
Hotels and restaurants	1.720308	4.571780	0.376288	0.773433	2.22425
Finance	6.690380	1.643168	4.071635	0.527516	12.6828
Real estate/housing	3.638895	2.027098	1.795125	20.90133	0.17410
Leasing and commercial services	2.454790	3.471094	0.707209	5.950867	0.41251
Scientific research, technology service, and geological survey	2.942352	2.817234	1.044412	0.313288	9.39186
Water conservation, environment, and public facility management	0.642506	0.885791	0.725348	0.090005	7.13854
# Public Facility Management					
Residents and other services	2.096437	5.478537	0.382664	1.337830	1.56704
Education	2.199026	3.996541	0.550232	0.940662	2.33774
Health, safety, and welfare	1.057814	1.983857	0.533211	0.280801	3.76714
Social, sports, and entertainment	0.505869	0.609309	0.830234	1.398810	0.36164
Public administration and social organizations	2.151180	2.868337	0.749975	1.265932	1.69929

Source: Data are from *China Statistics Abstract 2012* (XLS, DOC).

the levels of the eighth and sixth of the relevant values of the same indicator for the secondary and tertiary sectors. In terms of the comparative gains of capital, the investments in the primary and the tertiary economic sectors are in extreme excess, reaching 0.70572 and 0.83929, respectively, while the investment in the secondary sector is badly needed (1.21923).

Second, in terms of the employment situation, the relevant data of the 21 industries of Tianjin City should contain 42 data points. However, because of the lack of raw data, we only collected 30 of them. So, we can only base our discussion on these available statistics. Among all the industries, only one is basically in equilibrium, where the comparative gain of labor in "scientific research, technology service, and geological survey" is equal to 1.044412. All the rest (29/30 ≈ 97%) are situated in extremely imbalanced states. In particular, the comparative gains of labor of the following five industries are greater than 1, especially, the comparative gain of "finance" is even equal to 4.071635, which is about (4.071635/0.376288 ≈ 4.071635/0.382664 ≈) 10 times of the lowest values of this indicate of "Hotels and restaurants" and "Residents and other services."

Finance	Information transfer, computer, and software areas	Real estate/housing
Communication, transportation, storage, postal service	Scientific research, technology service, and geological survey	

There are 11 industries whose comparative gains of capital are greater than 1, see table below. In particular, the comparative gain of capital of "finance" reached beyond 10, while the value of this indicator of "real estate and housing" is too low, equal to merely 0.17410, which is only (0.17410/12.6828 ≈) 0.013 of "finance" (0.17410/9.39186 ≈) 0.019 of "residents and other services" (0.17410/8.02435 ≈) 0.021 of "wholesale and retail," and (0.17410/7.60908 ≈) 0.022 of "information transfer, computer, and software areas."

Finance	Scientific research, technology service, and geological survey	Construction	Wholesale and retail
Water conservation, environment, and public facility management	Health, safety, and welfare	Education	Hotels and restaurants
Public administration and social organizations	Information transfer, computer, and software areas	Residents and other services	

Therefore, in terms of the magnitudes of the comparative gains of labor, the 15 industries of Tianjin with available data can be ranked as follows:

1	2	3	4
Finance	Information transfer, computer, and software areas	Real estate/housing	Communication, transportation, storage, postal service
5	6	7	8
Scientific research, technology service, and geological survey	Wholesale and retail	Social, sports, and entertainment	Public administration and social organizations
9	10	11	12
Water conservation, environment, and public facility management	Leasing and commercial services	Education	Health, safety, and welfare

13	14	15	
Construction	Residents and other services	Hotels and restaurants	

In terms of the magnitudes of the comparative gains of capital, the 15 industries of Tianjin with available data can be ranked as follows:

1	2	3	4
Finance	Scientific research, technology service, and geological survey	Construction	Wholesale and retail
5	6	7	8
Water conservation, environment, and public facility management	Health, safety, and welfare	Education	Hotels and restaurants
9	10	11	12
Public administration and social organizations	Information transfer, computer, and software areas	Residents and other services	Communication, transportation, storage, postal service
13	14	15	
Leasing and commercial services	Social, sports, and entertainment	Real estate/ housing	

3.4.2 Investment and Employment in Tianjin's Economic Sectors and Industries

First, in terms of the comparative gains of labor, the values of this indicator of the following five industries:

Finance	Information transfer, computer, and software areas	Real estate and housing
Communication, transportation, storage, postal service	Scientific research, technology service, and geological survey	

are over 1, representing relatively optimal industries for employment. In all the other 10 industries, there has been excessive supply of labor so that appropriate control on the increase of labor into these industries should be imposed. Specifically, the excessive labor in the following industries should be transferred into the previous five industries.

Wholesale and retail	Social, sports, and entertainment	Public administration and social organizations	Water conservation, environment, and public facility	Leasing and commercial services
Education	Health, safety, and welfare	Construction	Residents and other services	Hotels and restaurants

In other words, no matter where the job seeker, who is looking for employment opportunities in Tianjin, is from, the eastern part of China, the other parts of China, or the rest of the world, his/her chance is potentially in the secondary economic sector with the industries ranked as follows:

1	2	3	4	5
Finance	Information transfer, computer, and software areas	Real estate/ housing	Communication, transportation, storage, postal service	Scientific research, technology service, and geological survey

Without special skills, do not waste time and effort in looking for employment opportunities in the primary economic sector and other industries, because those areas are experiencing excessive labor supplies.

That is the direction of employment.

Second, in terms of the comparative gains of capital, the following 11 industries have enjoyed the comparative gains of more than 1, representing relatively more optimal areas of additional investment.

Finance	Scientific research, technology service, and geological survey	Construction	Wholesale and retail
Water conservation, environment, and public facility management	Health, safety, and welfare	Education	Hotels and restaurants
Public administration and social organizations	Information transfer, computer, and software areas	Residents and other services	

However, for the following four industries, excessive investment of capital has been given. Further capital investment in these industries should be curbed and redirected to the previous 11 industries.

Communication, transportation, storage, postal service	Leasing and commercial services	Social, sports, and entertainment	Real estate/ housing

In other words, no matter whether the investors, who look for investment opportunities in Tianjin, are from eastern China, or other parts of the Mainland China, or the rest of the world, they should first consider the following industries, ranked from the most optimal to the least:

1	2	3	
Finance	Scientific research, technology service, and geological survey	Construction	Wholesale and retail
4	5	6	7
Water conservation, environment, and public facility management	Health, safety, and welfare	Education	Hotels and restaurants
8	9	10	
Public administration and social organizations	Information transfer, computer, and software areas	Residents and other services	

At the same time, try not to invest additional capital in the following four industries, because there has been a severe over supply of capital in these industries.

That is the direction of capital investment.

Communication, transportation, storage, postal service	Leasing and commercial services	Social, sports, and entertainment	Real estate/housing

Third, by combining the two criteria discussed above, it follows that the industries in Tianjin that are most optimal for investment and employment are the following, because the comparative gains of labor and capital of each of these industries are over 1.

Finance	Information transfer, computer, and software areas	Scientific research, technology service, and geological survey

The secondary choice is the industry of "wholesale and retail." And the last choice is the industry of "leasing and commercial services."

3.5 Beijing

Located in northern China, Beijing, also known as Peking (Cotterell, 2007), is the capital of the People's Republic of China, the second largest Chinese city in terms of urban population, and among the most populous cities in the world. The metropolis is governed as a directly controlled municipality under the national government. It is surrounded by the Hebei Province and neighbors with Tianjin Municipality in the southeast. For details, see Figure 3.5.

As China's political, cultural, and educational center, Beijing is home to the headquarters of most of China's largest state-owned companies, and is a major hub for the national highway, expressway, railway, and high-speed rail networks. The Beijing Capital International Airport is the second busiest in the world in terms of passenger traffic.

Historically, human habitation in the Beijing municipality area can be traced back 230,000–250,000 years ago when the well-known Peking man lived in the caves of Dragon Bone Hill near

Figure 3.5 Geographic location of Beijing.

the village of Zhoukoudian in the Fangshan District. Beijing became a walled city in Ji, a city state that existed from the eleventh to the seventh century BC. This city state was later conquered by the state of Yan and made its capital under the name Yanjing. After the time of Yan during the Warring States period, the early imperial dynasties continued to employ the city as their prefectural capital under various names. The area is connected by many canals constructed during the Sui dynasty. By the fifteenth century, Beijing had essentially taken its current shape. The city wall of the Ming Dynasty continued to serve its purpose until modern times, when it was removed and the second ring road was built. It is generally believed that Beijing was the largest city in the world for most of the fifteenth through eighteenth centuries.

During the republican era, the fomenters of the Xinhai Revolution of 1911 sought to replace Qing rule with a republic and intended to return the capital to Nanjing. After the Qing general Yuan Shikai forced the abdication of the last Qing emperor and ensured the success of the revolution, the revolutionaries accepted him as president of the new Republic of China. Yuan maintained his capital at Beijing and quickly consolidated power, declaring himself emperor in 1915. Following the success of the Nationalists' Northern Expedition, the capital was formally moved to Nanjing in 1928, while Beijing's name was changed back to Beiping. The city fell to Japan's occupation on July 29, 1937, during the Second Sino-Japanese War and was made the seat of the Provisional Government of the Republic of China, a puppet state that ruled the ethnic-Chinese portions of Japanese-occupied northern China. This government was later merged into the larger Wang Jingwei government based in Nanjing.

In the final phases of the Chinese Civil War, the People's Liberation Army seized control of the city peacefully on January 31, 1949; and on October 1 of that year, Mao Zedong announced the creation of the People's Republic of China. He restored the name of the city, as well as the new capital, to Beijing. In the 1950s, the city began to expand drastically into its surrounding neighborhoods with heavy industries in the west and residential neighborhoods in the north. During the so-called Cultural Revolution from 1966 to 1976, the Red Guard movement began in Beijing. By the fall of 1966, all schools in the city were closed and over a million Red Guards from across the country gathered in Beijing for eight rallies in Tian'anmen Square. In October 1976, the Gang of Four was arrested and the Cultural Revolution came to an end. In December 1978, under the leadership of Deng Xiaoping, the Third Plenum of the 11th Party Congress, held in Beijing, reversed the verdicts against victims of the Cultural Revolution and instituted the "policy of reform and opening up."

Since the early 1980s, the urban area of Beijing has expanded greatly. As of 2005, the size of newly developed Beijing was one-and-a-half times larger than before 1980, where Wangfujing and Xidan were developed into shopping districts, and Zhongguancun a major center of electronics in China.

Geographically, Beijing is located at the northern tip of North China Plain. Mountains in the north, northwest, and west help to shield the city and the agricultural heartland of northern China from the encroaching desert steppes. The northwestern part of the municipality is dominated by the Jundu Mountains, while the western part is framed by the Western Hills. The major rivers, flowing through the municipality, include the Chaobai, Yongding, Juma, which are all tributaries of the Hai River. Beijing is also the northern end of the Grand Canal which was built over 1400 years ago as a transportation route, and the recent South–North Water Transfer Project constructed in the past decade to bring water from the Yangtze River basin. The city spreads out with several concentric ring roads, where the second ring road traces the old city walls and the sixth ring road connects original satellite towns in the surrounding suburbs.

Economically, Beijing is one of the most developed cities in China with the tertiary industry accounting for 73.2% of its gross domestic product. It has been home to many of the Fortune Global 500 companies and the largest companies in China. Finance is one of the most important industries. As of the end of 2007, financial organizations in Beijing generated ¥128.6 billion in revenue, which amounted to 11.6% of the total financial industry revenue of the entire country. In 2010, Beijing's nominal GDP reached ¥1.37 trillion, while its per capita GDP reached ¥78,194. And, the real estate and automobile sectors have continued to boom in Beijing in recent years. Beijing has become increasingly known for its innovative entrepreneurs and high-growth startup companies, backed by a large community of Chinese and foreign venture capital firms.

For more related reading on Beijing, please consult Cotterell (2007), Li et al. (2007), Harper (2007), and MacKerras and Yorke (1991).

3.5.1 Current State of Investment and Employment Structures of Beijing

For Beijing, we first compute the total outputs of economic sectors and industries, the investment and employment situations (see Table 3.9), and the corresponding comparative gains. Then, by comparing the comparative gains, we obtain the current structural states of investment and employment for Beijing's economic sectors and industries. All the detailed calculations are similar to what has been described above. The results are summarized in Table 3.10.

Table 3.10 indicates that first, in terms of the three major economic sectors, among the six relevant data points of the three sectors, only one is basically in equilibrium, where the comparative gains of labor of the tertiary sector is equal to 1.028227, while the other five (5/6 = 83%) are in extremely imbalanced states. In particular, the comparative gain of labor of the primary sector is way too low, and is equal to only 0.151798, which is not even equal to one-seventh of the same indicator of the secondary sector. In terms of the capital investment, the comparative gain of

Table 3.9 GDPs, Total Investments in Fixed Properties, and Employment Situation of Beijing's Three Economic Sectors and Industries (2011)

Sectors (i)	*Total Output of Sector (0.1 Billion Yuan)*	*Total Investment on Fixed Properties (0.1 Billion Yuan)*	*Employment (10,000 People)*
Total	16251.9	5910.6	1069.7
Primary sector	136.3	47.2	59.1
Secondary sector	3752.5	762.2	219.2
Tertiary sector	12363.1	5101.3	791.4
Industries (*i*)			
Agriculture, forest, animal husbandry, fishery	363.1		2.3156
Mining		8.7	6.5967
Manufacturing		572	104.7638
Production/supply of electricity, gas, and water		171.2	8.9107
Construction	703.7	10.3	39.1999
Communication, transportation, storage, postal service	809	699.7	55.0454
Information transfer, computer, and software areas	1493.4	112.9	47.6871
# Information Transfer			
Wholesale and retail	2139.7	28.7	60.4255
Hotels and restaurants	348.4	38.1	26.2983
Finance	2215.4	86.7	26.6538
Real estate/housing	1074.9	3284.7	32.4241
Leasing and commercial services	1162.1	46.2	55.6866
Scientific research, technology service, and geological survey	1135.5	92	45.8786
Water conservation, environment, and public facility management	86.3	415.8	8.5461
# Public Facility Management			
Residents and other services	112.1	11.5	6.6259
Education	605.9	133	37.9183
Health, safety, and welfare	311.5	48.1	20.9738

Table 3.9 (continued) GDPs, Total Investments in Fixed Properties, and Employment Situation of Beijing's Three Economic Sectors and Industries (2011)

Sectors (i)	*Total Output of Sector (0.1 Billion Yuan)*	*Total Investment on Fixed Properties (0.1 Billion Yuan)*	*Employment (10,000 People)*
Social, sports, and entertainment	339.4	55	15.4674
Public administration and social organizations	529.5	48.8	38.9308

Source: Data are from *Beijing Statistical Yearbook 2012* (XLS, DOC).

capital of the secondary sector is way too low, equaling only 1.79052, while the capital investment in the tertiary sector is in extreme excess (0.88140).

Second, in terms of the employment situation, the relevant data of the 21 industries of Beijing should contain 42 data points. However, because of lack of raw data, we only collected 30 of them. So, we can only base our discussion on these available statistics. Among all the industries, only two are basically in equilibrium, where the comparative gains of labor in "health, safety, and welfare" and "communication, transportation, storage, postal service" are, respectively, equal to 0.977550 and 0.967354. All the rest ($28/30 \geq 93\%$) are situated in extremely imbalanced states. In particular, the comparative gains of the following 10 industries are over 1. Especially, the comparative gain of labor in "finance" even reached 5.470800, which is more than ($5.470800/0.664662 \approx$) 8 times of the lowest value of this indicator that is reached by "water conservation, environment, and public facility management."

Table 3.10 Structures and Comparative Gain Measures of Beijing's Three Economic Sectors and Industries (t = 2011)

Sectors (i)	*Demand Structure $d_i(t)$ (%)*	*Employment Structure $s_i^1(t)$ (%)*	*Comparative Gain of Labor $G_i^1(t)$*	*Investment Structure $s_i^2(t)$ (%)*	*Comparative Gain of Capital $G_i^2(t)$*
Total	100	100	–	100	–
Primary sector	0.838671	5.524914	0.151798	0.798565	1.05022
Secondary sector	23.08961	20.49173	1.126777	12.89548	1.79052
Tertiary sector	76.07172	73.98336	1.028227	86.30765	0.88140
Industries (*i*)					
Agriculture, forest, animal husbandry, fishery		0.216472			
Mining		0.616687		0.147193	
Manufacturing		9.793755		9.677529	
Production/supply of electricity, gas, and water		0.833009		2.896491	

continued

Table 3.10 (continued) Structures and Comparative Gain Measures of Beijing's Three Economic Sectors and Industries ($t = 2011$)

Sectors (i)	*Demand Structure* $d_i(t)$ *(%)*	*Employment Structure* $s_i^1(t)$ *(%)*	*Comparative Gain of Labor* $G_i^1(t)$	*Investment Structure* $s_i^2(t)$ *(%)*	*Comparative Gain of Capital* $G_i^2(t)$
Construction	4.329955	3.664570	1.181573	0.174263	24.8472
Communication, transportation, storage, postal service	4.977880	5.145873	0.967354	11.83805	0.42050
Information transfer, computer, and software areas	9.189079	4.457988	2.061261	1.910128	4.81072
# Information Transfer					
Wholesale and retail	13.16585	5.648827	2.330722	0.485568	27.1143
Hotels and restaurants	2.143749	2.458474	0.871984	0.644605	3.32568
Finance	13.63164	2.491708	5.470800	1.466856	9.29310
Real estate/housing	6.613996	3.031140	4.497199	55.57304	0.24529
Leasing and commercial services	7.150549	5.205815	1.373570	0.781647	9.14806
Scientific research, technology service, and geological survey	6.986875	4.288922	1.629052	1.556526	4.48876
Water conservation, environment, and public facility management	0.531015	0.798925	0.664662	7.034819	0.07548
# Public Facility Management					
Residents and other services	0.689766	0.619417	1.113573	0.194566	3.54515
Education	3.728179	3.544760	1.051744	2.250195	1.65683
Health, safety, and welfare	1.916699	1.960718	0.977550	0.813792	2.35527
Social, sports, and entertainment	2.088371	1.445957	1.444283	0.930532	2.24428
Public administration and social organizations	3.258081	3.639413	0.895221	0.825635	3.94615

Source: Data are from *China Statistics Abstract 2012* (XLS, DOC).

Finance	Real estate/ housing	Wholesale and retail	Information transfer, computer, and software areas	Scientific research, technology service, and geological survey
Social, sports, and entertainment	Leasing and commercial services	Construction	Residents and other services	Education

At the same time, the imbalance in terms of the comparative gains of capital seems to be way too severe. The comparative gains of capital in the following 12 industries are over 1. In particular, the comparative gains of capital of both "wholesale and retail" and "construction" are more than 10, while the value of this indicator of "water conservation, environment, and public facility management" is very low, and equal to only 0.07548. That value is merely (0.07548/27.1143 ≈) 0.0028 when compared to that of "wholesale and retail" (0.07548/24.8472 ≈) 0.00304 when compared to "construction" (0.07548/9.2931 ≈) 0.00818 when compared to that of "finance," and (0.07548/9.14806 ≈) 0.00825 when compared to that of "leasing and commercial services."

Wholesale and retail	Construction	Finance	Leasing and commercial services
Information transfer, computer, and software areas	Scientific research, technology service, and geological survey	Public administration and social organizations	Residents and other services
Hotels and restaurants	Health, safety, and welfare	Social, sports, and entertainment	Education

So, in terms of the magnitudes of the comparative gains of labor, the 15 industries of Beijing with available data can be ranked as follows:

In terms of the magnitudes of the comparative gains of capital, the 15 industries of Beijing with available data can be ranked as follows:

1	2	3	4
Finance	Real estate/ housing	Wholesale and retail	Information transfer, computer, and software areas
5	6	7	8
Scientific research, technology service, and geological survey	Social, sports, and entertainment	Leasing and commercial services	Construction
9	10	11	12
Residents and other services	Education	Health, safety, and welfare	Communication, transportation, storage, postal service

13	14	15	
Public administration and social organizations	Hotels and restaurants	Water conservation, environment, and public facility management	

1	2	3	4
Wholesale and retail	Construction	Finance	Leasing and commercial services
5	6	7	8
Information transfer, computer, and software areas	Scientific research, technology service, and geological survey	Public administration and social organizations	Residents and other services
9	10	11	12
Hotels and restaurants	Health, safety, and welfare	Social, sports, and entertainment	Education
13	14	15	
Communication, transportation, storage, postal service	Real estate/housing	Water conservation, environment, and public facility management	

3.5.2 Investment and Employment in Beijing's Economic Sectors and Industries

First, in terms of the comparative gains of labor, the values of this indicator of the following 10 industries:

Finance	Real estate/housing	Wholesale and retail	Information transfer, computer, and software areas	Scientific research, technology service, and geological survey
Social, sports, and entertainment	Leasing and commercial services	Construction	Residents and other services	Education

are over 1, representing relatively optimal industries for employment. In the other five industries, there has been excessive supply of labor so that appropriate control on the increase of labor into these industries should be imposed. Specifically, the excessive labor in the following industries should be transferred into the previous 10 industries.

Water conservation, environment, and public facility management	Hotels and restaurants	Public administration and social organizations

In other words, no matter where the job seeker, who is looking for employment opportunities in Beijing, is from, the eastern part of China, the other parts of China, or the rest of the world, his/her chance is potentially in the tertiary and the secondary economic sectors with the industries ranked as follows:

1	2	3	4	5
Finance	Real estate/ housing	Wholesale and retail	Information transfer, computer, and software areas	Scientific research, technology service, and geological survey
6	7	8	9	10
Social, sports, and entertainment	Leasing and commercial services	Construction	Residents and other services	Education

Without special skills, do not waste time and effort in looking for employment opportunities in the primary economic sector and other industries, because those areas are experiencing excessive labor supplies.

That is the direction of employment.

Second, in terms of the comparative gains of capital, the following 12 industries have enjoyed the comparative gains of more than 1, representing relatively more optimal areas of additional investment.

Wholesale and retail	Construction	Finance	Leasing and commercial services
Information transfer, computer, and software areas	Scientific research, technology service, and geological survey	Public administration and social organizations	Residents and other services
Hotels and restaurants	Health, safety, and welfare	Social, sports, and entertainment	Education

However, for the following three industries, excessive investment of capital has been given. Further capital investment in these industries should be curbed and redirected to the previous 12 industries.

Communication, transportation, storage, postal service	Real estate/housing	Water conservation, environment, and public facility management

In other words, no matter whether the investors, who look for investment opportunities in Beijing, are from eastern China, or other parts of the Mainland China, or the rest of the world, they should first consider the following industries, ranked from the most optimal to the least:

Wholesale and retail	Construction	Finance	Leasing and commercial services
Information transfer, computer, and software areas	Scientific research, technology service, and geological survey	Public administration and social organizations	Residents and other services
Hotels and restaurants	Health, safety, and welfare	Social, sports, and entertainment	Education

That is the direction of capital investment.

Third, by combining the two criteria discussed above, it follows that the industries in Beijing that are most optimal for investment and employment are the following, because the comparative gains of labor and capital of each of these industries are over 1.

Wholesale and retail	Finance	Information transfer, computer, and software areas	Scientific research, technology service, and geological survey	Leasing and commercial services
Construction	Residents and other services	Social, sports, and entertainment	Education	

The worst choice is the industry of "water conservation, environment, and public facility management."

3.6 Shandong Province

Shandong is a coastal province, a part of the East China region, and is located on the eastern edge of the North China Plain and in the lower reaches of the Yellow River. It extends out to sea in the form of the Shandong Peninsula. The province borders the Bohai Sea in the north, the Hebei Province in the northwest, the Henan Province in the west, the Jiangsu Province in the south, and the Yellow Sea in the southeast. It also shares a very short border with the Anhui Province between Henan and Jiangsu. It has been a major part of Chinese history since the beginning of Chinese civilization and served as a pivotal cultural and religious site of Taoism, Chinese Buddhism, and Confucianism. For details, see Figure 3.6.

Historically, due to its particular location, Shandong has been influenced by and played a major role in Chinese civilization since antiquity, where the earliest dynasties, such as the Shang Dynasty and Zhou Dynasty, exerted varying degrees of control over Shandong. During the Spring and

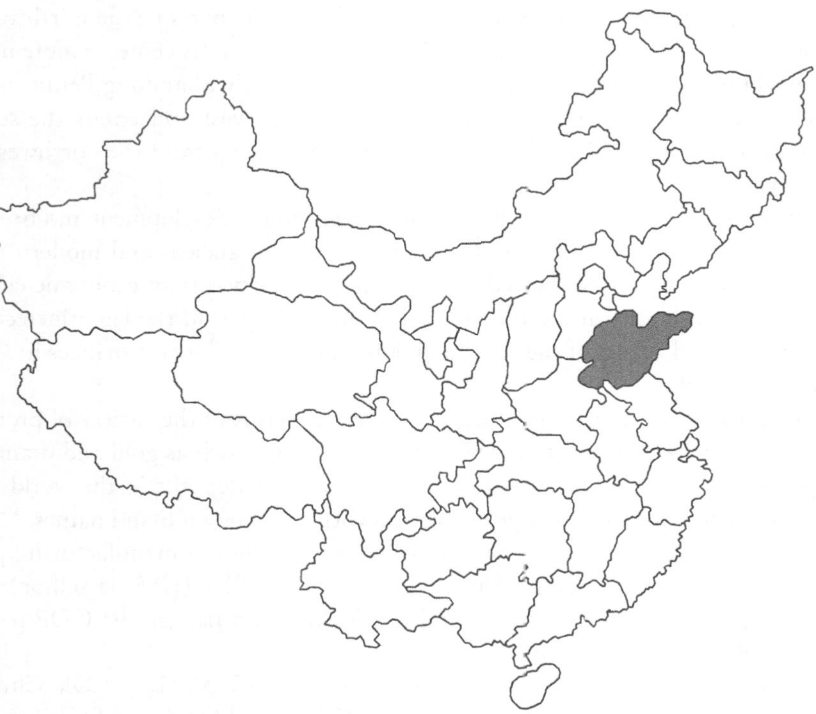

Figure 3.6 Geographic location of Shandong.

Autumn Period and the Warring States period, Shandong was home to two powerful states: Qi and Lu, which were eventually succumbed to the powerful state of Chu from the south. The Qin Dynasty conquered Qi and founded the first centralized Chinese state in 221 BCE. The following Han Dynasty created two "provinces" in what is now modern Shandong. In 589, the Sui Dynasty reestablished unity in China, and the Tang Dynasty (618–907) developed the next golden age of China. During the earlier part of this period, Shandong was ruled as part of Henan Circuit. Later on China splintered into warlord factions, resulting in the Five Dynasties and Ten Kingdoms period. Shandong was part of the Five Dynasties of the north. After Song Dynasty reunified China in the late tenth century, Song was forced to cede northern China to the Jurchen Jin Dynasty in 1142, where Shandong was administered by Jin. The modern Shandong was initially created by the Ming Dynasty. During the Qing Dynasty, Shandong acquired its current borders.

In the nineteenth century, China was increasingly exposed to Western influence. As a coastal province, Shandong was especially affected. For example, Qingdao was leased to Germany in 1897 and Weihai to Britain in 1898. Additionally, during the wave of Han Chinese immigration into Manchuria, Shandong was the main source of the ensuing tide of migrants. In World War I, Germany lost its influence in Shandong; and the Treaty of Versailles transferred the German concessions of the area to Japan. This outcome led to the May Fourth Movement, which helped to revert Shandong to Chinese control in 1922. In 1937 Japan began its invasion of China proper in the Second Sino-Japanese War; and Shandong was occupied in its entirety by Japan with Chinese resistance continuing in the countryside. By 1945, communist forces held parts of Shandong and they expanded their holdings over the next four years of the Chinese Civil War and gained the total control of the province in June 1949.

Geographically, the terrain of Shandong is mostly flat. Its parts in the northwestern, western, and southwestern are all part of the vast North China Plain. Its center is more mountainous with Mount Tai being the most prominent. Its east is the hilly Shandong Peninsula extending into the sea. The Yellow River passes through Shandong's west and enters the sea along the northern coast of Shandong. The Grand Canal enters Shandong from the northwest and leaves on the southwest.

In recent years, Shandong has enjoyed significant economic development and been one of the richest provinces of China. Its location at the intersection of ancient and modern north–south and east–west trading routes has helped to establish the province as an economic center. After a period of political instability and economic hardship that began in the late nineteenth century, Shandong has emerged as one of the most populous and most affluent provinces in China (GDP of ¥3.94 trillion in 2010).

Shandong ranks first among all Chinese provinces in terms of the variety of products it produces, including cotton and wheat as well as precious metals, such as gold and diamonds. It has extensive petroleum deposits and one of the biggest sapphire deposits in the world. Shandong's economic development focuses on large enterprises with well-known brand names.

It has been the biggest industrial producer and one of the top manufacturing provinces in China. In 2011, the nominal GDP for Shandong was ¥4.50 trillion (US$711 billion), making the province ranking third in the country behind Guangdong and Jiangsu. Its GDP per capita was ¥42,014 (US$6365), ranking eighth.

For more detailed reading, please consult Roberts (1837, pp. 122–123), Chinadaily.com (2011–2012), National Bureau of Statistics of China (2013), and Ministry of Commerce (2013).

3.6.1 Current State of Investment and Employment Structures of Shandong

For the Shandong Province, we first compute the total outputs of economic sectors and industries, the investment and employment situations (see Table 3.11), and the corresponding comparative gains. Then, by comparing the comparative gains, we obtain the current structural states of investment and employment for Tianjin's economic sectors and industries. All the detailed calculations are similar to that described above. The results are summarized in Table 3.12.

Table 3.12 indicates that first, in terms of the three major economic sectors, among the six relevant data points of the three sectors, all of them are in extremely imbalanced states. In particular, the comparative gain of labor of the primary sector is way too low, and is equal to only 0.256900, which is not even equal to one-sixth and one-fourth of those of the same indicator of the secondary and tertiary sectors. In terms of the capital investment, the comparative gains of capital of the primary and the secondary sectors are way too low, and are, respectively, 4.25896 and 1.10478, while the capital investment in the tertiary sector is in extreme excess (0.76559).

Second, in terms of the employment situation, the relevant data of the 21 industries of Shandong Province should contain 42 data points. However, because of lack of raw data, we only collected 30 of them. So, we can only base our discussion on these available statistics. Among all the industries, only one is basically in equilibrium, where the comparative gain of capital in "communication, transportation, storage, postal service" is 0.95352. All the rest (29/30 ≥ 97%) are situated in extremely imbalanced states. In particular, the comparative gains of labor of the following 13 industries are over 1. Especially, the comparative gain of labor in "real estate and housing" even

Table 3.11 GDPs, Total Investments in Fixed Properties, and Employment Situation of Shandong's Three Economic Sectors and Industries (2011)

Sectors (i)	*Total Output of Sector (0.1 Billion Yuan)*	*Total Investment on Fixed Properties (0.1 Billion Yuan)*	*Employment (10,000 People)*
Total	45361.85	25927.1	6485.6
Primary sector	3973.85	533.3	2211.6
Secondary sector	24017.11	12425.3	2185.6
Tertiary sector	17370.89	12968.5	2088.4
Industries (*i*)			
Agriculture, forest, animal husbandry, fishery		533.3392	2213.5
Mining		592.6163	82.9
Manufacturing		10765.7528	1362.3
Production/supply of electricity, gas, and water		628.5558	24.5
Construction	2741.22	438.3954	715.1
Communication, transportation, storage, postal service	2323.38	1395.6809	254.3
Information transfer, computer, and software areas	541.79	69.1253	63.5
# Information Transfer			
Wholesale and retail	5400.19	947.599	664.3
Hotels and restaurants	881.58	405.8822	195.5
Finance	1640.41	33.2947	40.4
Real estate/housing	1838.14	6032.0778	36.6
Leasing and commercial services	556.01	251.3748	61.8
Scientific research, technology service, and geological survey	317.99	238.4121	22.5
Water conservation, environment, and public facility management	163.51	1370.1989	15.4
# Public Facility Management			
Residents and other services	497.61	344.6163	48.9
Education	951.55	270.6188	120.3

continued

Table 3.11 (continued) GDPs, Total Investments in Fixed Properties, and Employment Situation of Shandong's Three Economic Sectors and Industries (2011)

Sectors (i)	*Total Output of Sector (0.1 Billion Yuan)*	*Total Investment on Fixed Properties (0.1 Billion Yuan)*	*Employment (10,000 People)*
Health, safety, and welfare	691.22	191.4104	53.2
Social, sports, and entertainment	144.82	542.8502	10.3
Public administration and social organizations	1417.69	875.3324	117.6

Source: Data are from *China Statistics Abstracts 2012* (XLS, DOC).

Table 3.12 Structures and Comparative Gain Measures of Shandong's Three Economic Sectors and Industries ($t = 2011$)

Sectors (i)	*Demand Structure $d_i(t)$ (%)*	*Employment Structure $s_i^1(t)$ (%)*	*Comparative Gain of Labor $G_i^1(t)$*	*Investment Structure $s_i^2(t)$ (%)*	*Comparative Gain of Capital $G_i^2(t)$*
Total	100	100	–	100	–
Primary sector	8.760335	34.10016	0.256900	2.056921	4.25896
Secondary sector	52.94561	33.69927	1.571120	47.92399	1.10478
Tertiary sector	38.29405	32.20057	1.189235	50.01909	0.76559
Industries (*i*)					
Agriculture, forest, animal husbandry, fishery		34.12946		2.057072	
Mining		1.278216		2.285702	
Manufacturing		21.00500		41.52317	
Production/supply of electricity, gas, and water		0.377760		2.424320	
Construction	6.043008	11.02597	0.548071	1.690877	3.57389
Communication, transportation, storage, postal service	5.132904	3.920994	1.309082	5.383097	0.95352
Information transfer, computer, and software areas	1.194374	0.979092	1.219879	0.266614	4.47978

Table 3.12 (continued) Structures and Comparative Gain Measures of Shandong's Three Economic Sectors and Industries ($t = 2011$)

Sectors (i)	*Demand Structure* $d_i(t)$ *(%)*	*Employment Structure* $s_i^1(t)$ *(%)*	*Comparative Gain of Labor* $G_i^1(t)$	*Investment Structure* $s_i^2(t)$ *(%)*	*Comparative Gain of Capital* $G_i^2(t)$
# Information Transfer					
Wholesale and retail	11.90470	10.24269	1.162262	3.654859	3.25722
Hotels and restaurants	1.943439	3.014370	0.644725	1.565475	1.24144
Finance	3.616277	0.622918	5.805377	0.128417	28.1605
Real estate/housing	4.052172	0.564327	7.180537	23.26553	0.17417
Leasing and commercial services	1.225722	0.952880	1.286333	0.969545	1.26422
Scientific research, technology service, and geological survey	0.701008	0.346922	2.020647	0.919548	0.76234
Water conservation, environment, and public facility management	0.360457	0.237449	1.518039	5.284814	0.06821
# Public Facility Management					
Residents and other services	1.096979	0.753978	1.454922	1.329174	0.82531
Education	2.097688	1.854879	1.130903	1.043768	2.00973
Health, safety, and welfare	1.523792	0.820279	1.857651	0.738264	2.06402
Social, sports, and entertainment	0.319255	0.158813	2.010253	2.093756	0.15248
Public administration and social organizations	3.125291	1.813248	1.723588	3.376129	0.92570

Source: Data are from *China Statistics Abstract 2012* (XLS, DOC).

reached 7.180537, which is more than (7.180537/0.548071 ≈) 13 times the lowest value of this indicator reached by residents and other services.

Real estate/ housing	Finance	Scientific research, technology service, and geological survey	Social, sports, and entertainment	Health, safety, and welfare
Public administration and social organizations	Water conservation, environment, and public facility management	Residents and other services	Communication, transportation, storage, postal service	Leasing and commercial services
Information transfer, computer, and software areas	Wholesale and retail	Education		

At the same time, the imbalance in terms of the comparative gains of capital seems to be way too severe. The comparative gains of capital in the following eight industries are over 1. In particular, the comparative gains of capital of "finance" reached beyond 10, equaling as high as 28.1605, while the value of this indicator of "water conservation, environment, and public facility management" is very low, equaling only 0.06821, which is (0.06821/28.1605 ≈) 0.0024 when compared to that of "finance" (0.06821/4.47978 ≈) 0.015 when compared to "information transfer, computer, and software areas" (0.06821/3.57389 ≈) 0.018 when compared to "construction," and (0.06821/3.25722 ≈) 0.021 when compared to "wholesale and retail."

Finance	Information transfer, computer, and software areas	Construction	Wholesale and retail
Health, safety, and welfare	Education	Leasing and commercial services	Hotels and restaurants

So, in terms of the magnitudes of the comparative gains of labor, the 15 industries of Shandong with available data can be ranked as follows:

1	2	3	4
Real estate/ housing	Finance	Scientific research, technology service, and geological survey	Social, sports, and entertainment

5	6	7	8
Health, safety, and welfare	Public administration and social organizations	Water conservation, environment, and public facility management	Residents and other services
9	10	11	12
Communication, transportation, storage, postal service	Leasing and commercial services	Information transfer, computer, and software areas	Wholesale and retail
13	14	15	
Education	Hotels and restaurants	Construction	

In terms of the magnitudes of the comparative gains of capital, the 15 industries of Shandong with available data can be ranked as follows:

1	2	3	4
Finance	Information transfer, computer, and software areas	Construction	Wholesale and retail
5	6	7	8
Health, safety, and welfare	Education	Leasing and commercial services	Hotels and restaurants
9	10	11	12
Communication, transportation, storage, postal service	Public administration and social organizations	Residents and other services	Scientific research, technology service, and geological survey
13	14	15	
Real estate/housing	Social, sports, and entertainment	Water conservation, environment, and public facility management	

3.6.2 Investment and Employment in Shandong's Economic Sectors and Industries

First, in terms of the comparative gains of labor, the values of this indicator of the following 13 industries

Real estate/ housing	Finance	Scientific research, technology service, and geological survey	Social, sports, and entertainment	Health, safety, and welfare
Public administration and social organizations	Water conservation, environment, and public facility management	Residents and other services	Communication, transportation, storage, postal service	Leasing and commercial services
Information transfer, computer, and software areas	Wholesale and retail	Education		

are over 1, representing relatively optimal industries for employment. In the other two industries, there has been excessive supply of labor so that appropriate control on the increase of labor into these industries should be imposed and the excessive labor in these industries should be transferred into the previous 13 industries.

In other words, no matter where the job seeker, who is looking for employment opportunities in Shandong, is from, the eastern part of China, the other parts of China, or the rest of the world, his/her chance is potentially in the tertiary and the secondary economic sectors with the industries ranked as follows:

1	2	3	4	5
Real estate/ housing	Finance	Scientific research, technology service, and geological survey	Social, sports, and entertainment	Health, safety, and welfare
6	7	8	9	10
Public administration and social organizations	Water conservation, environment, and public facility management	Residents and other services	Communication, transportation, storage, postal service	Leasing and commercial services
11	12	13		
Information transfer, computer, and software areas	Wholesale and retail	Education		

Without special skills, do not waste time and effort in looking for employment opportunities in the primary economic sector and other industries, because those areas are experiencing excessive labor supplies.

That is the direction of employment.

Second, in terms of the comparative gains of capital, the following seven industries have enjoyed the comparative gains of more than 1, representing relatively more optimal areas of additional investment.

Communication, transportation, storage, postal service	Public administration and social organizations	Residents and other services	Scientific research, technology service, and geological survey
Real estate/housing	Social, sports, and entertainment	Water conservation, environment, and public facility management	

However, for the following eight industries, excessive investment of capital has been given. Further capital investment in these industries should be curbed and redirected to the previous seven industries.

In other words, no matter whether the investors, who look for investment opportunities in Shandong, are from eastern China, or other parts of the Mainland China, or the rest of the world, they should first consider the following industries, ranked from the most optimal to the least:

1	2	3	4
Finance	Information transfer, computer, and software areas	Construction	Wholesale and retail
5	6	7	8
Health, safety, and welfare	Education	Leasing and commercial services	Hotels and restaurants

At the same time, try not to invest additional capital in the following seven industries, because there has been a severe over supply of capital in these industries.

That is the direction of capital investment.

Communication, transportation, storage, postal service	Public administration and social organizations	Residents and other services	Scientific research, technology service, and geological survey
Real estate/housing	Social, sports, and entertainment	Water conservation, environment, and public facility management	

Third, by combining the two criteria discussed above, it follows that the industries in Shandong that are most optimal for investment and employment are the following, because the comparative gains of labor and capital of each of these industries are over 1.

Finance	Health, safety, and welfare	Information transfer, computer, and software areas
Wholesale and retail	Leasing and commercial services	Education

The worst choice is the industry of "water conservation, environment, and public facility management."

3.7 Hebei Province

Hebei Province is located in the North China region. Both Beijing and Tianjin municipalities were carved out of Hebei. The province borders Liaoning Province in the northeast, Inner Mongolia to the north, Shanxi Province in the west, Henan Province in the south, and Shandong Province in the southeast. Bohai Bay of the Yellow Sea is located in the east of Hebei. And there is a small part of Hebei that is disjointed from the rest of the province and is wedged between the municipalities of Beijing and Tianjin. In terms of its geography, most of central and southern Hebei lies within the North China Plain. The western side of Hebei rises into the Taihang Mountains, while the Yan Mountains run through northern Hebei, beyond which lie the grasslands of Inner Mongolia. The Great Wall cuts through northern Hebei from east to west as well, and terminates at the seacoast of Shanhaiguan in northeastern Hebei. For details, see Figure 3.7.

Historically, Hebei was the home of Peking man that lived in the area around 200,000–700,000 years ago. During the Spring and Autumn period (722 BC–476 BC), Hebei was ruled by the states of Yan in the north and Jin in the south. Also, a nomadic people, known as Dí, invaded northern China and established the state Zhongshan in central Hebei. During the Warring States period (403–221 BC), Jin was partitioned, and much of its territory within Hebei went to Zhao. During the time of Three Kingdoms, Hebei was under the rule of the Kingdom of Wei. During

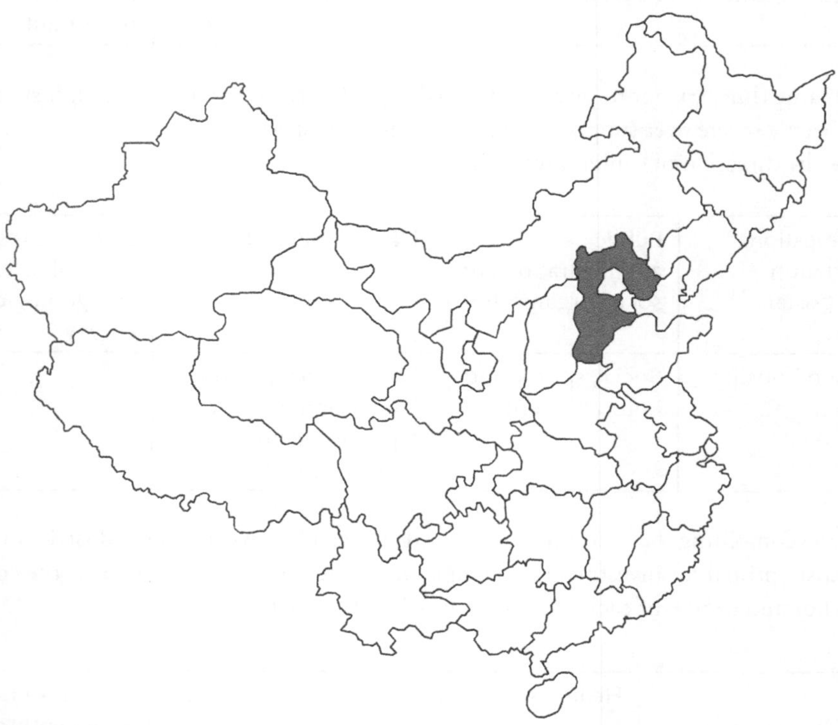

Figure 3.7 Geographic location of Hebei.

the Tang Dynasty (618–907), the area was formally designated as Hebei for the first time. The Yuan Dynasty divided China into provinces, but did not establish Hebei as a province. Rather, the area was directly administrated by the Secretariat at capital Dadu. In the Ming Dynasty, Hebei was directly ruled by the imperial capital, Beijing. When the Manchu Qing Dynasty came to power in 1644, Hebei was again directly ruled. The founding of the People's Republic of China in 1949 witnessed several changes: the region around Chengde and the region around Zhangjiakou were merged into Hebei, extending its borders northward beyond the Great Wall.

In 2011, Hebei's GDP reached ¥2.40 trillion (US$379 billion) and was ranked 6th in China. Its GDP per capita was ¥24,428. The primary, secondary, and tertiary sectors of industry contributed ¥203.46 billion, ¥877.74 billion, and ¥537.66 billion, respectively. The registered urban unemployment rate was 3.96%. Forty percent of Hebei's labor force works in the agriculture, forestry, and animal husbandry sectors. The majority of the production from these industries goes to Beijing and Tianjin. The main agricultural products of Hebei are cereal crops, such as wheat, maize, millet, and sorghum; and cash crops, such as cotton, peanut, soya bean, and sesame, are also produced. Kailuan is one of China's first modern coal mines with over 100 years of history, producing over 20 million metric tons of coal annually. Much of the North China Oilfield is located in Hebei along with major iron mines. Iron and steel manufacturing are the largest industries in Hebei. The province has continued its growth as a manufacturing and transportation center for the region.

For detailed reading, please consult Moore (2011) and Pomeranz (2010).

3.7.1 Current State of Investment and Employment Structures of Hebei

For the Hebei Province, we first compute the total outputs of economic sectors and industries, the investment and employment situations (see Table 3.13), and the corresponding comparative gains. Then, by comparing the comparative gains, we obtain the current structural states of investment and employment for Tianjin's economic sectors and industries. All the detailed calculations are similar to what has been described above. The results are summarized in Table 3.14.

Because there are too many missing data for this province, the relevant comparative gains of investment and employment can be computed only partially. So, the investment and employment problems can only be addressed partially.

Table 3.14 indicates that first, in terms of the three major economic sectors, all of the six relevant data points of the three sectors are in extremely imbalanced states. In particular, the comparative gain of labor of the primary sector is way too low, and is equal to only 0.326227, which

Table 3.13 GDPs, Total Investments in Fixed Properties, and Employment Situation of Hebei's Three Economic Sectors and Industries (2011)

Sectors (i)	*Total Output of Sector (0.1 Billion Yuan)*	*Total Investment on Fixed Properties (0.1 Billion Yuan)*	*Employment (10,000 People)*
Total	24515.76	16389.3254	3962.42
Primary sector	2905.73	590.3648	1439.63
Secondary sector	13126.86	7462.5206	1319.83
Tertiary sector	8483.17	8336.4401	1202.96

continued

Table 3.13 (continued) GDPs, Total Investments in Fixed Properties, and Employment Situation of Hebei's Three Economic Sectors and Industries (2011)

Sectors (i)	*Total Output of Sector (0.1 Billion Yuan)*	*Total Investment on Fixed Properties (0.1 Billion Yuan)*	*Employment (10,000 People)*
Industries (*i*)			
Agriculture, forest, animal husbandry, fishery	2905.73	559.6269	1439.63
Mining		439.7102	96.65
Manufacturing		5468.1058	791.02
Production/supply of electricity, gas, and water		668.9598	35.94
Construction	1356.48	53.7937	396.22
Communication, transportation, storage, postal service	2046.22	1521.2226	183.93
Information transfer, computer, and software areas		41.2495	17.93
# Information Transfer			
Wholesale and retail	1780.63	464.3961	326.44
Hotels and restaurants	338.91	147.7126	151.92
Finance	746.01	12.8319	29.53
Real estate/housing	918.02	3549.5818	10.42
Leasing and commercial services		109.7529	37.79
Scientific research, technology service, and geological survey		60.8801	12.21
Water conservation, environment, and public facility management		1396.5308	17.23
# Public Facility Management			
Residents and other services		51.3029	126.17
Education		149.1112	121.68
Health, safety, and welfare		112.2103	51.81
Social, sports, and entertainment		128.9332	17.39
Public administration and social organizations		147.4409	98.51

Source: All data underlined is from *China Statistics Yearbook 2012*, edited by National Bureau of Statistics of China, published by China Statistics Press and the rest from *Hebei Province's Statistics Yearbook 2012*.

Table 3.14 Structures and Comparative Gain Measures of Hebei's Three Economic Sectors and Industries ($t = 2011$)

Sectors (i)	*Demand Structure* $d_i(t)$ *(%)*	*Employment Structure* $s_i^1(t)$ *(%)*	*Comparative Gain of Labor* $C_i^1(t)$	*Investment Structure* $s_i^2(t)$ *(%)*	*Comparative Gain of Capital* $C_i^2(t)$
Total	100	100	–	100	–
Primary sector	11.85250	36.33209	0.326227	3.60213	3.29041
Secondary sector	53.54458	33.30869	1.607526	45.53281	1.17596
Tertiary sector	34.60292	30.35922	1.139783	50.86506	0.68029
Industries (*i*)					
Agriculture, forest, animal husbandry, fishery	11.85250	36.33209	0.326227	3.414582	3.47114
Mining		2.439166		2.682906	
Manufacturing		19.96305		33.36382	
Production/supply of electricity, gas, and water		0.907021		4.08168	
Construction	5.533094	9.999445	0.553340	0.328224	16.8577
Communication, transportation, storage, postal service	8.346549	4.641860	1.798104	9.281789	0.89924
Information transfer, computer, and software areas		0.452501		0.251685	
# Information Transfer					
Wholesale and retail	7.263205	8.238400	0.881628	2.833528	2.56331
Hotels and restaurants	1.382417	3.834021	0.360566	0.901273	1.53385
Finance	3.042981	0.745252	4.083160	0.078294	38.8660
Real estate/housing	3.744612	0.26291	14.23966	21.65789	0.17290
Leasing and commercial services		0.953710		0.669661	
Scientific research, technology service, and geological survey		0.308145		0.371462	

continued

Table 3.14 (continued) Structures and Comparative Gain Measures of Hebei's Three Economic Sectors and Industries (t = 2011)

Sectors (i)	*Demand Structure $d_i(t)$ (%)*	*Employment Structure $s_i^1(t)$ (%)*	*Comparative Gain of Labor $G_i^1(t)$*	*Investment Structure $s_i^2(t)$ (%)*	*Comparative Gain of Capital $G_i^2(t)$*
Water conservation, environment, and public facility management		0.434835		8.520978	
# Public Facility Management					
Residents and other services		3.184165		0.313026	
Education		3.070851		0.909807	
Health, safety, and welfare		1.307534		0.684655	
Social, sports, and entertainment		0.438873		0.78669	
Public administration and social organizations		2.486107		0.899615	

Source: Data are from *China Statistics Abstract 2012* (XLS, DOC).

is not even equal to one-fifth, and one-third of those of the same indicator of the secondary and tertiary sectors. In terms of the capital investment, the comparative gains of capital of the primary and the secondary sectors are way too low, and are, respectively, 3.29041 and 1.17596, while the capital investment in the tertiary sector is in extreme excess (0.68029).

Second, in terms of the employment situation, the comparative gains of labor and investment can only be calculated partially based on the available data. So, we can only address the investment and employment problems partially.

The relevant data of the 21 industries of Hebei Province should contain 42 data points. However, because of lack of raw data, we only collected 14 of them. So, we can only base our discussion on these available statistics. All the industries are in extremely imbalanced states. In particular, the comparative gains of labor of the following three industries are over 1. Especially, the comparative gain of labor in "real estate and housing" even reached as high as 14.23966, which is more than (14.23966/0.326227 ≈) 44 times of the lowest value of this indicator that is reached by "agriculture, forest, animal husbandry, fishery."

Real estate/housing	Finance	Communication, transportation, storage, postal service

At the same time, the imbalance in terms of the comparative gains of capital seems to be very severe. The comparative gains of capital of the following five industries are over 1. In particular, the comparative gains of capital of both "finance" and "construction" reached beyond 10, arriving as high as at 38.8660 and 16.8577, respectively, while the value of this indicator of "real estate and housing" is very low, equaling merely 0.17290, which is merely (0.17290/38.8660 ≈) 0.0044 when compared to "finance," (0.17290/16.8577 ≈) 0.010 when compared to "construction," (0.17290/3.47114 ≈) 0.049 when compared to "agriculture, forest, animal husbandry, fishery," and (0.17290/2.56331 ≈) 0.069 when compared to "wholesale and retail."

Finance	Construction	Agriculture, forest, animal husbandry, fishery	Wholesale and retail	Hotels and restaurants

So, in terms of the magnitudes of the comparative gains of labor, the seven industries of Hebei with available data can be ranked as follows:

1	2	3	4
Real estate/ housing	Finance	Communication, transportation, storage, postal service	Wholesale and retail
5	6	7	
Construction	Hotels and restaurants	Agriculture, forest, animal husbandry, fishery	

In terms of the magnitudes of the comparative gains of capital, the seven industries of Hebei with available data can be ranked as follows:

1	2	3	4
Finance	Construction	Agriculture, forest, animal husbandry, fishery	Wholesale and retail
5	6	7	
Hotels and restaurants	Communication, transportation, storage, postal service	Real estate/housing	

3.7.2 Investment and Employment in Hebei's Economic Sectors and Industries

First, in terms of the comparative gains of labor, the available data suggest that the values of this indicator of the following three industries:

Real estate/housing	Finance	Communication, transportation, storage, postal service

are over 1, representing relatively optimal industries for employment. In the other four industries, there has been excessive supply of labor so that appropriate control on the increase of labor into these industries should be imposed. The excessive labor in these industries should be transferred into the previous three industries.

In other words, no matter where the job seeker, who is looking for employment opportunities in Hebei, is from, the eastern part of China, the other parts of China, or the rest of the world, his/her chance is potentially in the tertiary and the secondary economic sectors with the industries ranked as follows:

1	2	3
Real estate/housing	Finance	Communication, transportation, storage, postal service

Without special skills, do not waste time and effort in looking for employment opportunities in the primary economic sector and other industries, because those areas are experiencing excessive labor supplies.

That is the direction of employment.

Second, in terms of the comparative gains of capital, based on the available data, only the following five industries have enjoyed the comparative gains of more than 1, representing relatively more optimal areas of additional investment.

Finance	Construction	Agriculture, forest, animal husbandry, fishery	Wholesale and retail	Hotels and restaurants

However, for the following two industries, excessive investment of capital has been given. Further capital investment in these industries should be curbed and redirected to the previous five industries.

Communication, transportation, storage, postal service	Real estate/housing

In other words, no matter whether the investors, who look for investment opportunities in Hebei, are from eastern China, or other parts of the Mainland China, or the rest of the world, they should first consider the following industries, ranked from the most optimal to the least:

1	2	3	4	5
Finance	Construction	Agriculture, forest, animal husbandry, fishery	Wholesale and retail	Hotels and restaurants

That is the direction of capital investment.

Communication, transportation, storage, postal service	Real estate/housing

Third, by combining the two criteria discussed above, it follows that the only industry in Hebei that is most optimal for investment and employment is "finance," because its comparative gains of labor of capital are over 1.

3.8 Zhejiang Province

Zhejiang is an eastern coastal province. It borders Jiangsu Province and Shanghai municipality in the north, Anhui Province in the northwest, Jiangxi Province in the west, and Fujian Province in the south. To the east is the East China Sea. The province consists mostly (about 70% of its total area) of hills with the highest altitudes in the south and west. Valleys and plains are found along the coastline and rivers. The province lies just south of the Yangtze Delta, and consists of plains around the cities of Hangzhou, Jiaxing, and Huzhou, where the Grand Canal enters from the northern border and ends in Hangzhou. For details, see Figure 3.8.

Historically, the area of modern Zhejiang was outside the major sphere of early Chinese civilization during the Shang Dynasty (sixteenth to the eleventh century BC). In the Spring and Autumn Period, the state of Yue emerged in northern Zhejiang. This state was heavily influenced by the Chinese civilization. Under King Goujian, Yue reached its pinnacle and wiped out the powerful state of Wu in 473 BC that was located to its immediate north. In 221 BC, the state of Qin completed the conquest of the last states of China, making what is now Zhejiang part of a unified Chinese empire.

Although Zhejiang was under the control of a unified China throughout the Qin Dynasty (221–206 BC) and Han Dynasty (206 BC to 220 AD), it was still on the fringe of the empire. During the Three Kingdoms era, Zhejiang was part of the Kingdom of Wu (222–280). The Sui Dynasty reestablished unity of China, rebuilt, and expanded the Grand Canal, which linked Hangzhou to the North China Plain and Zhejiang to other centers of Chinese civilization. In the Tang Dynasty (618–907), Zhejiang was part of the Jiangnan East Circuit and was cited in history

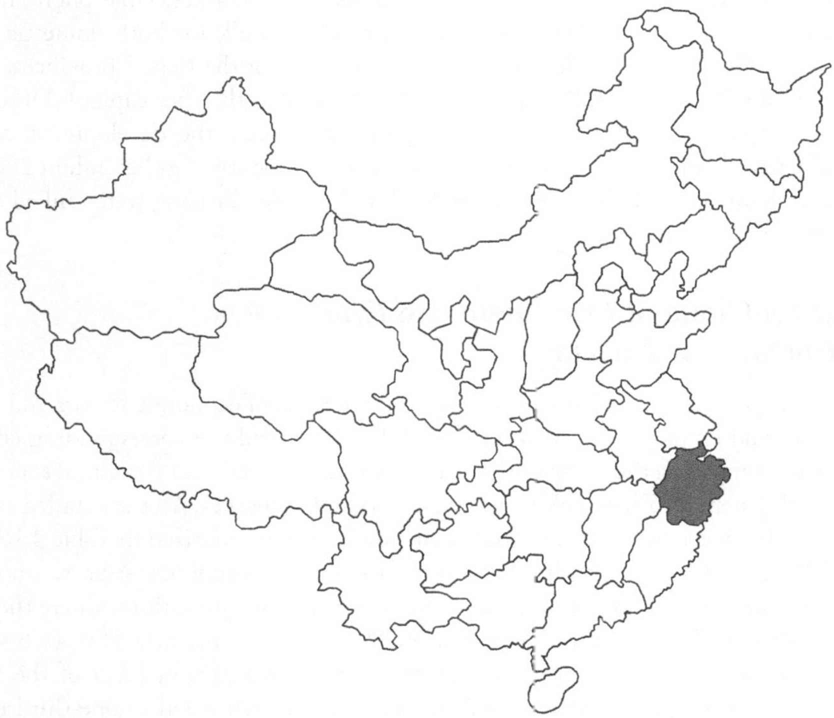

Figure 3.8 Geographic location of Zhejiang.

books for its prosperity. When the Northern Song Dynasty reestablished unity of China in 960, the prosperity of South China overtook that of North China. Since 1127, Hongzhou, a city of Zhejiang that might have been the largest city in the world at the time (Geography.about.com, 2013), was renowned for its prosperity and beauty. And, ever since then, north Zhejiang has been synonymous with luxury and affluence in Chinese culture. When Marco Polo visited the city of Hongzhou in the Yuan Dynasty, he called it the "finest and noblest city in the world" (Fordham.edu, 2014). The Ming Dynasty, which defeated the Mongols in 1368, created the province of Zhejiang, whose borders have not changed much since then. During the Ming dynasty and the following Qing dynasty, Zhejiang's ports represented important centers of international trade.

After the First Opium War, Ningbo became one of the five Chinese treaty ports opened to virtually unrestricted foreign trade under the terms of the Treaty of Nanjing, signed in 1843. During the Second Sino-Japanese War, Japan occupied much of Zhejiang province, placing it under the control of the Japanese puppet state known as the Reorganized National Government of China. During the Cultural Revolution (1966–1976), Zhejiang was in chaos with a stagnant economy. Although Zhejiang benefited less from central government investment than some other provinces, it has been an epicenter of capitalist development in China, and has led the nation in the development of a market economy and private enterprises (Interscience.wiley.com, 2013). By taking the opportunities provided by the economic reforms of Deng Xiaoping, south Zhejiang has become a major center of export. That, together with the traditional prosperity of north Zhejiang, has allowed Zhejiang to leapfrog forward and become one of the richest provinces of China.

Electromechanical industries, textiles, chemical industries, food, and construction materials are some of the main manufacturing sectors of Zhejiang. In recent years, the province has established its own economic model, which prioritizes and encourages entrepreneurship with an emphasis on small businesses responsive to the whims of the market, large public investments on infrastructure, and the production of low-cost goods in bulk for both domestic consumption and export. Consequently, Zhejiang has made itself one of the richest provinces. Zhejiang's nominal GDP for 2011 was ¥3.20 trillion (US$506 billion) with a per capita GDP of ¥44,335 (US$6490) (People.com, 2013). Compared to other provinces, the development of different regions in Zhejiang is more balanced. Although the countryside still lags behind, in 2006, the per capita disposable incomes of all 11 prefecture-level divisions in Zhejiang were ranked among the top 30 among all Chinese cities.

3.8.1 Current State of Investment and Employment Structures of Zhejiang

For the Zhejiang Province, we first compute the total outputs of economic sectors and industries, the investment and employment situations (see Table 3.15), and the corresponding comparative gains. Then, by comparing the comparative gains, we obtain the current structural states of investment and employment for Zhejiang's economic sectors and industries. All the detailed calculations are similar to what has been described above. The results are summarized in Table 3.16.

Table 3.16 indicates that first, in terms of the three major economic sectors, among the six relevant data points of the three sectors, only one is basically in equilibrium, where the comparative gain of labor of the secondary sector is 1.007096, while all the other five (5/6 = 83%) are in extremely imbalanced states. In particular, the comparative gain of labor of the first sector is way too low, and is equal to only 0.336214, which is not even equal to one-third of those of the same indicator of the secondary and tertiary sectors. In terms of the capital investment, the comparative gains of capital of the primary and the secondary sectors are way too low, and are,

Table 3.15 GDPs, Total Investments in Fixed Properties, and Employment Situation of Zhejiang's Three Economic Sectors and Industries (2011)

Sectors (i)	*Total Output of Sector (0.1 Billion Yuan)*	*Total Investment on Fixed Properties (0.1 Billion Yuan)*	*Employment (10,000 People)*
Total	32318.85	14077.249	3674.11
Primary sector	1583.04	96.7743	535.27
Secondary sector	16555.58	5223.1092	1868.83
Tertiary sector	14180.23	8757.3655	1270.01
Industries (*i*)			
Agriculture, forest, animal husbandry, fishery		96.7743	535.27
Mining		25.2886	6.34
Manufacturing		4538.6735	1500.92
Production/supply of electricity, gas, and water		620.3767	14.94
Construction	1872.55	38.7704	346.63
Communication, transportation, storage, postal service	1206.95	1102.9276	145.6
Information transfer, computer, and software areas	756.1	138.4257	43.54
# Information Transfer			
Wholesale and retail	3288.53	226.0317	456.96
Hotels and restaurants	620.25	164.1069	149.3
Finance	2730.29	25.4264	32.16
Real estate/housing	1677.13	5183.8971	33.17
Leasing and commercial services	575.38	120.3249	75.74
Scientific research, technology service, and geological survey	289.59	49.8061	23.85
Water conservation, environment, and public facility management	138.45	1193.4475	14.38
# Public Facility Management			
Residents and other services	407.78	22.6827	117.29
Education	774.64	150.0622	66.06

continued

Table 3.15 (continued) GDPs, Total Investments in Fixed Properties, and Employment Situation of Zhejiang's Three Economic Sectors and Industries (2011)

Sectors (i)	*Total Output of Sector (0.1 Billion Yuan)*	*Total Investment on Fixed Properties (0.1 Billion Yuan)*	*Employment (10,000 People)*
Health, safety, and welfare	493.07	94.9974	37.98
Social, sports, and entertainment	218.56	115.4769	15.29
Public administration and social organizations	1003.51	169.7524	58.7

Source: Data are from *China Statistics Abstracts 2012* (XLS, DOC).

respectively, 7.12515 and 1.38063, while the capital investment in the tertiary sector is in extreme excess (0.70530).

Second, in terms of the employment situation, the relevant data of the 21 industries of Zhejiang Province should contain 42 data points. However, because of lack of raw data, we only collected 30 of them. So, we can only base our discussion on these available statistics. All of the industries are situated in extremely imbalanced states. In particular, the comparative gains of labor of the following nine industries are over 1. Especially, the comparative gain of labor in "finance" even

Table 3.16 Structures and Comparative Gain Measures of Zhejiang's Three Economic Sectors and Industries ($t = 2011$)

Sectors (i)	*Demand Structure $d_i(t)$ (%)*	*Employment Structure $s_i^1(t)$ (%)*	*Comparative Gain of Labor $G_i^1(t)$*	*Investment Structure $s_i^2(t)$ (%)*	*Comparative Gain of Capital $G_i^2(t)$*
Total	100	100	–	100	–
Primary sector	4.898194	14.56870	0.336214	0.687452	7.12515
Secondary sector	51.22577	50.86484	1.007096	37.10320	1.38063
Tertiary sector	43.87604	34.56647	1.269324	62.20935	0.70530
Industries (*i*)					
Agriculture, forest, animal husbandry, fishery		14.56870		0.687452	
Mining		0.172559		0.179642	
Manufacturing		40.85125		32.24120	
Production/supply of electricity, gas, and water		0.406629		4.406946	
Construction	5.793987	9.434394	0.614135	0.275412	21.0375

Table 3.16 (continued) Structures and Comparative Gain Measures of Zhejiang's Three Economic Sectors and Industries ($t = 2011$)

Sectors (i)	*Demand Structure* $d_i(t)$ *(%)*	*Employment Structure* $s_i^1(t)$ *(%)*	*Comparative Gain of Labor* $G_i^1(t)$	*Investment Structure* $s_i^2(t)$ *(%)*	*Comparative Gain of Capital* $G_i^2(t)$
Communication, transportation, storage, postal service	3.734508	3.962864	0.942376	7.834823	0.47666
Information transfer, computer, and software areas	2.339502	1.185049	1.974181	0.983329	2.37916
# Information Transfer					
Wholesale and retail	10.17527	12.43730	0.818125	1.605652	6.33716
Hotels and restaurants	1.919159	4.063569	0.472284	1.165760	1.64627
Finance	8.447980	0.875314	9.651371	0.180621	46.7720
Real estate/housing	5.189324	0.902804	5.748010	36.82465	0.14092
Leasing and commercial services	1.780323	2.061452	0.863626	0.854747	2.08287
Scientific research, technology service, and geological survey	0.896041	0.649137	1.380357	0.353806	2.53258
Water conservation, environment, and public facility management	0.428388	0.391387	1.094537	8.477846	0.05053
# Public Facility Management					
Residents and other services	1.261740	3.192338	0.395240	0.161130	7.83056
Education	2.396867	1.797986	1.333084	1.065991	2.24849
Health, safety, and welfare	1.525642	1.033720	1.475876	0.674829	2.26078
Social, sports, and entertainment	0.676262	0.416155	1.625023	0.820309	0.82440
Public administration and social organizations	3.105030	1.597666	1.943479	1.205863	2.57494

Source: Data are from *China Statistics Abstract 2012* (XLS, DOC).

reaches 9.651371, which is more than (9.651371/0.395240≈) 24 times of the lowest value of this indicator that is reached by "residents and other services."

Finance	Real estate/ housing	Information transfer, computer, and software areas	Public administration and social organizations	Social, sports, and entertainment
Health, safety, and welfare	Scientific research, technology service, and geological survey	Education	Water conservation, environment, and public facility management	

At the same time, the imbalance in terms of the comparative gains of capital seems to be way too severe. The comparative gains in the following 11 industries are over 1. In particular, the comparative gains of capital of both "finance" and "construction" are more than 10, while the value of this indicator of "water conservation, environment, and public facility management" is very low, equaling only 0.05053, which is merely (0.05053/46.7720 ≈) 0.0017 when compared to that of "finance" (0.05053/21.0375 ≈) 0.0024 when compared to that of "construction" (0.05053/7.83056 ≈) 0.0064 when compared to that of "residents and other services," and (0.05053/6.33716 ≈) 0.008 when compared to that of "wholesale and retail."

Finance	Construction	Residents and other services	Wholesale and retail
Public administration and social organizations	Scientific research, technology service, and geological survey	Information transfer, computer, and software areas	Health, safety, and welfare
Education	Leasing and commercial services	Hotels and restaurants	

So, in terms of the magnitudes of the comparative gains of labor, the 15 industries of Zhejiang with available data can be ranked as follows:

1	2	3	4
Finance	Real estate/housing	Information transfer, computer, and software areas	Public administration and social organizations
5	6	7	8
Social, sports, and entertainment	Health, safety, and welfare	Scientific research, technology service, and geological survey	Education

9	10	11	12
Water conservation, environment, and public facility management	Communication, transportation, storage, postal service	Leasing and commercial services	Wholesale and retail
13	14	15	
Construction	Hotels and restaurants	Residents and other services	

In terms of the magnitudes of the comparative gains of capital, the 15 industries of Zhejiang with available data can be ranked as follows:

1	2	3	4
Finance	Construction	Residents and other services	Wholesale and retail
5	6	7	8
Public administration and social organizations	Scientific research, technology service, and geological survey	Information transfer, computer, and software areas	Health, safety, and welfare
9	10	11	12
Education	Leasing and commercial services	Hotels and restaurants	Social, sports, and entertainment
13	14	15	
Communication, transportation, storage, postal service	Real estate/housing	Water conservation, environment, and public facility management	

3.8.2 Investment and Employment in Zhejiang's Economic Sectors and Industries

First, in terms of the comparative gains of labor, the values of this indicator of the following nine industries

Finance	Real estate/housing	Information transfer, computer, and software areas	Public administration and social organizations	Social, sports, and entertainment
Health, safety, and welfare	Scientific research, technology service, and geological survey	Education	Water conservation, environment, and public facility management	

are over 1, representing relatively optimal industries for employment. In all the other six industries, there has been excessive supply of labor so that appropriate control on the increase of labor into these industries should be imposed. Specifically, the excessive labor in the following industries should be transferred into the previous nine industries.

Residents and other services	Hotels and restaurants	Construction
Wholesale and retail	Leasing and commercial services	Communication, transportation, storage, postal service

In other words, no matter where the job seeker, who is looking for employment opportunities in Zhejiang, is from, the eastern part of China, the other parts of China, or the rest of the world, his/her chance is potentially in the tertiary and the secondary economic sector with the industries ranked as follows:

1	2	3	4	5
Finance	Real estate/housing	Information transfer, computer, and software areas	Public administration and social organizations	Social, sports, and entertainment
6	7	8	9	
Health, safety, and welfare	Scientific research, technology service, and geological survey	Education	Water conservation, environment, and public facility management	

Without special skills, do not waste time and effort in looking for employment opportunities in the primary economic sector and other industries, because those areas are experiencing excessive labor supplies.

That is the direction of employment.

Second, in terms of the comparative gains of capital, the following 11 industries have enjoyed the comparative gains of more than 1, representing relatively more optimal areas of additional investment.

Finance	Construction	Residents and other services	Wholesale and retail
Public administration and social organizations	Scientific research, technology service, and geological survey	Information transfer, computer, and software areas	Health, safety, and welfare
Education	Leasing and commercial services	Hotels and restaurants	

However, for the following four industries, excessive investment of capital has been given. Further capital investment in these industries should be curbed and redirected to the previous 11 industries.

Social, sports, and entertainment	Communication, transportation, storage, postal service	Real estate/ housing	Water conservation, environment, and public facility management

In other words, no matter whether the investors, who look for investment opportunities in Zhejiang, are from eastern China, or other parts of the Mainland China, or the rest of the world, they should first consider the following industries, ranked from the most optimal to the least:

1	2	3	4
Finance	Construction	Residents and other services	Wholesale and retail
5	6	7	8
Public administration and social organizations	Scientific research, technology service, and geological survey	Information transfer, computer, and software areas	Health, safety, and welfare
9	10	11	
Education	Leasing and commercial services	Hotels and restaurants	

That is the direction of capital investment.

Third, by combining the two criteria discussed above, it follows that the industries in Zhejiang that are most optimal for investment and employment are the following, because the comparative gains of labor and capital of each of these industries are over 1.

Finance	Information transfer, computer, and software areas	Public administration and social organizations
Health, safety, and welfare	Scientific research, technology service, and geological survey	Education

The worst choice is the industry of "communication, transportation, storage, postal service."

3.9 Fujian Province

Located along the southeast coast of China, the Fujian Province borders with Zhejiang Province in the north, Jiangxi Province in the west, and Guangdong Province in the south (Figure 3.9). Across the Taiwan Strait lies Taiwan in the east. The province faces the East China Sea in the east, the South China Sea in the south, and the Taiwan Strait in the southeast. Although the majority of the population consists of Han Chinese, Fujian is one of the most culturally and linguistically diverse provinces in China.

Figure 3.9 Geographic location of Fujian.

The province is mostly mountainous with higher altitude in the northwest, where the Wuyi Mountains are located and forms the border between Fujian and Jiangxi. The highest point of the Wuyi Mountains reaches an altitude of 2157 m. The Min River and its tributaries run through much of northern and central Fujian. Among other rivers are the Jin River and the Jiulong River.

The area of the present day Fuzhou, a major city in Fujian, was the home to the kingdom of Minyue until the emperor of Qin Dynasty abolished the status in his effort to unify China. In 202 BC Liu Bang, the founding emperor of the Han Dynasty, restored Minyue's status as a tributary independent kingdom, which extended beyond the borders of contemporary Fujian into eastern Guangdong, eastern Jiangxi, and southern Zhejiang. When the Han emperor decided to wipe out the potential threat of Minyue, the rulers in Fuzhou surrendered, which brought an abrupt end to the first kingdom in Fujian history.

The rugged terrain contributed to Fujian's relatively backward economy and level of development, despite major population boost from northern China during the invasions of nomadic peoples from the north and civil war in the early fourth century. Like other southern provinces, Fujian often served as a destination for exiled prisoners and dissidents at that time. After the Tang Dynasty (618–907) ended, another wave of immigrants arrived who set up an independent Kingdom of Min with its capital in Fuzhou. This kingdom was soon taken by Southern Tang, another southern kingdom. In 1689, the Qing dynasty officially merged Taiwan into Fujian province. After Taiwan was separated into its own province in 1885 and taken by Japan in 1895, the current shape of Fujian was formed.

Because of its mountainous landscape, Fujian was the most secluded province in eastern China due to the underdeveloped network of transportation systems before the 1950s. Even so, Fujian has had a strong academic tradition since the Southern Song Dynasty.

Starting in the late 1970s, Fujian has economically benefited from its geographic and cultural proximity to Taiwan. In 2003, for example, Xiamen ranked number eight in terms of per capita GDP among 659 Chinese cities, which placed the city ahead of Shanghai and Beijing, while Fuzhou ranked number 21 (number 4 among 30 provincial capitals). Currently, Fujian is one of the affluent provinces. In terms of agricultural production, rice is the main crop, supplemented by sweet potatoes and wheat and barley, while cash crops include sugar cane, rapeseed, longan, lychees, tea, and seafood. In 2011, Fujian's nominal GDP reached ¥1.74 trillion (US$276.3 billion). It's GDP per capita was ¥46,802 (US$7246). By 2015, Fujian expects to have at least 50 enterprises that have over 10 billion RMB in annual revenues. The government also expects 55% of GDP growth to come from the industrial sector.

For more detailed reading, please consult China Network (2010), Fujian Economic News (2014), Fukien (2008), Ministry of Commerce (2013), National Bureau of Statistics (2013), and Roberts (1837).

3.9.1 Current State of Investment and Employment Structures of Fujian

For Fujian Province, we first compute the total outputs of economic sectors and industries, the investment and employment situations (see Table 3.17), and the corresponding comparative gains. Then, by comparing the comparative gains, we obtain the current structural states of investment and employment for Fujian's economic sectors and industries. All the detailed calculations are similar to what has been described above. The results are summarized in Table 3.18.

Table 3.18 indicates that first, in terms of the three major economic sectors, all of the six relevant data points of the three sectors are in extremely imbalanced states. In particular, the comparative gain of labor of the first sector is way too low, and is equal to only 0.348798, which is only equivalent to one-fourth, and one-third of those of the same indicator of the secondary and tertiary sectors. In terms of the capital investment, the comparative gains of capital of the primary and the secondary sectors are way too low, and are, respectively, 5.28794 and 1.39583, while the capital investment in the tertiary sector is in extreme excess (0.63941).

Second, in terms of the employment situation, the relevant data of the 21 industries of Fujian Province should contain 42 data points. However, because of lack of raw data, we only collected 38 of them, where the data in bold or underlined come from the urban employment statistics of *China Statistics Yearbook* (2012). These data are not comparable with other data. Especially, these data might contain errors. In fact, on one hand, the provincial statistics of employment numbers in the tertiary economic sector indicated 88,366,000, while the national statistics yearbook is less than 2 million. And on the other hand, the overall provincial number of urban residents working in the industry of "residents and other services" is merely 13,000 and the number in "hotels and restaurants" only 87,000. How can these be possible? So, there are only 19 effective data points in terms of the comparative gains of capital. Based on these 19 effective data, our results indicate that all of them are situated in extremely imbalanced states. In particular, the comparative gains of capital of the following 14 industries are over 1. Especially, the comparative gain of capital in "construction" even reached 26.5495, which is more than (26.5495/0.04068≈) 664 times of the lowest value of this indicator that is reached by "water conservation, environment, and public facility management," is (26.5495/0.18271≈) 145 times of the value of this indicator reached by "real estate and housing," is (26.5495/0.38444≈) 70 times of the value of this indicator

Table 3.17 GDPs, Total Investments in Fixed Properties, and Employment Situation of Fujian's Three Economic Sectors and Industries (2011)

Sectors (i)	*Total Output of Sector (0.1 Billion Yuan)*	*Total Investment on Fixed Properties (0.1 Billion Yuan)*	*Employment (10,000 People)*
Total	17560.18	10119.47	2459.99
Primary sector	1612.24	175.7	647.53
Secondary sector	9069.2	3744.26	928.81
Tertiary sector	6878.74	6199.5	883.66
Industries (*i*)			
Agriculture, forest, animal husbandry, fishery	1612.24	175.7	620.37
Mining	403.91	111.09	6.9
Manufacturing	6912.22	3064.84	283
Production/supply of electricity, gas, and water	358.96	538.08	9.5
Construction	1394.11	30.26	99.9
Communication, transportation, storage, postal service	963.85	1384.74	17.9
Information transfer, computer, and software areas	373.73	142.49	5.4
# Information Transfer			
Wholesale and retail	1511.29	172.15	20.1
Hotels and restaurants	300.35	142.7	8.7
Finance	862.41	23.25	13.6
Real estate/housing	911.16	2873.87	9.1
Leasing and commercial services	342.97	86.95	6.6
Scientific research, technology service, and geological survey	121.89	25.96	5.7
Water conservation, environment, and public facility management	57.12	809.12	4.7
# Public Facility Management			
Residents and other services	314.65	16.24	1.3
Education	293.96	147.81	46.1
Health, safety, and welfare	260.68	59.99	17.1

Table 3.17 (continued) GDPs, Total Investments in Fixed Properties, and Employment Situation of Fujian's Three Economic Sectors and Industries (2011)

Sectors (i)	*Total Output of Sector (0.1 Billion Yuan)*	*Total Investment on Fixed Properties (0.1 Billion Yuan)*	*Employment (10,000 People)*
Social, sports, and entertainment	152.38	138.97	3.8
Public administration and social organizations	412.3	175.27	30.6

Source: The data underlined are from *China Statistics Yearbook 2012*, edited by National Bureau of Statistics of China, published by China Statistics Press and the rest from *Fujian Province's Statistics Yearbook 2012.*

Table 3.18 The Structures and Comparative Gain Measures of Fujian's Three Economic Sectors and Industries ($t = 2011$)

Sectors (i)	*Demand Structure $d_i(t)$ (%)*	*Employment Structure $s_i^1(t)$ (%)*	*Comparative Gain of Labor $G_i^1(t)$*	*Investment Structure $s_i^2(t)$ (%)*	*Comparative Gain of Capital $G_i^2(t)$*
Total	100	100	–	100	–
Primary sector	9.181227	26.32246	0.348798	1.736257	5.28794
Secondary sector	51.64640	37.75666	1.367875	37.00055	1.39583
Tertiary sector	39.17238	35.92128	1.090506	61.26309	0.63941
Industries (*i*)					
Agriculture, forest, animal husbandry, fishery	9.181227	25.21840	**0.364069**	1.736257	5.28794
Mining	2.300147	0.280489	**8.200492**	1.097785	2.09526
Manufacturing	39.36304	11.50411	**3.421649**	30.28657	1.29969
Production/supply of electricity, gas, and water	2.044170	0.386180	**5.293304**	5.317275	0.38444
Construction	7.939042	4.060992	**1.954951**	0.299028	26.5495
Communication, transportation, storage, postal service	5.488839	0.727645	**7.543290**	13.68392	0.401116
Information transfer, computer, and software areas	2.128281	0.219513	**9.695464**	1.408078	1.51148
# Information Transfer					
Wholesale and retail	8.606347	0.817076	**10.53310**	1.701176	5.05906

continued

Table 3.18 (continued) The Structures and Comparative Gain Measures of Fujian's Three Economic Sectors and Industries (*t* = 2011)

Sectors (i)	*Demand Structure $d_i(t)$ (%)*	*Employment Structure $s_i^1(t)$ (%)*	*Comparative Gain of Labor $G_i^1(t)$*	*Investment Structure $s_i^2(t)$ (%)*	*Comparative Gain of Capital $G_i^2(t)$*
Hotels and restaurants	1.710404	0.353660	**4.836295**	1.410153	1.21292
Finance	4.911168	0.552848	**8.883401**	0.229755	21.3757
Real estate/housing	5.188785	0.369920	**14.02677**	28.39941	0.18271
Leasing and commercial services	1.953112	0.268294	**7.279752**	0.859235	2.27308
Scientific research, technology service, and geological survey	0.694127	0.231708	**2.995695**	0.256535	2.70578
Water conservation, environment, and public facility management	0.325281	0.191058	**1.702530**	7.995676	0.04068
# Public Facility Management					
Residents and other services	1.791838	0.052846	**33.90695**	0.160483	11.1653
Education	1.674015	1.873991	**0.893288**	1.46065	1.14608
Health, safety, and welfare	1.484495	0.695128	**2.135581**	0.592818	2.50413
Social, sports, and entertainment	0.867759	0.154472	**5.617573**	1.373293	0.63188
Public administration and social organizations	2.347926	1.243907	**1.887541**	1.732008	1.35561

Source: Data are from *China Statistics Abstract 2012* (XLS, DOC).

reached by "production/supply of electricity, gas, and water," is (26.5495/0.401116≈) 66 times of the value of this indicator reached by "communication, transportation, storage, postal service," and is (26.5495/0.63188≈) 42 times of the value of this indicator reached by "social, sports, and entertainment."

Construction	Finance	Residents and other services	Agriculture, forest, animal husbandry, fishery	Wholesale and retail
Scientific research, technology service, and geological survey	Health, safety, and welfare	Leasing and commercial services	Mining	Information transfer, computer, and software areas

Public administration and social organizations	Manufacturing	Hotels and restaurants	Education	

So, in terms of the magnitudes of the comparative gains of capital, the 19 industries of Fujian with available data can be ranked as follows:

1	2	3	4
Construction	Finance	Residents and other services	Agriculture, forest, animal husbandry, fishery
5	6	7	8
Wholesale and retail	Scientific research, technology service, and geological survey	Health, safety, and welfare	Leasing and commercial services
9	10	11	12
Mining	Information transfer, computer, and software areas	Public administration and social organizations	Manufacturing
13	14	15	16
Hotels and restaurants	Education	Social, sports, and entertainment	Communication, transportation, storage, postal service
17	18	19	
Production/ supply of electricity, gas, and water	Real estate/housing	Water conservation, environment, and public facility management	

3.9.2 Investment and Employment in Fujian's Economic Sectors and Industries

First, in terms of the comparative gains of labor, no matter where the job seeker, who is looking for employment opportunities in Fujian, is from, the east part of China, other parts of China, or the rest of the world, his opportunity is in the tertiary and the secondary economic sector. Try not to waste time and effort in the primary sector.

That is the direction of employment.

Second, in terms of the comparative gains of capital, the following 14 industries have enjoyed the comparative gains of more than 1, representing relatively more optimal areas of additional investment.

Construction	Finance	Residents and other services	Agriculture, forest, animal husbandry, fishery	Wholesale and retail
Scientific research, technology service, and geological survey	Health, safety, and welfare	Leasing and commercial services	Mining	Information transfer, computer, and software areas
Public administration and social organizations	Manufacturing	Hotels and restaurants	Education	

However, for the following five industries, excessive investment of capital has been given. Further capital investment in these industries should be curbed and redirected to the previous 14 industries.

Social, sports, and entertainment	Communication, transportation, storage, postal service	Production/supply of electricity, gas, and water
Real estate/housing	Water conservation, environment, and public facility management	

In other words, no matter whether the investors, who look for investment opportunities in Fujian, are from eastern China, or other parts of the Mainland China, or the rest of the world, they should first consider the following industries, ranked from the most optimal to the least:

1	2	3	4	5
Construction	Finance	Residents and other services	Agriculture, forest, animal husbandry, fishery	Wholesale and retail
6	7	8	9	10
Scientific research, technology service, and geological survey	Health, safety, and welfare	Leasing and commercial services	Mining	Information transfer, computer, and software areas
11	12	13	14	
Public administration and social organizations	Manufacturing	Hotels and restaurants	Education	

That is the direction of capital investment.

Social, sports, and entertainment	Communication, transportation, storage, postal service	Production/supply of electricity, gas, and water	Real estate/housing	Water conservation, environment, and public facility management

3.10 Hainan Province

Located in the South China Sea, Hainan is the smallest and southernmost province. It is separated from the Leizhou Peninsula of the Guangdong Province in the north by the shallow and narrow Qiongzhou Strait. Hainan Island had traditionally been part of Guangdong until 1988 when it became part of the newly created Hainan Province. The area if the island is similar to that of Belgium; In the west is the Gulf of Tonkin. Wuzhi Mountain is the highest mountain on the island with an elevation of 1840 m. Most of the rivers of the province originate in the central area of the island and flow radially in different directions. The Wanning River in the southern part of the island is the largest river surrounding Hainan. It is 350 km (220 mi) long. The Nandu River in the northern part of the island is 314 km (195 mi) long. The Changhua River in the west is 230 km (140 mi) long and the Wanquan River in the east is 162 km (101 mi) long. For details, see Figure 3.10.

When compared to other parts of Mainland China, Hainan has enjoyed better air quality. For example, throughout 2012, Hainan had the highest air quality in the country for 351 days (Chinadaily.com, 2013).

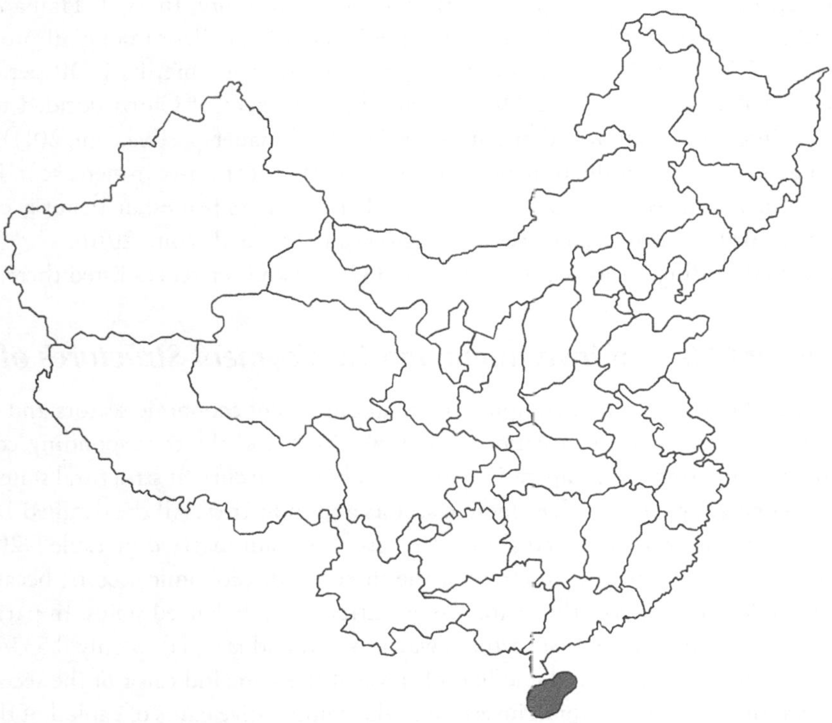

Figure 3.10 Geographic location of Hainan.

Historically, the Hainan Island first entered written Chinese history in 110 BC, when the Han Dynasty established a military garrison there following the arrival of General Lu Bode. In 46 BC due to the high cost, the Han Dynasty abandoned the island. As descendants of the ancient Yue tribes, Li people are the original inhabitants of Hainan who settled on the island about 7000 years ago (BMC Evolutionary Biology, 2011). They mainly reside in the nine cities and counties in the middle and southern part of Hainan.

During the Three Kingdoms Period (184–280), Hainan was under the control of Eastern Wu. At the time of the Song Dynasty (980–1279), Hainan became part of Guangxi province. Under the Yuan Dynasty (1206–1368) the island became an independent province; then in 1370 was placed under the administration of Guangdong by the ruling Ming Dynasty. In 1921, Hainan was planned to become a Special Administrative Region. In 1944, it officially became Hainan Special Administrative Region with 16 counties containing the South China Sea Islands. In 1988, when the island was made a separate province, it was designated a Special Economic Zone in an effort to increase investment.

Economically, Hainan is predominantly agricultural with agricultural products accounting for more than a half of the island's exports. Hainan's elevation to the status of a province was accompanied by its designation as China's largest "special economic zone" with the intent of hastening the development of the island's plentiful resources. Prior to that, the province had been largely untouched by industrialization. Even today, tourism plays an important part of Hainan's economy because of its tropical beaches and lush forests. The central government has encouraged foreign investment in Hainan and has allowed the island to rely to a large extent on market forces.

The industrial development of the province has been largely limited to the processing of its mineral and agricultural products, particularly rubber and iron ore. In 2011, Hainan's nominal GDP was ¥251.5 billion (US$39.9 billion), making it the 4th smallest among all provinces and contributes just 0.53% to the entire country's economy. At that time, its GDP per capita was ¥19,166 (US$2805). In December 2009, the central government of China decided to establish Hainan as an "international tourist destination" by 2020 (Chinaperspective.com, 2011). This decision contributed to a surge in the province's economy, with a year-on-year increase in investment of 136.9% in the first three months of 2010. Since then Hainan's real estate sector accounted for more than one third of the province's economic growth (Chinadaily.com, 2010).

For additional reading, please consult Thurgood (2006) and references listed there.

3.10.1 Current State of Investment and Employment Structures of Hainan

For the Hainan Province, we first compute the total outputs of economic sectors and industries, the investment and employment situations (see Table 3.19), and the corresponding comparative gains. Then, by comparing the comparative gains, we obtain the current structural states of investment and employment for Hainan's economic sectors and industries. All the detailed calculations are similar to what has been described above. The results are summarized in Table 3.20.

Table 3.20 indicates that first, in terms of the three major economic sectors, because of lack of data, all the relevant results of the sectors are in extremely imbalanced states. In particular, the comparative gain of labor of the first sector is way too low, and is equal to only 0.533403, which is not even equal to one-fourth, and one-half of those of the same indicator of the secondary and tertiary sectors. In terms of the capital investment, the comparative gains of capital of the primary and the secondary sectors are way too low, and are respectively 36.3016 and 1.49749, while the capital investment in the tertiary sector is in extreme excess (0.56671).

Table 3.19 GDPs, Total Investments in Fixed Properties, and Employment Situation of Hainan's Three Economic Sectors and Industries (2011)

Sectors (i)	*Total Output of Sector (0.1 Billion Yuan)*	*Total Investment on Fixed Properties (0.1 Billion Yuan)*	*Employment (10,000 People)*
Total	2522.66	*1611.41*	459.22
Primary sector	659.23	*11.6*	224.98
Secondary sector	714.5	*304.78*	54.73
Tertiary sector	1148.93	*1295.03*	179.51
Industries (*i*)			
Agriculture, forest, animal husbandry, fishery	106.6	**24.2**	224.9799
Mining		**14.9**	1.4492
Manufacturing		**132.6**	25.7243
Production/supply of electricity, gas, and water		**119.6**	2.309
Construction	239.46	**30.7**	25.2473
Communication, transportation, storage, postal service	119.74	**100.3**	16.4421
Information transfer, computer, and software areas	59.29	**29.1**	4.7767
# Information Transfer			
Wholesale and retail	258.06	**7**	49.9354
Hotels and restaurants	89.75	**148.4**	23.3937
Finance	105.24	**6.4**	2.548
Real estate/housing	208.71	**779.2**	8.5661
Leasing and commercial services	21.74	**4.4**	8.8784
Scientific research, technology service, and geological survey	17.95	**3.7**	2.5978
Water conservation, environment, and public facility management	11.5	**108**	2.4602
# Public Facility Management			
Residents and other services	31.47	**0.5**	7.3736
Education	70.84	**29.2**	12.5991

continued

Table 3.19 (continued) GDPs, Total Investments in Fixed Properties, and Employment Situation of Hainan's Three Economic Sectors and Industries (2011)

Sectors (i)	*Total Output of Sector (0.1 Billion Yuan)*	*Total Investment on Fixed Properties (0.1 Billion Yuan)*	*Employment (10,000 People)*
Health, safety, and welfare	37.86	**36.7**	4.7997
Social, sports, and entertainment	19.98	**46.2**	2.4636
Public administration and social organizations	96.8	**36.1**	10.595

Source: The data in italics is from Hainan Province's statistical report of gross domestic products 2011, the data in bold is from *China Statistics Yearbook 2012*, edited by National Bureau of Statistics of China, published by China Statistics Press and the rest from *Hainan Province's Statistical Yearbook* 2012.

Second, in terms of the industries, the relevant data of the 21 industries of the Hainan Province should contain 42 data points. However, because of the lack of raw data, we only collected 32 of them. So, we can only base our discussion on these 32 available statistics. Among all the industries, only one is basically in equilibrium, where the comparative gain of labor in "education" is 1.023531. All the rest (31/32 ≥ 96%) are situated in extremely imbalanced states. In particular, the comparative gains of labor of the following 10 industries are over 1. Especially, the comparative gain of labor in "finance" even reached 7.518717, which is more than (7.518713/0.086253≈) 84 times of the lowest value of this indicator reached by "agriculture, forest, animal husbandry, fishery," is more than (7.518717/0.445745≈) 84 times of the value of this indicator reached by "leasing and commercial services," is equal to (7.518717/0.683775≈) 11 times of the value of this indicator reached by "hotels and restaurants," is equal to (7.518717/0.776925≈) 10 times of the

Table 3.20 Structures and Comparative Gain Measures of Hainan's Three Economic Sectors and Industries ($t = 2011$)

Sectors (i)	*Demand Structure $d_i(t)$ (%)*	*Employment Structure $s_i^1(t)$ (%)*	*Comparative Gain of Labor $G_i^1(t)$*	*Investment Structure $s_i^2(t)$ (%)*	*Comparative Gain of Capital $G_i^2(t)$*
Total	100	100	–	100	–
Primary sector	26.13234	48.99177	0.533403	0.719866	36.3016
Secondary sector	28.32328	11.91803	2.376506	18.91387	1.49749
Tertiary sector	45.54439	39.09020	1.165110	80.36626	0.56671
Industries (*i*)					
Agriculture, forest, animal husbandry, fishery	4.225698	48.99175	0.086253	1.50179	2.81377
Mining		0.315579		0.924656	
Manufacturing		5.601738		8.228818	

Table 3.20 (continued) Structures and Comparative Gain Measures of Hainan's Three Economic Sectors and Industries ($t = 2011$)

Sectors (i)	*Demand Structure* $d_i(t)$ *(%)*	*Employment Structure* $s_i^1(t)$ *(%)*	*Comparative Gain of Labor* $G_i^1(t)$	*Investment Structure* $s_i^2(t)$ *(%)*	*Comparative Gain of Capital* $G_i^2(t)$
Production/supply of electricity, gas, and water		0.502809		7.422071	
Construction	9.492361	5.497866	1.725554	1.905164	4.98244
Communication, transportation, storage, postal service	4.746577	3.580441	1.325696	6.224363	0.76258
Information transfer, computer, and software areas	2.350297	1.040177	2.259517	1.805872	1.30147
# Information Transfer					
Wholesale and retail	10.22968	10.87396	0.940750	0.434402	23.5489
Hotels and restaurants	3.557753	5.203105	0.683775	9.209326	0.38632
Finance	4.171787	0.554854	7.518717	0.397168	10.5038
Real estate/housing	8.273410	1.887135	4.384112	48.35517	0.17110
Leasing and commercial services	0.861789	1.933365	0.445745	0.273053	3.15613
Scientific research, technology service, and geological survey	0.711551	0.565698	1.257827	0.229613	3.09892
Water conservation, environment, and public facility management	0.455868	0.535736	0.850922	6.702205	0.06802
# Public Facility Management					
Residents and other services	1.247493	1.605679	0.776925	0.031029	40.2044
Education	2.808147	2.74357	1.023531	1.812078	1.54968
Health, safety, and welfare	1.500797	1.045185	1.435915	2.277509	0.65896
Social, sports, and entertainment	0.792021	0.536475	1.476343	2.867054	0.27625
Public administration and social organizations	3.837219	2.307173	1.663169	2.240274	1.71283

Source: Data are from *China Statistics Abstract 2012* (XLS, DOC).

value of this indicator reached by "residents and other services," is equal to (7.518717/0.850922≈) 9 times of the value of this indicator reached by "water conservation, environment, and public facility management," and is equal to (7.518717/0.940750≈) 8 times of the value of this indicator reached by "wholesale and retail."

Finance	Real estate/ housing	Information transfer, computer, and software areas	Construction	Public administration and social organizations
Social, sports, and entertainment	Health, safety, and welfare	Communication, transportation, storage, postal service	Scientific research, technology service, and geological survey	Education

At the same time, the imbalance in terms of the comparative gains of capital seems to be way too severe. The comparative gains of this indicator of the following 10 industries are greater than 1. In particular, the comparative gains of capital of "residents and other services," "wholesale and retail," and "Finance" are all over 10, while the value of this indicator for "Water conservation, environment, and public facility management" is way too low, equaling only 0.06802, which is merely (0.0680/40.2044≈) 0.0012 when compared to that of "Residents and other services," is (0.0680/23.5489≈) 0.0028 when compared to that of "Wholesale and retail," and (0.0680/10.5038≈) 0.0065 when compared to that of "Finance."

Residents and other services	Wholesale and retail	Finance	Construction	Leasing and commercial services
Scientific research, technology service, and geological survey	Agriculture, forest, animal husbandry, fishery	Public administration and social organizations	Education	Information transfer, computer, and software areas

So, in terms of the magnitudes of the comparative gains of labor, the 16 industries of Hainan with available data can be ranked as follows:

1	2	3	4
Finance	Real estate/ housing	Information transfer, computer, and software areas	Construction
5	6	7	8
Public administration and social organizations	Social, sports, and entertainment	Health, safety, and welfare	Communication, transportation, storage, postal service

9	10	11	12
Scientific research, technology service, and geological survey	Education	Wholesale and retail	Water conservation, environment, and public facility management
13	14	15	16
Residents and other services	Hotels and restaurants	Leasing and commercial services	Agriculture, forest, Animal Husbandry, fishery

In terms of the magnitudes of the comparative gains of capital, the 16 industries of Hainan with available data can be ranked as follows:

1	2	3	4
Residents and other services	Wholesale and retail	Finance	Construction
5	6	7	8
Leasing and commercial services	Scientific research, technology service, and geological survey	Agriculture, forest, Animal Husbandry, fishery	Public administration and social organizations
9	10	11	12
Education	Information transfer, computer, and software areas	Communication, transportation, storage, postal service	Health, safety, and welfare
13	14	15	16
Hotels and restaurants	Social, sports, and entertainment	Real estate/housing	Water conservation, environment, and public facility management

3.10.2 *Investment and Employment in Hainan's Economic Sectors and Industries*

First, in terms of the comparative gains of labor, the values of this indicator of the following 10 industries

Finance	Real estate/ housing	Information transfer, computer, and software areas	Construction	Public administration and social organizations
Social, sports, and entertainment	Health, safety, and welfare	Communication, transportation, storage, postal service	Scientific research, technology service, and geological survey	Education

are over 1, representing relatively optimal industries for employment. In all the other six industries, there has been excessive supply of labor so that appropriate control on the increase of labor into these industries should be imposed. Specifically, the excessive labor in the following industries should be transferred into the previous 10 industries.

Agriculture, forest, animal husbandry, fishery	Leasing and commercial services	Hotels and restaurants
Residents and other services	Water conservation, environment, and public facility management	Wholesale and retail

In other words, no matter where the job seeker, who is looking for employment opportunities in Hainan, is from, the eastern part of China, the other parts of China, or the rest of the world, his/her chance is potentially in the secondary and the tertiary economic sectors with the industries ranked as follows:

Finance	Real estate/ housing	Information transfer, computer, and software areas	Construction	Public administration and social organizations
Social, sports, and entertainment	Health, safety, and welfare	Communication, transportation, storage, postal service	Scientific research, technology service, and geological survey	Education

Without special skills, do not waste time and effort in looking for employment opportunities in the primary economic sector and other industries, because those areas are experiencing excessive labor supplies.

That is the direction of employment.

Second, in terms of the comparative gains of capital, the following 10 industries have enjoyed the comparative gains of more than 1, representing relatively more optimal areas of additional investment.

Residents and other services	Wholesale and retail	Finance	Construction	Leasing and commercial services
Scientific research, technology service, and geological survey	Agriculture, forest, animal husbandry, fishery	Public administration and social organizations	Education	Information transfer, computer, and software areas

However, for the following six industries, excessive investment of capital has been given. Further capital investment in these industries should be curbed and redirected to the previous 10 industries.

Water conservation, environment, and public facility management	Real estate/housing	Social, sports, and entertainment
Hotels and restaurants	Health, safety, and welfare	Communication, transportation, storage, postal service

In other words, no matter whether the investors, who look for investment opportunities in Hainan, are from eastern China, or other parts of the Mainland China, or the rest of the world, they should first consider the following industries, ranked from the most optimal to the least:

Residents and other services	Wholesale and retail	Finance	Construction	Leasing and commercial services
Scientific research, technology service, and geological survey	Agriculture, forest, animal husbandry, fishery	Public administration and social organizations	Education	Information transfer, computer, and software areas

That is the direction of capital investment.

Third, by combining the two criteria discussed above, it follows that the industries in Shanghai that are most optimal for investment and employment are the following, because the comparative gains of labor and capital of each of these industries are over 1.

Finance	Construction	Public administration and social organizations
Scientific research, technology service, and geological survey	Information transfer, computer, and software areas	Education

The worst choice is the industry of "water conservation, environment, and public facility management."

3.11 Jiangsu Province

Jiangsu is an eastern coastal province with the highest population density in China. It borders Shandong Province in the north, Anhui Province in the west, and Zhejiang Province and Shanghai in the south. The province has a coastline of over 1000 km (620 miles) along the Yellow Sea, and the Yangtze River passes through the southern part of the province. Geographically, Jiangsu is very flat with plains covering 68% of its total area while water bodies cover another 18%. And it is laced with a well-developed irrigation system. For example, the southern city of Suzhou is so crisscrossed with canals that it has been dubbed "Venice of the East" (Uvista.com, 2013). The Grand Canal cuts through Jiangsu from north to south, traversing all the east–west river systems. The Huai He, a major river in central China and the traditional border between

Figure 3.11 Geographic location of Jiangsu.

North China and South China, cuts through north Jiangsu before reaching the Yellow Sea. For details, see Figure 3.11.

Historically, during the earliest Chinese dynasties, the area in what is now Jiangsu was far removed from the center of the Chinese civilization, which was in northwest Henan. In the Zhou Dynasty more contacts were made, and eventually the state of Wu (centered at the present Suzhou) appeared as a vassal to the Zhou Dynasty in south Jiangsu. Near the end of the Spring and Autumn Period, Wu became a great power and was able to defeat the state of Qi in 484 BC, a major power in the north located in modern-day Shandong province, and contest for the position of overlord over all states of China. The state of Wu was eventually conquered in 473 BC by the state of Yue, another state that had emerged in the south in modern-day Zhejiang province. Yue was in turn conquered by the powerful state of Chu from the west in 333 BC. Eventually, the state of Qin swept away all the other states, and established China as a unified nation in 221 BC. Dunring the Han Dynasty (206 BC to 220 AD), Jiangsu was a relative backwater, far removed from the centers of civilization in the North China Plain. It became significant when the northern nomads invaded during the Western Jin Dynasty, starting in the fourth century. In the Tang Dynasty (618–907), Jiangsu was still not significant among the different parts of China; it was during the Song Dynasty (960–1279), which saw the development of a market economy, that south Jiangsu emerged as a center of trade. From then on, south Jiangsu has been synonymous with opulence and luxury in China. Today south Jiangsu remains one of the richest parts of China with Shanghai as a direct extension of south Jiangsu culture.

The Ming Dynasty, which was established in 1368, initially put its capital in Nanjing. However, the Yongle Emperor moved the capital to Beijing. During the Qing Dynasty in

1666 Jiangsu and Anhui were split apart as separate provinces, and Jiangsu was given borders approximately the same as today. With the Western incursion into China in the 1840s, the rich and mercantile south Jiangsu was increasingly exposed to Western influence, where Shanghai, originally a little town of Jiangsu, quickly developed into a metropolis of trade, banking, and cosmopolitanism, and was split later as an independent municipality. The Taiping Rebellion (1851–1864) swept through much of South China and established Nanjing by 1853 as its capital. The Republic of China was established in 1912 and used Nanjing as its capital. However, this was interrupted by the second Sino-Japanese War. On December 13, 1937, Nanjing fell and soon became the seat of the collaborationist government of East China under Wang Jingwei. Jiangsu remained under Japanese occupation until 1945. After the war, Nanjing was once again the capital of the Republic of China. After Kuomintang was defeated by the communist forces and retreated to Taiwan, the Republic of China government continues to claim Nanjing as its rightful capital. After communist takeover of Mainland China, Beijing was made the capital of the People's Republic.

The economic reforms of Deng Xiaoping initially focused on the south coast of China, in Guangdong Province. Starting in the 1990s they were applied more evenly to the rest of China. Suzhou and Wuxi, two southern cities of Jiangsu, have since become particularly prosperous, ranking among the top 10 cities in China in terms of GDP. As of this writing, Jiangsu enjoys the highest GDP per capita of all Chinese provinces. Its nominal GDP in 2011 was about half the size that of the GDP of India.

With an extensive irrigation system, Jiangsu's agriculture is based primarily on rice and wheat, followed by maize and sorghum. Its cash crops include cotton, soybeans, peanuts, rapeseed, sesame, ambry hemp, and tea. Silkworms represent an important part of the province's agriculture with the Lake Taihu region being a major base of silk production in China.

Jiangsu is traditionally oriented toward light industries, such as textiles and food. However, since 1949, heavy industries have developed in the province (Peopledaily.com, 2013). Currently, Jiangsu is very wealthy among the provinces of China with the GDP per capita of ¥52,448 (US$7945).

3.11.1 Current State of Investment and Employment Structures of Jiangsu

For the Jiangsu Province, we first compute the total outputs of economic sectors and industries, the investment and employment situations (see Table 3.21), and the corresponding comparative gains. Then, by comparing the comparative gains, we obtain the current structural states of investment and employment for Jiangsu's economic sectors and industries. All the detailed calculations are similar to what has been described above. The results are summarized in Table 3.22.

Table 3.22 indicates that first, in terms of the three major economic sectors, all the relevant data of the three sectors are in extremely imbalanced states. In particular, the comparative gain of labor of the first sector is way too low, and is equal to only 0.290261, which is not even equal to one-fourth, and one-third of those of the same indicator of the secondary and tertiary sectors. In terms of the capital investment, the comparative gains of capital of the primary and the secondary sectors are way too low, and are, respectively, 2.36981 and 1.21988, while the capital investment in the tertiary sector is in extreme excess (0.79128).

Second, in terms of the employment situation, the relevant data of the 21 industries of Jiangsu Province should contain 42 data points. However, because of lack of raw data, we only collected 38 of them, where the data in bold are from the urban employment statistics of *China Statistics Yearbook* (2012). Because these data are not comparable with the other data, we have only 19

Table 3.21 GDPs, Total Investments in Fixed Properties, and Employment Situation of Jiangsu's Three Economic Sectors and Industries (2011)

Sectors (i)	*Total Output of Sector (0.1 Billion Yuan)*	*Total Investment on Fixed Properties (0.1 Billion Yuan)*	*Employment (10,000 People)*
Total	49110.27	*17726.30*	4758.23
Primary sector	3064.78	*466.8*	1023.02
Secondary sector	25203.28	*7457.4*	2017.49
Tertiary sector	20842.21	*9507.3*	1717.72
Industries (*i*)			
Agriculture, forest, animal husbandry, fishery	3064.78	155.20	**9.4**
Mining	296.27	68.02	**12.9**
Manufacturing	20978.51	13073.81	**353.7**
Production/supply of electricity, gas, and water	1005.83	629.30	**13.2**
Construction	2922.67	156.08	**65.2**
Communication, transportation, storage, postal service	2127.93	1190.32	**31.4**
Information transfer, computer, and software areas	910.86	180.43	**9.9**
# Information Transfer			
Wholesale and retail	5341.39	588.04	**32.1**
Hotels and restaurants	919.13	335.50	**11.7**
Finance	2600.11	55.41	**28.9**
Real estate/housing	2747.89	6587.09	**7.9**
Leasing and commercial services	1191.29	394.19	**11.2**
Scientific research, technology service, and geological survey	496.42	226.12	**12**
Water conservation, environment, and public facility management	280.76	1737.39	**13**
# Public Facility Management			
Residents and other services	568.78	105.34	**1.1**
Education	1217.21	218.82	**87.1**

Table 3.21 (continued) GDPs, Total Investments in Fixed Properties, and Employment Situation of Jiangsu's Three Economic Sectors and Industries (2011)

Sectors (i)	*Total Output of Sector (0.1 Billion Yuan)*	*Total Investment on Fixed Properties (0.1 Billion Yuan)*	*Employment (10,000 People)*
Health, safety, and welfare	664.54	145.66	**39.7**
Social, sports, and entertainment	268.01	209.76	**6**
Public administration and social organizations	1507.89	258.19	**65**

Source: The data in italic is from Statistics Report of Jiangsu Province's Domestic Economy and Social Development 2011, the data in bold is from *China Statistics Yearbook 2012*, edited by National Bureau of Statistics of China, published by China Statistics Press and the rest from *Jiangsu Province's Statistics Yearbook 2012.*

effective value points (regarding the comparative gains of capital). Among all these 19 effective data points, only one is basically in equilibrium, where the comparative gain of capital in "hotels and restaurants" is equal to 0.98885. All the rest (18/19 ≥ 94%) are situated in extremely imbalanced states. In particular, the comparative gains of capital of the following 11 industries are over 1. Especially, the comparative gain of labor in the industry "finance" even reached 16.9375. That value is equal to (16.9375/0.05833≈) 292 times of the lowest value of this indicator that is reached by "water conservation, environment, and public facility management" (16.9375/0.15057≈) 112 times

Table 3.22 Structures and Comparative Gain Measures of Jiangsu's Three Economic Sectors and Industries (t = 2011)

Sectors (i)	*Demand Structure* $d_i(t)$ *(%)*	*Employment Structure* $s_i^1(t)$ *(%)*	*Comparative Gain of Labor* $C_i^1(t)$	*Investment Structure* $s_i^2(t)$ *(%)*	*Comparative Gain of Capital* $G_i^2(t)$
Total	100	100	–	100	–
Primary sector	6.240609	21.50001	0.290261	2.633375	2.36981
Secondary sector	51.31977	42.40001	1.210372	42.06969	1.21988
Tertiary sector	42.43962	36.09998	1.175613	53.63387	0.79128
Industries (*i*)					
Agriculture, forest, animal husbandry, fishery	6.240609	0.197552	**31.58963**	0.875535	7.12776
Mining	0.603275	0.271109	**2.225210**	0.383724	1.57216
Manufacturing	42.71715	7.433436	**5.746623**	73.75374	0.57919
Production/supply of electricity, gas, and water	2.048105	0.277414	**7.382845**	3.550092	0.57692

continued

Table 3.22 (continued) Structures and Comparative Gain Measures of Jiangsu's Three Economic Sectors and Industries ($t = 2011$)

Sectors (i)	*Demand Structure* $d_i(t)$ *(%)*	*Employment Structure* $s_i^1(t)$ *(%)*	*Comparative Gain of Labor* $G_i^1(t)$	*Investment Structure* $s_i^2(t)$ *(%)*	*Comparative Gain of Capital* $G_i^2(t)$
Construction	5.951240	1.370257	**4.343155**	0.8805	6.75894
Communication, transportation, storage, postal service	4.332963	0.659909	**6.565999**	6.714994	0.64527
Information transfer, computer, and software areas	1.854724	0.208061	**8.914347**	1.017866	1.82217
# Information Transfer					
Wholesale and retail	10.87632	0.674621	**16.12213**	3.317331	3.27864
Hotels and restaurants	1.871564	0.245890	**7.611394**	1.892668	0.98885
Finance	5.294432	0.607369	**8.716999**	0.312586	16.9375
Real estate/housing	5.595347	0.166028	**33.70120**	37.15998	0.15057
Leasing and commercial services	2.425745	0.235382	**10.30558**	2.223758	1.09083
Scientific research, technology service, and geological survey	1.010827	0.252195	**4.008124**	1.275619	0.79242
Water conservation, environment, and public facility management	0.571693	0.273211	**2.092498**	9.801199	0.05833
# Public Facility Management					
Residents and other services	1.158169	0.023118	**50.09850**	0.594258	1.94893
Education	2.478524	1.830513	**1.354006**	1.234437	2.00782
Health, safety, and welfare	1.353159	0.834344	**1.621824**	0.821717	1.64675
Social, sports, and entertainment	0.545731	0.126097	**4.327857**	1.183326	0.46118
Public administration and social organizations	3.070417	1.366054	**2.247654**	1.456536	2.10803

Source: Data are from *China Statistics Abstract 2012* (XLS, DOC).

of the value of this indicator reached by "real estate and housing" (16.9375/0.46118≈) 37 times of the value of this indicator reached by "social, sports, and entertainment" (16.9375/0.57692≈) 29 times of the value of this indicator reached by "production/supply of electricity, gas, and water" (16.9375/0.57919≈) 29 times of the value of this indicator reached by "manufacturing," and (16.9375/0.64527≈) 26 times of the value of this indicator reached by "communication, transportation, storage, postal service."

Finance	Agriculture, forest, animal husbandry, fishery	Construction	Wholesale and retail
Public administration and social organizations	Education	Residents and other services	Information transfer, computer, and software areas
Health, safety, and welfare	Mining	Leasing and commercial services	

So, in terms of the magnitudes of the comparative gains of capital, the 19 industries of Jiangsu with available data can be ranked as follows:

1	2	3	4
Finance	Agriculture, forest, animal husbandry, fishery	Construction	Wholesale and retail
5	6	7	8
Public administration and social organizations	Education	Residents and other services	Information transfer, computer, and software areas
9	10	11	12
Health, safety, and welfare	Mining	Leasing and commercial services	Hotels and restaurants
13	14	15	16
Scientific research, technology service, and geological survey	Communication, transportation, storage, postal service	Manufacturing	Production/supply of electricity, gas, and water
17	18	19	
Social, sports, and entertainment	Real estate/housing	Water conservation, environment, and public facility management	

3.11.2 Investment and Employment in Jiangsu's Economic Sectors and Industries

First, in terms of the comparative gains of labor, because of lack of data, no meaningful discussion can be produced.

Second, in terms of the comparative gains of capital, the following 11 industries have enjoyed the comparative gains of more than 1, representing relatively more optimal areas of additional investment.

Finance	Agriculture, forest, animal husbandry, fishery	Construction	Wholesale and retail
Public administration and social organizations	Education	Residents and other services	Information transfer, computer, and software areas
Health, safety, and welfare	Mining	Leasing and commercial services	

However, for the following eight industries, excessive investment of capital has been given. Further capital investment in these industries should be curbed and redirected to the previous 11 industries.

Water conservation, environment, and public facility management	Real estate/ housing	Social, sports, and entertainment	Production/supply of electricity, gas, and water
Manufacturing	Communication, transportation, storage, postal service	Scientific research, technology service, and geological survey	Hotels and restaurants

In other words, no matter whether the investors, who look for investment opportunities in Jiangsu, are from eastern China, or other parts of the Mainland China, or the rest of the world, they should first consider the following industries, ranked from the most optimal to the least:

1	2	3	4
Finance	Agriculture, forest, animal husbandry, fishery	Construction	Wholesale and retail
5	6	7	8
Public administration and social organizations	Education	Residents and other services	Information transfer, computer, and software areas
9	10	11	
Health, safety, and welfare	Mining	Leasing and commercial services	

At the same time, try not to invest additional capital in other industries, because there has been a severe over supply of capital in these industries.

That is the direction of capital investment.

3.12 Liaoning Province

Liaoning is located in the southern part of the Northeast. The modern province was established in 1907 with the current name in use since 1929. The province consists of three approximate geographical regions: highlands in the west, plains in the middle, and hills in the east. The highlands are dominated by the Nulu'erhu Mountains of low hills, which roughly border along Inner Mongolia. The central part of Liaoning consists of the watersheds of such rivers as the Liao, Daliao, and their tributaries. This region is mostly flat and at low altitudes. And the eastern part of Liaoning is dominated by the Changbai Shan and Qianshan ranges, which extends into the sea to form the Liaodong Peninsula. The highest point in Liaoning of about 1336 m is found in this region on Mount Huabozi. For details, see Figure 3.12.

Historically, the governments of Liaoning were headed by various peoples, such as the Korean kingdoms of Gojoseon, Goguryeo, Balhae, the Chinese as the Yan state, Han Dynasty, and the nomadic states as Xiongnu, Donghu, Xianbei, Khitan, Jurchen, Mongol Empire, and Northern Yuan.

The Ming Empire took control of Liaoning in 1371, three years after having expulsed the Mongols from Beijing. The Manchus conquered Liaoning in the early seventeenth century,

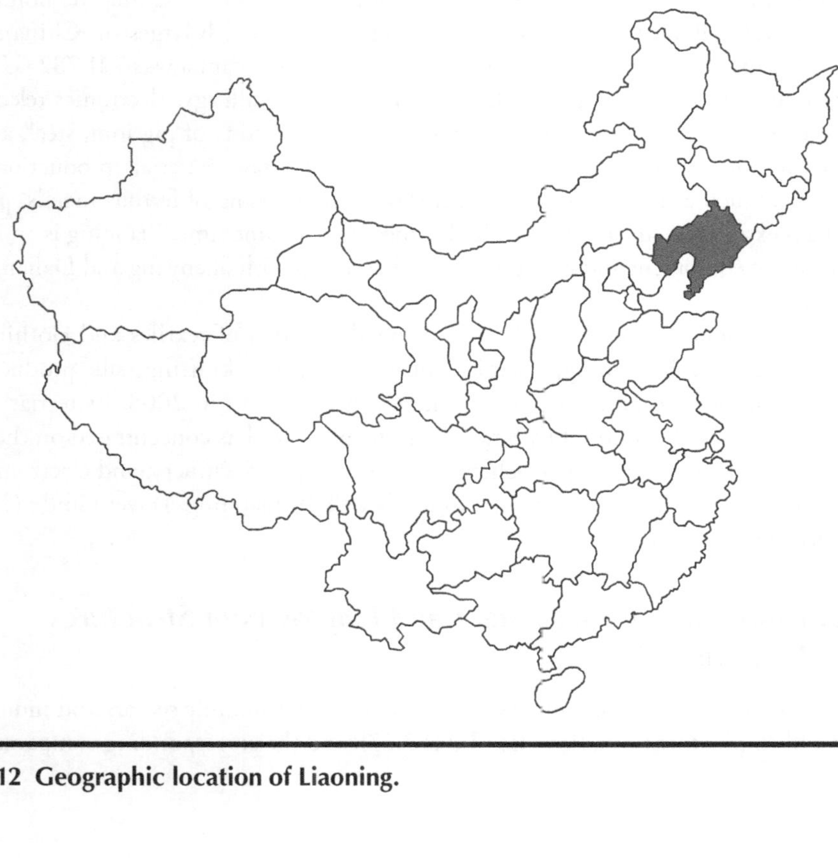

Figure 3.12 Geographic location of Liaoning.

decades before the rest of China fell to them. The Qing conquest of Liaodong resulted in a significant population loss in the area, as many local Chinese residents were either killed or fled south, and many cities were destroyed by the retreating Ming forces themselves (Edmonds, 1985). In the latter half of the seventeenth century, starting with laws issued in 1651 and 1653, the Qing government recruited migrants from the south, notably from Shandong, to settle the relatively sparsely populated area of Fengtian Province, which corresponds roughly to today's Liaoning. However, the rest of China's northeast remained officially off limits to Han Chinese for most of the Manchu era. In 1860, the Manchu government began to reopen the region to migration, which quickly resulted in Han Chinese becoming the dominant ethnic group in the region.

In 1904–1905 many key battles of the Russo-Japanese War, such as the Battle of Port Arthur and the Battle of Mukden, took place in Liaoning. To that point in time, these battles represented the largest land conflicts ever fought. During the Warlord Era in the early twentieth century, Liaoning was under the control of Fengtian Clique, including Zhang Zuolin and his son Zhang Xueliang. In 1931, Japan invaded and the area came under the rule of the Japanese-controlled puppet state of Manchukuo. The Chinese Civil War, following Japanese defeat in 1945, had its first major battles (the Liaoshen Campaign) in and around Liaoning. In 1949 when the People's Republic of China was founded, Liaoning did not exist. Instead there were two provinces, known as Liaodong and Liaoxi, as well as five municipalities, Shenyang, Luda, Anshan, Fushun, and Benxi, which were all merged into Liaoning in 1954.

Economically, Liaoning was one of the early provinces in China to industrialize, first under Japanese occupation, and then even more in the 1950s and 1960s. Recognizing the special difficulties faced by Liaoning and other provinces in Northeast China because of their heritage of heavy industry, the central government recently launched a "Revitalize the Northeast" Campaign. In comparison, Liaoning represents the largest economy of Northeast China. Its nominal GDP for 2011 was ¥2.20 trillion (US$348 billion), making it the seventh largest in China among 31 provinces, municipalities, and autonomous regions. Its GDP per capita was ¥41,782 (US$6172).

Leading industries in Liaoning include petrochemicals, metallurgy, electronics telecommunications, and machinery. The province is a major producer in China of pig iron, steel, and metal-cutting machine tools. It is also one of the most important raw materials production bases in China. Industries such as mining, quarrying, smelting and pressing of ferrous metals, petroleum, and natural gas extraction, are all of great significance. At the same time, Liaoning is an important production base of equipment and machinery manufacturing with Shenyang and Dalian being the industrial centers.

The province's light industry mainly focuses on the sectors of textiles and clothing, which include cotton and wool spinning, chemical fiber production, knitting, silk production, and the manufacturing of both garments and textile machinery. And in 2008, its tertiary industry accounted for 34.5% of the total GDP. Since then, the province has concentrated on the development of its four pillar industries—petrochemicals, metallurgy, machinery, and electronics.

For additional reading, please consult Edmonds (1985), Liaoning Travel Guide (2010), and People's Daily Online (2012).

3.12.1 Current State of Investment and Employment Structures of Liaoning

For Liaoning Province, we first compute the total outputs of economic sectors and industries, the investment and employment situations (see Table 3.23), and the corresponding comparative gains.

Table 3.23 GDPs, Total Investments in Fixed Properties, and Employment Situation of Liaoning's Three Economic Sectors and Industries (2011)

Sectors (i)	*Total Output of Sector (0.1 Billion Yuan)*	*Total Investment on Fixed Properties (0.1 Billion Yuan)*	*Employment (10,000 People)*
Total	22226.7	17726.3	2364.8785
Primary sector	1915.57	537	699.8727
Secondary sector	12152.1	7468.9	645.0823
Tertiary sector	8158.98	9720.4	1019.9235
Industries (*i*)			
Agriculture, forest, animal husbandry, fishery	1915.57	536.9476	**26.5**
Mining		547.4414	**33.1**
Manufacturing		5975.6094	**167.6**
Production/supply of electricity, gas, and water		698.2815	**17.1**
Construction	1455.61	247.5799	**47.9**
Communication, transportation, storage, postal service	1143.17	907.5333	**32.9**
Information transfer, computer, and software areas		105.457	**8.5**
# Information Transfer			
Wholesale and retail	1960.33	532.033	**22**
Hotels and restaurants	436.13	323.7418	**7.3**
Finance	755.57	56.6805	**22.4**
Real estate/housing	876.12	4884.5258	**11.5**
Leasing and commercial services		410.4312	**12.6**
Scientific research, technology service, and geological survey		107.776	**13.9**
Water conservation, environment, and public facility management		1489.2846	**14.2**

continued

Table 3.23 (continued) GDPs, Total Investments in Fixed Properties, and Employment Situation of Liaoning's Three Economic Sectors and Industries (2011)

Sectors (i)	*Total Output of Sector (0.1 Billion Yuan)*	*Total Investment on Fixed Properties (0.1 Billion Yuan)*	*Employment (10,000 People)*
# Public Facility Management			
Residents and other services		156.5292	**3**
Education		178.0523	**55.1**
Health, safety, and welfare		92.9871	**28.1**
Social, sports, and entertainment		184.2577	**5.2**
Public administration and social organizations		291.1372	**50.6**

Source: The data in bold is from *China Statistics Yearbook 2012,* edited by National Bureau of Statistics of China, published by China Statistics Press and the rest from *Liaoning Province's Statistics Yearbook 2012.*

Then, by comparing the comparative gains, we obtain the current structural states of investment and employment for Liaoning's economic sectors and industries. All the detailed calculations are similar to what has been described above. The results are summarized in Table 3.24.

Table 3.24 indicates that first, in terms of the three major sectors, all of the six relevant data of the three sectors are in extremely imbalanced states. In particular, the comparative gain of labor of the first sector is way too low, and is equal to only 0.291214, which is roughly equal to one-seventh of that of the same indicator of the secondary sector. In terms of the capital investment, the comparative gains of capital of the primary and the secondary sectors are way too low, and are, respectively, 2.84490 and 1.29759, while the investment of labor and capital in

Table 3.24 Structures and Comparative Gain Measures of Liaoning's Three Economic Sectors and Industries ($t = 2011$)

Sectors (i)	*Demand Structure $d_i(t)$ (%)*	*Employment Structure $s_i^1(t)$ (%)*	*Comparative Gain of Labor $G_i^1(t)$*	*Investment Structure $s_i^2(t)$ (%)*	*Comparative Gain of Capital $G_i^2(t)$*
Total	100	100	–	100	–
Primary sector	8.618328	29.59445	0.291214	3.029397	2.84490
Secondary sector	54.67343	27.27761	2.004334	42.13457	1.29759
Tertiary sector	36.70801	43.12795	0.851142	54.83603	0.66941
Industries (*i*)					
Agriculture, forest, animal husbandry, fishery	8.618328	1.120565	**7.691056**	3.029101	2.84518
Mining		1.399649		3.0883	

Table 3.24 (continued) Structures and Comparative Gain Measures of Liaoning's Three Economic Sectors and Industries ($t = 2011$)

Sectors (i)	*Demand Structure* $d_i(t)$ *(%)*	*Employment Structure* $s_i^1(t)$ *(%)*	*Comparative Gain of Labor* $G_i^1(t)$	*Investment Structure* $s_i^2(t)$ *(%)*	*Comparative Gain of Capital* $G_i^2(t)$
Manufacturing		7.087045		33.71042	
Production/supply of electricity, gas, and water		0.723086		3.93924	
Construction	6.548925	2.025474	**3.233280**	1.396681	4.63892
Communication, transportation, storage, postal service	5.143229	1.391192	**3.696994**	5.1197	1.00460
Information transfer, computer, and software areas		0.359426		0.594918	
# Information Transfer					
Wholesale and retail	8.819708	0.930280	**9.480699**	3.001376	2.93855
Hotels and restaurants	1.962190	0.308684	**6.356630**	1.826336	1.07439
Finance	3.399380	0.947195	**3.588893**	0.319754	10.6312
Real estate/housing	3.941746	0.486283	**8.105869**	27.55525	0.14305
Leasing and commercial services		0.532797		2.31538	
Scientific research, technology service, and geological survey		0.587768		0.608001	
Water conservation, environment, and public facility management		0.600457		8.401554	
# Public Facility Management					
Residents and other services		0.126856		0.883034	
Education		2.329929		1.004453	
Health, safety, and welfare		1.188222		0.524571	
Social, sports, and entertainment		0.219884		1.039459	
Public administration and social organizations		2.139645		1.642403	

Source: Data are from *China Statistics Abstract 2012* (XLS, DOC).

the tertiary sector is in extreme excess with their comparative gains equal to respective 0.851142 and 0.66941.

Second, in terms of the industries, the relevant data of the 21 industries of Liaoning Province should contain 42 data points. However, because of lack of raw data, we only collected 14 of them, where the data in bold are from the urban employment data of 2012 *China Statistics Yearbook*, and are not comparable with the rest of the data. So, there are only seven effective data points in terms of the comparative gains of capital. Among all these seven effective data values, only one is basically in equilibrium, where the comparative gain of capital in "communication, transportation, storage, postal service" is equal to 1.00460. All the rest six data points (6/7 ≥ 86%) are situated in extremely imbalanced states. In particular, the comparative gains of capital of the following six industries are over 1. Especially, the comparative gain of capital in "finance" is equal to 10.6312, which is more than (10.6312/0.14305 ≈) 74 times of the lowest value of this indicator that is reached by "real estate and housing."

Finance	Construction	Wholesale and retail
Agriculture, forest, animal husbandry, fishery	Hotels and restaurants	Communication, transportation, storage, postal service

Therefore, in terms of the magnitudes of the comparative gains of capital, the seven industries of Liaoning Province about which we have data are ranked as follows:

1	2	3	4
Finance	Construction	Wholesale and retail	Agriculture, forest, animal husbandry, fishery
5	6	7	
Hotels and restaurants	Communication, transportation, storage, postal service	Real estate and housing	

3.12.2 Investment and Employment in Liaoning's Economic Sectors and Industries

First, in terms of the comparative gains of labor, the excessive supplies of labor in the primary and the tertiary economic sectors should be relocated into the secondary sector as much as possible. As for detailed analysis, because of lack of the employment data of the industries, the problem of adjusting labor forces cannot be meaningfully addressed.

Second, in terms of the comparative gains of capital, the following six industries have enjoyed the comparative gains of more than 1, representing relatively more optimal areas of additional investment.

Finance	Construction	Wholesale and retail	Agriculture, forest, animal husbandry, fishery
Hotels and restaurants	Communication, transportation, storage, postal service		

However, for "real estate and housing," excessive investment of capital has been given. Further capital investment in this industry should be curbed and redirected to the previous six industries.

In other words, Liaoning is the worst province in East China in terms of opportunities of investment and employment. Even so, within the province, there are still the problems of investment and employment. According to our computations, the industries of Liaoning that are suitable for investment are ranked as follows:

1	2	3
Finance	Construction	Wholesale and retail
4	5	6
Agriculture, forest, animal husbandry, fishery	Hotels and restaurants	Communication, transportation, storage, postal service

Chapter 4

Central China

A General Area

From Chapter 3, it follows that if China is divided into three parts: the east, center, and west, then Central China represents the general area for investment and employment. Although this area is general in terms of the opportunities for investment and employment, the regions in this part of China can still be ranked from the most optimal to the least optimal based on their respective characteristics. On the basis of the theoretical foundation, models, and methods, provided in Chapter 2, this chapter calculates and analyzes the statistics available for the year 2011, and then provides the consequent guidelines for investment and employment opportunities in this region.

The geographic location of Central China is shown in Figure 4.1.

4.1 Industries and Economic Sectors of the Central Plane

4.1.1 Current State of the Regional Economic Structure of Central China

From Table 2.1, the 2011 GDP of Central China and the relevant investment and employment are shown in Table 4.1.

From these data in Table 4.1, we can obtain the values in Table 4.2 by using the relevant formulas given in Chapter 1.

The previous table indicates that the regional economic structure of this part of China experiences severe imbalances, no matter whether it is in terms of the supply and demand of labor or the supply and demand of capital.

In fact, from the fifth column (the comparative gains of labor) and the seventh column (the comparative gains of capital) of the table, it follows that among the 20 relevant data of the 10 provinces, municipalities, and regions of this area, only six of them are basically in equilibrium: the comparative gains of labor of Jinlin, Hunan, and Guangxi, and the comparative gains of capital of Guangxi, Henan, and Jinlin are, respectively, equal to 1.0069, 0.9967, 0.9624, and 0.992, 1.026, 0.962. And all others, amounting to (14/20≥) 70%, are in various states of imbalance or severe imbalance. Here, the most severe imbalance of supply and demand appears to such areas as Jiangxi, Inner Mongolia, Anhui, and Shanxi. In fact, Jiangxi's comparative gain of labor is the

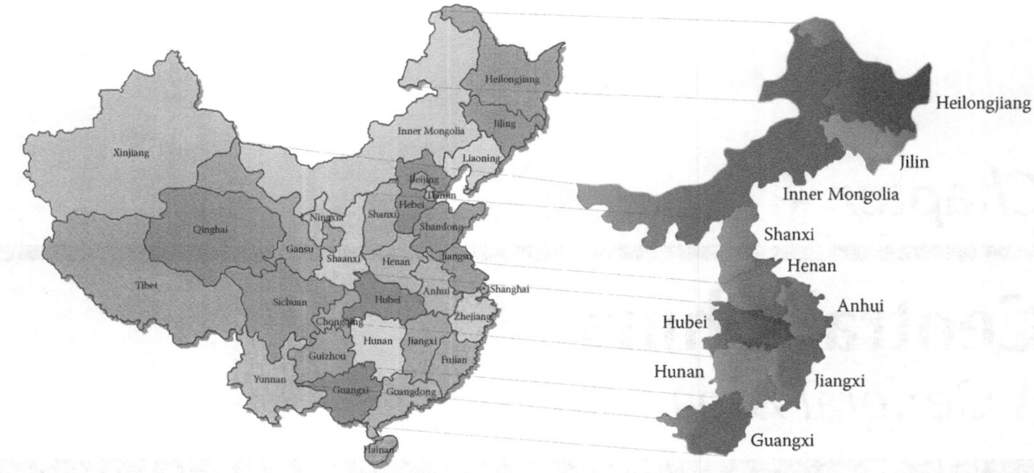

Figure 4.1 The geographic location of Central China.

lowest (= 0.7368), Anhui's comparative gain of capital is the lowest (= 0.832); and the comparative gain of labor of Inner Mongolia and the comparative gain of capital of Heilongjiang are, respectively, the highest, equaling 1.5677 and 1.140.

So, in terms of the magnitudes of comparative gains of labor, the 10 provinces, municipalities, and regions of this area can be ordered as follows:

1	2	3	4	5
Inner Mongolia	Henan	Shanxi	Jilin	Hunan
6	7	8	9	10
Guangxi	Anhui	Heilongjiang	Hubei	Jiangxi

Table 4.1 GDPs, Values of Fixed Properties, and Employments of Various Provinces, Municipalities, and Autonomous Regions of Central China (2011)

Area	*Province/Municipality/ Autonomous Region (i)*	*Regional GDP (0.1 Billion Yuan)*	*Value of Fixed Properties (0.1 Billion Yuan)*	*Employment (10,000 people)*
Central China	12. Shanxi	11237.55	7073.1	1225.2
	13. Inner Mongolia	14359.88	10366.2	1120.1
	14. Jilin	10568.83	7441.7	1283.7
	15. Heilongjiang	12582	7475.4	1672.1
	16. Anhui	15300.65	12455.7	2028.8
	17. Jiangxi	11702.82	9087.6	1942.4
	18. Henan	26931.03	17769.0	2892.4
	19. Hubei	19632.26	12557.3	2701.7
	20. Hunan	19669.56	11880.9	2413.3
	21. Guangxi	11720.87	7990.7	1489.4
	Total	153705.45	104097.6	18769.1

Source: Data are from *China Statistical Abstract 2012* (XLS, DOC).

Table 4.2 The Regional Economic Structures and Measures of Comparative Gains of the Provinces, Municipalities, and Autonomous Regions of Central China (2011)

Area	*Province/ Municipality/ Autonomous Region (i)*	*Demand Structure $d_i(t)$ (%)*	*Employment Structure $s_i^1(t)$ (%)*	*Comparative Gain of Labor $G_i^1(t)$*	*Investment Structure $s_i^2(t)$ (%)*	*Comparative Gain of Capital $G_i^2(t)$*
Central China	12. Shanxi	7.311	6.518	1.1217	6.795	1.076
	13. Inner Mongolia	9.342	5.959	1.5677	9.958	0.938
	14. Jilin	6.876	6.829	1.0069	7.147	0.962
	15. Heilongjiang	8.186	8.896	0.9202	7.181	1.140
	16. Anhui	9.955	10.794	0.9223	11.965	0.832
	17. Jiangxi	7.614	10.334	0.7368	8.730	0.872
	18. Henan	17.521	15.388	1.1386	17.070	1.026
	19. Hubei	12.773	14.374	0.8886	12.063	1.059
	20. Hunan	12.797	12.839	0.9967	11.413	1.121
	21. Guangxi	7.626	7.924	0.9624	7.676	0.993
	Total	100	100	—	100	—

In terms of the magnitudes of the comparative gains of capital, these 10 provinces, municipalities, and regions of this area can be ranked as follows:

1	2	3	4	5
Heilongjiang	Hunan	Shanxi	Hubei	Henan
6	7	8	9	10
Guangxi	Jilin	Inner Mongolia	Jiangxi	Anhui

4.1.2 Investment and Employment Opportunities in Central China

First, in terms of labor, the comparative gains of the first four provinces and regions are over 1, representing the most ideal places for employment. The other six provinces and regions experience excessive labor so that limitations should be imposed to control the import of labors. In particular, for the two provinces of Jiangxi and Hubei, they should try different methods to transport their excessive labors to the first four provinces and regions.

In other words, no matter whether the job seekers, who look for employment opportunities in the central area of China, are from Central China, or other parts of the mainland China, or the rest of the world, the order of the provinces and municipalities in terms of their employment opportunities is given as follows:

1	2	3	4
Inner Mongolia	Henan	Shanxi	Jilin

Without special skills, don't look for employment opportunities in the other six provinces and regions, because these provinces and regions are experiencing excessive labor supplies.

This is the direction of employment.

Second, in terms of capital, the comparative gains of the first five provinces and regions are more than 1, representing ideal places for potential investments. The other five provinces and regions experience excessive investment capital so that the injection of capital should be controlled and reduced. Specifically, the five provinces and regions, such as Anhui, Jiangxi, Inner Mongolia, Jilin, and Guangxi, have suffered from excessive capital which should be relocated to the five previously ranked provinces and municipalities.

In other words, no matter whether the investors, who look for investment opportunities in the central area of China, are from Central China, or other parts of the mainland China, or the rest of the world, the order of the provinces and regions in terms of their investment opportunities is given as follows:

1	2	3	4	5
Heilongjiang	Hunan	Shanxi	Hubei	Henan

Unless there are particular niches, don't go to the other five provinces and regions.

This is the direction of investment.

Third, by combining the two criteria discussed above, it follows that the most optimal provinces and regions in this area in terms of investment and employment opportunities are Henan and Shanxi. It is because their comparative gains of labor and capital are all more than 1. The secondary choices are Inner Mongolia, Hunan, and Jilin. And the last choices are Jiangxi and Anhui.

In the following sections, we will analyze computationally the investment and employment problems of these 10 provinces and regions of Central China by making use of the recent statistics (2011) and produce our best conclusions for the reader to make his/her appropriate decisions.

Note: Since we have no access to relevant statistics of regions internal to individual provinces, municipalities, and autonomous regions, we have not been able to produce any computational conclusions on investment and employment regarding these finer divisions of China. Instead, we can only focus on the investigation of the corresponding investment and employment problems in terms of economic sectors of the individual provinces, municipalities, and autonomous regions.

4.2 Henan Province

Henan, often referred to as Zhongyuan or Zhongzhou which literally means "central plains" or "midland," is located in the central part of the country. It is the birthplace of Chinese civilization with over 5000 years of written history, and remained China's cultural, economic, and political center until approximately 1000 years ago. It covers a large part of the fertile and densely populated North China Plain. It borders with six other provinces, such as Shaanxi in the west, Hubei in the south, Shanxi in the northwest, Hebei in the northeast, Shandong in the east (northeast), and Anhui in the southeast. The province is ranked third among all Chinese provinces in terms of its population. If it were a country by itself, Henan would be the 12th most populous country

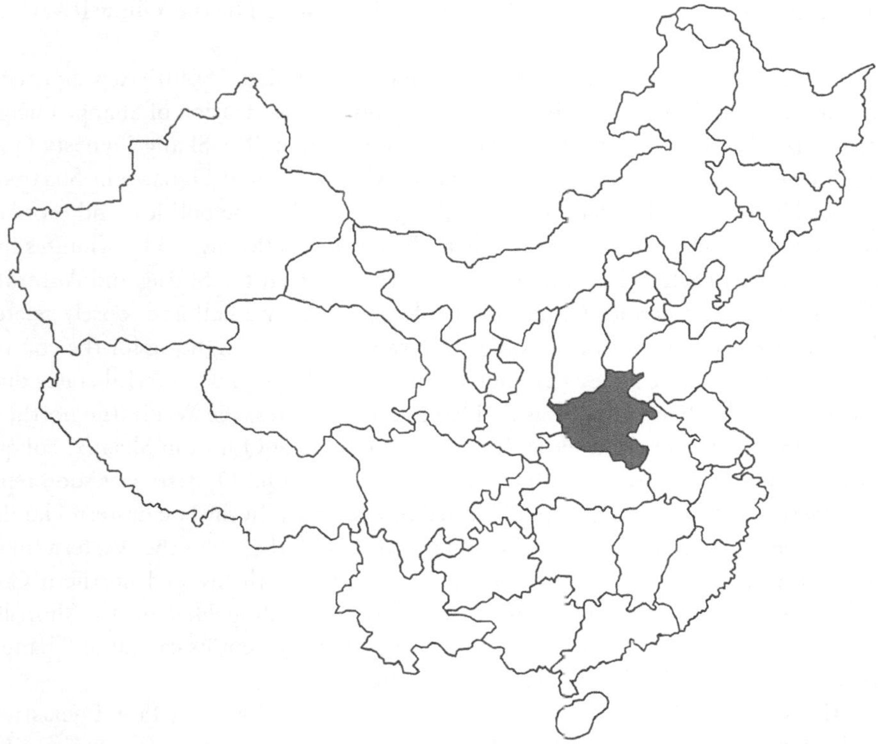

Figure 4.2 The geographic location of Henan.

in the world. At the same time, Henan is the fifth largest provincial economy in China and the largest among inland provinces. Even so, its GDP per capita is low compared to other eastern and central provinces, making the province one of the less developed areas in China. For details, see Figure 4.2.

Geographically, Henan consists of a diverse landscape with floodplains in the east and mountains in the west. The Taihang Mountains intrude partially into Henan's northwestern borders from Shanxi, forming the eastern edge of Loess Plateau. In the west the Xionger and Funiu Mountains form an extensive network of mountain ranges and plateaus, supporting one of the few remaining temperate deciduous forests which once covered all of Henan. In the far south, the Dabie Mountains divides Hubei from Henan. Different from the rest of northern China, desertification is not a problem for Henan. The Yellow River passes through central Henan. It enters from the northwest. Excessive amounts of sediment are formed due to the silt it picks up from the Loess Plateau that raises the riverbed and causes frequent floods. However, this problem of flooding has been resolved recently by the construction of dams, levees, and reservoirs, as well as the depletion of water resources. The Huai River in southern Henan is important as it divides northern and southern Chinese climate and culture.

Historically, Henan is widely regarded as the Cradle of Chinese civilization along with Shanxi and Shaanxi provinces. Although its economic prosperity has resulted from its fertile plains and location, its very location has also made it suffer from nearly all of major wars in Chinese history. Additionally, the frequent floods of the Yellow River have caused significant damage from time to time. For example, Kaifeng has been buried by the Yellow River's silt seven times due to flooding.

The Xia Dynasty, the first Chinese dynasty that existed about 5000 years ago within east Henan, collapsed around the sixteenth century BC following the invasion of Shang, a neighboring vassal state centered around today's Shangqiu in eastern Henan. The Shang Dynasty lasted from the sixteenth century BC to the eleventh century BC when the Zhou Dynasty of Shaanxi arrived from the west. Zhou established its capital in Chang'an, causing the political and economic center to move away from Henan. In 722 BC, when Chang'an was threatened by Xionites invasions, the capital was moved back to Luoyang in Henan. That is when the Spring and Autumn period started. During this time period, China was made up of many small and loosely related states (under the rule of Zhou emperor in Luoyang) that were constantly fighting for the control of the central plain. Later on, these states were pooled into seven large and powerful states during the Warring States period, when Henan was divided into three states, the Wei in the north, the Chu in the south, and the Han in the middle. In 221 BC, the state of Qin from Shaanxi conquered all of the other six states and ended the 800 years of warfare. The Qin Dynasty was soon replaced by the Han Dynasty in 206 BC with its capital located in Chang'an. In the late Eastern Han Dynasty, war and rivalry between regional warlords were rampant. With the fall of the Western Jin Dynasty in the fourth and fifth century, nomadic peoples from the north invaded northern China and established many successive regimes in northern China, including Henan. The short-lived Sui Dynasty reunified China in 589. The succeeding Tang Dynasty kept its capital in Chang'an with Henan being one of the wealthiest places in the empire.

During the period of Five Dynasties and Ten Kingdoms that followed Tang Dynasty, Kaifeng in eastern Henan was the capital of four dynasties. The Song Dynasty that reunified China in 982 also selected Kaifeng as its capital. In 1127, the Song Dynasty succumbed to Jurchen (Jin Dynasty) invaders from the north, and in 1142 abandoned all of northern China, including Henan. Kaifeng served as the Jurchen's southern capital from 1157 while the Jurchen kept their main capital further north until 1214 when they moved the imperial court to Kaifeng in order to flee the Mongol onslaught. In 1234, they succumbed to combined forces of Mongol and Song Dynasty while Mongols took control. In 1279, the Mongols conquered entire China, establishing the Yuan Dynasty and setting up the equivalent of modern Henan province. Henan remained important in the Ming Dynasty and Qing Dynasty that followed. However, its economy slowly deteriorated due to frequent natural disasters.

In 1938, when the Imperial Japanese Army captured Kaifeng, the government led by Chiang Kai-shek bombed the Huayuankow dam in Zhengzhou in order to prevent the Japanese forces from advancing further, which caused millions of deaths because of the consequent flooding and famine in Henan, Anhui, and Jiangsu.

In 1978, when the communist leader Deng Xiaoping initiated the open door policy and embraced capitalism, China entered an economic boom that continues today. The boom did not reach inland provinces, such as Henan, initially. But by the 1990s, Henan's economy was expanding at an even faster rate than national average.

Henan has rapidly grown its economy over the past two decades, expanding at a rate faster than the national average of 10%. This rapid growth has transformed Henan into one province that matches other central provinces, though still relatively impoverished on a national scale. In 2011, Henan's nominal GDP was ¥3.20 trillion (US$427 billion), making it the fifth largest economy in China, while its GDP per capita was ranked the 19th. In particular, Henan is a semi-industrialized economy with an underdeveloped service sector. In 2009, Henan's primary, secondary, and tertiary industries were, respectively, worth ¥277 billion (US$40 billion), ¥1.097 trillion (US$160 billion), and ¥563 billion (US$82 billion). Agriculture has traditionally been a core of Henan's economy with the nation's highest outputs of wheat and sesame and second highest output of rice. Food production and processing make up more than 14% of the output from the province's secondary industry.

Even though the traditional industry in Henan has been in light textiles and food processing, recent developments have expanded into metallurgy, petrol, cement, chemical industry, machinery, and electronics. The province has the second largest molybdenum reserves in the world with large amounts of coal, aluminum, alkaline metals, and tungsten discovered in western Henan. Export and processing of these materials has been one of the main sources of revenues. In 2008, the total trade volume of both import and export amounted to US$17.5 billion. Since 2002, over 7000 foreign enterprises have been approved, and foreign fund investments of US$10.64 billion have been used in contracts with a realized foreign fund investment of US$5.3 billion. In contrast, Henan's service sector is rather underdeveloped. Finance and commerce have been mainly concentrated in urban centers, such as Zhengzhou and Luoyang. To make the economy more knowledge and technology based, the government has established various development zones to promote industries, such as software, information technologies, new materials, bio-pharmaceutical, and photo-machinery-electronics. (http://en.wikipedia.org/wiki/Henan-cite_note-19.) The province is also a major destination for tourists with places such as the Shaolin Temple and Longmen Grottoes attracting millions of tourists each year.

For additionally relevant readings, refer to Ministry of Commerce (2013); National Bureau of Statistics of China (2013); and Osnos (2011).

4.2.1 Current State of Investment and Employment Structures of Henan

All that is necessary for us to do is the computations of the gross domestic outputs of all the economic sectors and industries and the investment and employment situations (Table 4.3) of Henan and the relevant comparative gains. By comparing the resultant comparative gains, we can then obtain the current states of investment and employment for the three economic sectors and industries. All the detailed computations are similar to what has been described earlier. The results are given in Table 4.4.

First, in terms of the three major economic sectors, among the six relevant data points of the three sectors, all of them are situated in extreme instability, where the comparative gain of labor of the primary sector is way too low and equal to merely 0.302683. That is not even reaching one-sixth, and one-third of that same indicator of the secondary and the tertiary sectors, respectively. In terms of the capital investment, the comparative gains of the primary and the secondary sectors are much too low, and are respectively equal to 3.21486 and 1.11526, while the capital investment in the tertiary sector is in extreme excess (0.66566).

Second, in terms of the employment situation, the relevant data of the 21 industries of Henan province should contain 42 data points. However, because of lack of raw data, we only collected 38 of them. So, we can only base our discussion on these available statistics. Among all the industries, only two are basically in equilibrium, where the comparative gains of labor in "communication, transportation, storage, postal service" and "hotels and restaurants" are, respectively, 1.015974 and 1.040394. All the rest (36/38 ≥ 94%) are situated in extremely imbalanced states. In particular, the comparative gains of labor of the following 15 industries are all over 1. Especially, the comparative gain of labor in "real estate/housing" is even equal to 10.33234, which is more than

Table 4.3 GDPs, Total Investments in Fixed Properties, and Employment Situation of Henan's Three Economic Sectors and Industries (2011)

Sectors (i)	*Total Output of Sector (0.1 Billion Yuan)*	*Total Investment on Fixed Properties (0.1 Billion Yuan)*	*Employment (10,000 People)*
Total	26931.0316	17770.5099	6197.85
Primary sector	3512.24	720.8906	2670.45
Secondary sector	15427.07633	9127.5748	1852.5
Tertiary sector	7991.715275	7922.0445	1674.9
Industries (*i*)			
Agriculture, forest, animal husbandry, fishery	3440.39	720.8906	2670.4482
Mining	2278.07	**717.4**	65.1709
Manufacturing	11116.14	**7903.6**	1109.32174
Production/supply of electricity, gas, and water	555.11	**492.3**	22.0114
Construction	1477.756328	14.4259	656.000802
Communication, transportation, storage, postal service	961.5	812.5448	217.7984518
Information transfer, computer, and software areas	325.9	31.91	35.7358793
# Information Transfer			
Wholesale and retail	1586.09	410.1412	535.438508
Hotels and restaurants	797.99	175.1811	176.517521
Finance	868.2	16.57	25.1507
Real estate/housing	987	4456.3293	21.984
Leasing and commercial services	265.04	56.39	32.1518
Scientific research, technology service, and geological survey	166.04	41.26	16.3037
Water conservation, environment, and public facility management	63.65	1283.85	13.5285
# Public Facility Management			
Residents and other services	220.66	68.7181	323.857299
Education	739.02	230.38	117.5861
Health, safety, and welfare	263.97	157.39	43.7514
Social, sports, and entertainment	74.33	110.44	8.9907
Public administration and social organizations	672.325275	70.95	106.1064

Source: Data are from *Henan Province's Statistical Yearbook 2012.*

Table 4.4 The Structures and Comparative Gain Measures of Henan's Three Economic Sectors and Industries ($t = 2011$)

Sectors (i)	*Demand Structure $d_i(t)$ (%)*	*Employment Structure $s_i^1(t)$ (%)*	*Comparative Gain of Labor $G_i^1(t)$*	*Investment Structure $s_i^2(t)$ (%)*	*Comparative Gain of Capital $G_i^2(t)$*
Total	100	100	—	100	—
Primary sector	13.04161	43.08672	0.302683	4.056668	3.21486
Secondary sector	57.28364	29.88940	1.916521	51.36361	1.11526
Tertiary sector	29.67475	27.02389	1.098093	44.57973	0.66566
Industries (*i*)				**0**	
Agriculture, forest, animal husbandry, fishery	12.77482	43.08669	0.296491	4.056668	3.14909
Mining	8.458904	1.051508	8.044544	4.037025	2.09533
Manufacturing	41.27632	17.89849	2.306134	44.47593	0.92806
Production/supply of electricity, gas, and water	2.061228	0.355146	5.803894	2.77032	0.74404
Construction	5.487188	10.58433	0.518426	0.081179	67.5938
Communication, transportation, storage, postal service	3.570231	3.514097	1.015974	4.572434	0.78082
Information transfer, computer, and software areas	1.210128	0.576585	2.098785	0.179567	6.73914
# Information Transfer					
Wholesale and retail	5.889451	8.639101	0.681720	2.307988	2.55177
Hotels and restaurants	2.963087	2.848044	1.040394	0.985797	3.00578
Finance	3.223790	0.405797	7.944339	0.093244	34.5736
Real estate/housing	3.664917	0.354704	10.33234	25.0771	0.14615
Leasing and commercial services	0.984144	0.518757	1.897117	0.317323	3.10139

continued

Table 4.4 (continued) The Structures and Comparative Gain Measures of Henan's Three Economic Sectors and Industries ($t = 2011$)

Sectors (i)	*Demand Structure $d_i(t)$ (%)*	*Employment Structure $s_i^1(t)$ (%)*	*Comparative Gain of Labor $G_i^1(t)$*	*Investment Structure $s_i^2(t)$ (%)*	*Comparative Gain of Capital $G_i^2(t)$*
Scientific research, technology service, and geological survey	0.616538	0.263054	2.343768	0.232182	2.65540
Water conservation, environment, and public facility management	0.236344	0.218277	1.082772	7.22461	0.03271
# Public Facility Management					
Residents and other services	0.819352	5.225317	0.156804	0.386697	2.11885
Education	2.744121	1.897208	1.446400	1.296417	2.11670
Health, safety, and welfare	0.980170	0.705913	1.388515	0.885681	1.10669
Social, sports, and entertainment	0.276001	0.145062	1.902649	0.621479	0.44410
Public administration and social organizations	2.496471	1.711987	1.458230	0.399257	6.25279

(10.33234/0.156804≈) 66 times of the lowest value of this indicator that is reached by "residents and other services."

Real estate/ housing	Mining	Finance	Production/ supply of electricity, gas, and water	Scientific research, technology service, and geological survey
Manufacturing	Information transfer, computer, and software areas	Social, sports, and entertainment	Leasing and commercial services	Public administration and social organizations
Education	Health, safety, and welfare	Water conservation, environment, and public facility management	Hotels and restaurants	Communication, transportation, storage, postal service

At the same time, the imbalance in terms of the comparative gains of capital seems to be way too severe. The comparative gains of capital in the following 13 industries are all over 1. In particular, the comparative gains of capital of "construction" and "finance" are over 10, while the value of this indicator of "water conservation, environment, and public facility management" is too low, equaling to merely 0.03271, which is only (0.03271/67.5938≈) 0.00049 when compared to that of "construction," (0.03271/34.5736≈) 0.00095 when compared to that of "finance," (0.03271/6.73914≈) 0.0049 when compared to that of "information transfer, computer, and software areas," (0.03271/6.25279≈) 0.0052 when compared to that of "public administration and social organizations," and (0.03271/**3.14909**≈) 0.010 when compared to that of "agriculture, forest, animal husbandry, fishery."

Construction	Finance	Information transfer, computer, and software areas	Public administration and social organizations	Agriculture, forest, animal husbandry, fishery
Leasing and commercial services	Hotels and restaurants	Scientific research, technology service, and geological survey	Wholesale and retail	Residents and other services
Education	Mining	Health, safety, and welfare		

So, in terms of the magnitudes of the comparative gains of labor, the industries of the Henan Province with available data can be ranked as follows:

1	2	3	4	5
Real estate/housing	Mining	Finance	Production/supply of electricity, gas, and water	Scientific research, technology service, and geological survey
6	7	8	9	10
Manufacturing	Information transfer, computer, and software areas	Social, sports, and entertainment	Leasing and commercial services	Public administration and social organizations
11	12	13	14	15
Education	Health, safety, and welfare	Water conservation, environment, and public facility management	Hotels and restaurants	Communication, transportation, storage, postal service
16	17	18	19	20
Wholesale and retail	Construction	Agriculture, forest, animal husbandry, fishery	Residents and other services	

In terms of the magnitudes of the comparative gains of capital, the industries of the Henan Province with available data can be ranked as follows:

1	2	3	4	5
Construction	Finance	Information transfer, computer, and software areas	Public administration and social organizations	Agriculture, forest, animal husbandry, fishery
6	7	8	9	10
Leasing and commercial services	Hotels and restaurants	Scientific research, technology service, and geological survey	Wholesale and retail	Residents and other services
11	12	13	14	15
Education	Mining	Health, safety, and welfare	Manufacturing	Communication, transportation, storage, postal service
16	17	18		
Social, sports, and entertainment	Real estate/ housing	Water conservation, environment, and public facility management		

4.2.2 Investment and Employment in Henan's Economic Sectors and Industries

First, in terms of the comparative gains of labor, the values of this indicator for the following industries:

Real estate/housing	Mining	Finance	Production/supply of electricity, gas, and water
Scientific research, technology service, and geological survey	Manufacturing	Information transfer, computer, and software areas	Social, sports, and entertainment
Leasing and commercial services	Public administration and social organizations	Education	Health, safety, and welfare
Water conservation, environment, and public facility management	Hotels and restaurants	Communication, transportation, storage, postal service	

are over 1, representing relatively optimal industries for employment. In the other four industries, there has been an excessive supply of labor so that appropriate control on the increase of labor into these industries should be imposed. Specifically, the excessive labor in the following industries should be transferred into the previous 15 industries.

Residents and other services	Agriculture, forest, animal husbandry, fishery	Construction	Wholesale and retail

In other words, no matter whether the job seekers, who look for employment opportunities in the Henan Province, are from Central China, or other parts of the mainland China, or the rest of the world, they should first consider the secondary and the tertiary economic sectors with the relevant industries ranked as follows:

1	2	3	4
Real estate/housing	Mining	Finance	Production/supply of electricity, gas, and water
5	6	7	8
Scientific research, technology service, and geological survey	Manufacturing	Information transfer, computer, and software areas	Social, sports, and entertainment
9	10	11	12
Leasing and commercial services	Public administration and social organizations	Education	Health, safety, and welfare
13	14	15	
Water conservation, environment, and public facility management	Hotels and restaurants	Communication, transportation, storage, postal service	

Without special skills, don't look for employment opportunities in the primary economic sector and other industries, because those areas are experiencing excessive labor supplies.

That is the direction of employment.

Second, in terms of capital, the following 13 industries have enjoyed the comparative gains of more than 1, representing relatively more optimal areas of additional investment.

Construction	Finance	Information transfer, computer, and software areas	Public administration and social organizations	Agriculture, forest, animal husbandry, fishery
Leasing and commercial services	Hotels and restaurants	Scientific research, technology service, and geological survey	Wholesale and retail	Residents and other services
Education	Mining	Health, safety, and welfare		

However, for the following six industries, excessive investment of capital has been given. Further capital investment in these industries should be curbed and redirected to the previous 13 industries.

Water conservation, environment, and public facility management	Real estate/housing	Social, sports, and entertainment
Production/supply of electricity, gas, and water	Communication, transportation, storage, postal service	Manufacturing

In other words, no matter whether the investors, who look for investment opportunities in the Henan Province, are from Central China, or other parts of the mainland China, or the rest of the world, they should consider the following industries ranked from the most optimal to the least:

Construction	Finance	Information transfer, computer, and software areas	Public administration and social organizations	Agriculture, forest, animal husbandry, fishery
Leasing and commercial services	Hotels and restaurants	Scientific research, technology service, and geological survey	Wholesale and retail	Residents and other services
Education	Mining	Health, safety, and welfare		

That is the direction of capital investment.

Third, by combining the two criteria discussed above, it follows that the industries in the Henan Province that are most optimal for both investment and employment are the following, because the comparative gains of labor and capital of each of these industries are over 1.

Finance	Information transfer, computer, and software areas	Mining	Scientific research, technology service, and geological survey	Public administration and social organizations
Leasing and commercial services	Hotels and restaurants	Education	Health, safety, and welfare	

The secondary choice is the industry of "wholesale and retail." And the worst choice is the industry of "water conservation, environment, and public facility management."

4.3 Shanxi Province

Shanxi is located in the North China region. It borders Hebei in the east, Henan in the south, Shaanxi in the west, and Inner Mongolia in the north. The capital of the province is Taiyuan.

Geographically, Shanxi is located on a plateau with higher ground to the east, the Taihang Mountains, the west, the Lüliang Mountains, and the highest peak, the Mount Wutai, in the northeast, and a series of valleys in the center through which the Fen River runs. The Great Wall forms most of the northern border with Inner Mongolia. The Yellow River forms the western border with Shaanxi. The Fen and Qin rivers, two tributaries of the Yellow River, which drain much of the area, run north-to-south through the province; and Sanggan and Hutuo rivers, tributaries of the Hai River, drain the north of the province. In terms of the climate, other than that of continental monsoons, Shanxi is rather arid with long, dry, and cold winters and warm and humid summers. For more details, see Figure 4.3.

Historically, Shanxi was the territory of state of Jin during the Spring and Autumn period (722–403 BC). Then it underwent a three-way split into the states of Han, Zhao, and Wei in 403 BC, which was the start of the Warring States period (403–221 BC). The Han Dynasty (206 BC–AD 220) ruled Shanxi as the province. The Tang Dynasty (618–907) was originated in Taiyuan. During the first part of the Five Dynasties and Ten Kingdoms period (907–960), Shanxi supplied rule for three of the Five Dynasties, and was the only one of the Ten Kingdoms located in northern China. During the Northern Song Dynasty (960–1127), the 16 ceded prefectures continued to be an area of contention between Song China and the Liao Dynasty. Later the Southern Song Dynasty abandoned all of North China to the Jurchen Jin Dynasty (1115–1234) in 1127, including Shanxi. The Mongol Yuan Dynasty divided China into provinces but did not establish Shanxi as a province. Shanxi only gained its present name and approximate borders in the Ming Dynasty (1368–1644). During the Qing Dynasty (1644–1911), Shanxi extended northwards beyond the

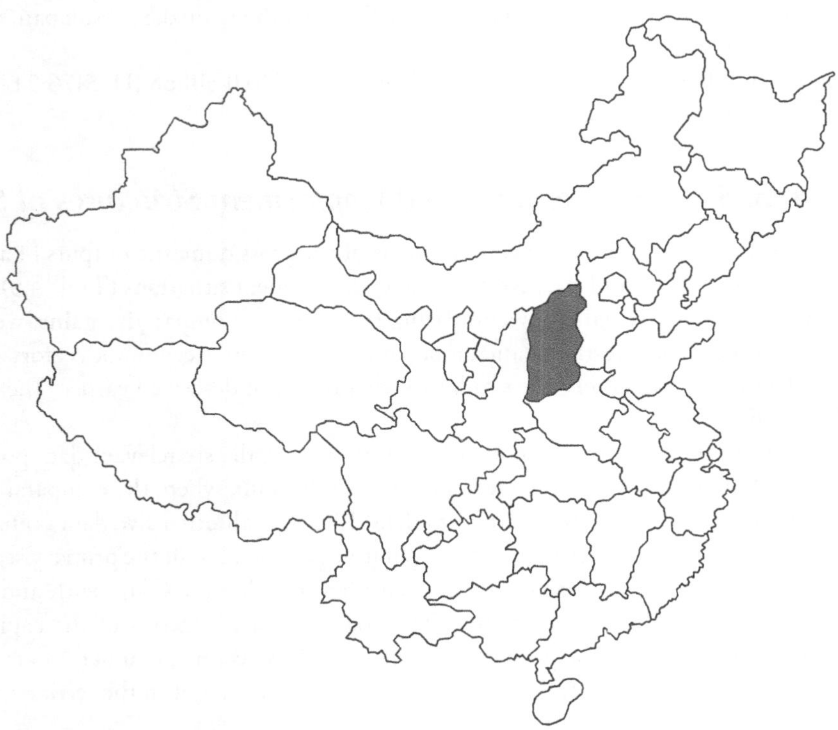

Figure 4.3 The geographic location of Shanxi.

Great Wall to include parts of Inner Mongolia. During the rule of the Republic of China over mainland China (1912–1949), the warlord Yen Hsi-shan held Shanxi, regardless of the frequent political upheavals shaking the rest of China. During the Second Sino-Japanese War, Japan occupied much of the province after defeating China in the Battle of Taiyuan. Shanxi was also a major battlefield between the Japanese and the Chinese communist guerrillas during the war. After the defeat of Japan, the countryside of Shanxi became important bases for the communist People's Liberation Army in the ensuing Chinese Civil War. Although Yen had incorporated thousands of former Japanese soldiers into his own forces, he was not successful in defending Taiyuan against the People's Liberation Army in early 1949.

For centuries Shanxi was the center of trade and banking. Due to its geographic location in the Greater China and its natural environment, Shanxi was the richest province in Central China.

Economically, the GDP per capita of Shanxi has been below the national average. Compared to the provinces in East China, Shanxi is less developed due to its geographic location that limits the province to participate in international trades, while the local climate and dwindling water resources have limited the development of agriculture in Shanxi. Since Shanxi possesses about one-third of China's total coal deposits, the province has been a leading producer of coal in China with an annual production exceeding 300 million metric tons. Some of the most important coalfields are located in Datong, Ningwu, Xishan, Hedong, Qinshui, and Huoxi. The province also contains about one-third of the total Chinese bauxite reserves.

The economy in Shanxi has been concentrated on heavy industries, such as coal and chemical production, power generation, metal refining, and military-related industries. The satellite launch center in Taiyuan is one of China's three satellite launch centers. Many private and state-owned mining corporations have invested billions of dollars in the province's mining industry. Foreign investors in Shanxi include mining companies from Canada, the United States, Japan, the United Kingdom, Germany, and Italy.

The nominal GDP of Shanxi province in 2011 reached ¥1110.0 billion (US$176.2 billion), and per-capita GDP ¥21,544 (US$3154).

4.3.1 Current State of Investment and Employment Structures of Shanxi

All that is necessary for us to do is the computations of the gross domestic outputs of all the economic sectors and industries and the investment and employment situations (Table 4.5) of Jiangxi and the relevant comparative gains. By comparing the resultant comparative gains, we can then obtain the current states of investment and employment for the three economic sectors and industries. All the detailed computations are similar to what has been described earlier. The results are given in Table 4.6.

First, in terms of the three major economic sectors, among the six relevant data points of the three sectors of Shanxi Province, only one is in basic equilibrium, where the comparative gain of labor of the tertiary economic sector is equal to 0.986171. And all other five data points are situated in extreme instability. In particular, the comparative gain of labor of the primary sector is way too low and equal to merely 0.152839, which is not even reaching one-fourteenth and one-sixth of that same indicator of the secondary and the tertiary sectors. In terms of the capital investment, the comparative gains of the primary and the secondary sectors are much too low, and are respectively equal to 1.55175 and 1.30009, while the capital investment in the tertiary sector is in extreme excess (0.69241).

Second, in terms of the employment situation, the relevant data of the 21 industries of Shanxi Province should contain 42 data points. However, because of lack of raw data, we only collected

Table 4.5 The GDPs, Total Investments in Fixed Properties, and Employment Situation of Shanxi's Three Economic Sectors and Industries (2011)

Sectors (i)	*Total Output of Sector (0.1 Billion Yuan)*	*Total Investment on Fixed Properties (0.1 Billion Yuan)*	*Employment (10,000 People)*
Total	11237.55	7373.0582	1738.9
Primary sector	641.42	271.2048	649.4
Secondary sector	6635.26	3348.5814	468
Tertiary sector	3960.87	3753.272	621.5
Industries (*i*)			
Agriculture, forest, animal husbandry, fishery	641.42	271.2048	649.4
Mining		1425.9307	162.7
Manufacturing		1365.4424	160.9
Production/supply of electricity, gas, and water		547.6398	10.7
Construction	675.3	9.5685	133.3
Communication, transportation, storage, postal service	756.29	1235.8054	91.5
Information transfer, computer, and software areas	238.32	25.9861	14.4
# Information Transfer			
Wholesale and retail	846.65	148.2459	183.1
Hotels and restaurants	261.33	51.7264	55.4
Finance	519.32	1.6722	15.9
Real estate/housing	224.91	1478.6894	5.1
Leasing and commercial services	99.25	15.787	11.6
Scientific research, technology service, and geological survey	58.88	24.1779	7.3
Water conservation, environment, and public facility management	28.18	490.2643	7.5
# Public Facility Management			
Residents and other services	126.1	9.5516	23.4
Education	225.54	125.4947	49.3
Health, safety, and welfare	100.85	52.0071	17.6
Social, sports, and entertainment	59.35	77.691	6.1
Public administration and social organizations	415.9	16.173	58.4

Source: Data are from *China Statistics Yearbook* (2012).

Table 4.6 The Structures and Comparative Gain Measures of Shanxi's Three Economic Sectors and Industries ($t = 2011$)

Sectors (i)	*Demand Structure $d_i(t)$ (%)*	*Employment Structure $s_i^1(t)$ (%)*	*Comparative Gain of Labor $G_i^1(t)$*	*Investment Structure $s_i^2(t)$ (%)*	*Comparative Gain of Capital $G_i^2(t)$*
Total	100	100	—	100	—
Primary sector	5.707828	37.34545	0.152839	3.678322	1.55175
Secondary sector	59.04543	26.91357	2.193891	45.41645	1.30009
Tertiary sector	35.24674	35.74099	0.986171	50.90523	0.69240
Industries (i)					
Agriculture, forest, animal husbandry, fishery	5.707828	37.34545	0.152839	3.678322	1.55175
Mining		9.356490		19.33975	
Manufacturing		9.252976		18.51935	
Production/supply of electricity, gas, and water		0.615332		7.427580	
Construction	6.009317	7.694520	0.780987	0.129777	46.3051
Communication, transportation, storage, postal service	6.730026	5.267698	1.277603	16.76110	0.40153
Information transfer, computer, and software areas	2.120747	0.828110	2.560949	0.352447	6.01721
# Information Transfer					
Wholesale and retail	7.534116	10.52965	0.715515	2.010643	3.74712
Hotels and restaurants	2.325507	3.185922	0.729932	0.701560	3.31477
Finance	4.621292	0.914371	5.054066	0.022680	203.762
Real estate/housing	2.001415	0.293289	6.824040	20.05531	0.09979

continued

Table 4.6 (continued) The Structures and Comparative Gain Measures of Shanxi's Three Economic Sectors and Industries ($t = 2011$)

Sectors (i)	*Demand Structure $d_i(t)$ (%)*	*Employment Structure $s_i^1(t)$ (%)*	*Comparative Gain of Labor $G_i^1(t)$*	*Investment Structure $s_i^2(t)$ (%)*	*Comparative Gain of Capital $G_i^2(t)$*
Leasing and commercial services	0.883200	0.667088	1.323962	0.214117	4.12484
Scientific research, technology service, and geological survey	0.523958	0.419806	1.248096	0.327922	1.59781
Water conservation, environment, and public facility management	0.250766	0.431307	0.581410	6.649402	0.03771
# Public Facility Management					
Residents and other services	1.122131	1.345678	0.833877	0.129547	8.66194
Education	2.007021	2.835126	0.707913	1.702071	1.17916
Health, safety, and welfare	0.897438	1.012134	0.886679	0.705367	1.27230
Social, sports, and entertainment	0.528140	0.350796	1.505545	1.053715	0.50122
Public administration and social organizations	3.700985	3.358445	1.101994	0.219353	16.8723

32 of them. So, we can only base our discussion on these available statistics. All the industries are situated in extremely imbalanced states. In particular, the comparative gains of labor of the following eight industries are all over 1. Especially, the comparative gain of labor in "real estate/housing" is even equal to 6.824040, which is equal to (6.824040/0152839≈) 45 times of the lowest value of this indicator that is reached by "Agriculture, forest, animal husbandry, fishery."

Real estate/housing	Finance	Information transfer, computer, and software areas	Social, sports, and entertainment
Leasing and commercial services	Communication, transportation, storage, postal service	Scientific research, technology service, and geological survey	Public administration and social organizations

At the same time, the imbalance in terms of the comparative gains of capital seems to be way too severe. The comparative gains in the following 12 industries are over 1. In particular, the comparative gains of capital of "finance," "construction," and "public administration and social organizations" are even over 10, while the value of this indicator for "water conservation, environment, and public facility management" is too low, equaling only 0.03771. That is merely (0.03771/203.762≈) 0.00019 when compared to the value of this indicator of "finance," (0.03771/46.3051≈) 0.00082 when compared to the value of this indicator of "construction," (0.03771/16.8723≈) 0.00225 when compared to the value of this indicator of "public administration and social organizations," and (0.03771/8.6619≈) 0.00439 when compared to the value of this indicator of "residents and other services."

Finance	Construction	Public administration and social organizations	Residents and other services
Information transfer, computer, and software areas	Leasing and commercial services	Wholesale and retail	Hotels and restaurants
Scientific research, technology service, and geological survey	Agriculture, forest, animal husbandry, fishery	Health, safety, and welfare	Education

So, in terms of the magnitudes of the comparative gains of labor, the industries of Shanxi Province with available data can be ranked as follows:

1	2	3	4
Real estate/ housing	Finance	Information transfer, computer, and software areas	Social, sports, and entertainment
5	6	7	8
Leasing and commercial services	Communication, transportation, storage, postal service	Scientific research, technology service, and geological survey	Public administration and social organizations
9	10	11	12
Health, safety, and welfare	Residents and other services	Construction	Hotels and restaurants
13	14	15	16
Wholesale and retail	Education	Water conservation, environment, and public facility management	Agriculture, forest, animal husbandry, fishery

In terms of the magnitudes of the comparative gains of capital, the industries of Shanxi Province with available data can be ranked as follows:

1	2	3	4
Finance	Construction	Public administration and social organizations	Residents and other services
5	6	7	8
Information transfer, computer, and software areas	Leasing and commercial services	Wholesale and retail	Hotels and restaurants
9	10	11	12
Scientific research, technology service, and geological survey	Agriculture, forest, animal husbandry, fishery	Health, safety, and welfare	Education
13	14	15	16
Social, sports, and entertainment	Communication, transportation, storage, postal service	Real estate/housing	Water conservation, environment, and public facility management

4.3.2 Investment and Employment in Shanxi's Economic Sectors and Industries

First, in terms of the comparative gains of labor, the values of this indicator for the following eight industries:

Real estate/housing	Finance	Information transfer, computer, and software areas	Social, sports, and entertainment
Leasing and commercial services	Communication, transportation, storage, postal service	Scientific research, technology service, and geological survey	Public administration and social organizations

are over 1, representing relatively optimal industries for employment. In the other eight industries, there has been an excessive supply of labor so that appropriate control on the increase of labor into these industries should be imposed. Specifically, the excessive labor in the following industries should be transferred into the previous eight industries.

Agriculture, forest, animal husbandry, fishery	Water conservation, environment, and public facility management	Education	Wholesale and retail
Hotels and restaurants	Construction	Residents and other services	Health, safety, and welfare

In other words, no matter whether the job seekers, who look for employment opportunities in Shanxi Province, are from Central China, or other parts of the mainland China, or the rest of the world, they should first consider the secondary economic sector with the relevant industries ranked as follows:

1	2	3	4
Real estate/ housing	Finance	Information transfer, computer, and software areas	Social, Sports, and entertainment
5	6	7	8
Leasing and commercial services	Communication, transportation, storage, postal service	Scientific research, technology service, and geological survey	Public administration and social organizations

Without special skills, don't look for employment opportunities in the primary economic and tertiary sectors and other industries, because those areas are experiencing excessive labor supplies.

That is the direction of employment.

Second, in terms of capital, the following 12 industries have enjoyed the comparative gains of more than 1, representing relatively more optimal areas of additional investment.

Finance	Construction	Public administration and social organizations	Residents and other services
Information transfer, computer, and software areas	Leasing and commercial services	Wholesale and retail	Hotels and restaurants
Scientific research, technology service, and geological survey	Agriculture, forest, animal husbandry, fishery	Health, safety, and welfare	Education

However, for the following four industries, excessive investment of capital has been given. Further capital investment in these industries should be curbed and redirected to the previous 12 industries.

Social, sports, and entertainment	Communication, transportation, storage, postal service	Real estate/ housing	Water conservation, environment, and public facility management

In other words, no matter whether the investors, who look for investment opportunities in Shanxi Province, are from Central China, or other parts of the mainland China, or the rest of the world, they should consider the following industries ranked from the most optimal to the least:

1	2	3	4
Finance	Construction	Public administration and social organizations	Residents and other services

5	6	7	8
Information transfer, computer, and software areas	Leasing and commercial services	Wholesale and retail	Hotels and restaurants
9	10	11	12
Scientific research, technology service, and geological survey	Agriculture, forest, animal husbandry, fishery	Health, safety, and welfare	Education

That is the direction of capital investment.

Third, by combining the two criteria discussed above, it follows that the industries in Shanxi province that are most optimal for both investment and employment are the following, because the comparative gains of labor and capital of each of these industries are over 1.

Finance	Information transfer, computer, and software areas	Leasing and commercial services	Public administration and social organizations	Scientific research, technology service, and geological survey

The secondary choice is the industries of "social, sports, and entertainment" and "communication, transportation, storage, postal service." And the worst choice is the industry of "water conservation, environment, and public facility management."

4.4 Hunan Province

Hunan is located in the south-central part of China to the south of the middle reaches of the Yangtze River and south of Lake Dongting. The province borders Hubei in the north, Jiangxi in the east, Guangdong in the southeast, Guangxi in the southwest, Guizhou in the west, and Chongqing in the northwest. The capital of the province is Changsha. For more details, see Figure 4.4.

Geographically, Hunan is located on the south bank of the Yangtze River. It covers an area of 211,800 square kilometers (81,800 sq mi), making it the 10th largest province in China. The east, south, and west sides of the province are surrounded by mountains, such as the Wuling Mountains in the northwest, the Xuefeng Mountains in the west, the Nanling Mountains in the south, and the Luoxiao Mountains in the east. These mountains occupy more than 80% of the land mass of the province. The Xiang, Zi, Yuan, and Lishui Rivers converge into the Yangtze River at Lake Dongting in north Hunan. Most of Hunan lies in the basins of four major tributaries of the Yangtze River.

The climate of Hunan is humid subtropical with short, cool, damp winters, very hot and humid summers, and plenty of rainfall.

Historically, Hunan entered the written history of China around 350 BC. During the Zhou Dynasty, it became part of the State of Chu. From this time on for hundreds of years, the area was a destination of migration of Han Chinese from the north. The migration was especially prevalent during the periods of Eastern Jin Dynasty and the Southern and Northern Dynasties when nomadic invaders pushed these people southward. Hunan and Hubei together was a part of the province named Huguang until the Qing dynasty, which created the Hunan province in 1664 and renamed to its current name in 1723. Due to its particular position along the Yangzi River and on the Imperial Highway constructed between northern and southern China, Hunan became

Figure 4.4 The geographic location of Hunan.

an important center of communications. The province produced grain abundantly that had fed many parts of China.

In 1910, there were uprisings against the crumbling Qing dynasty and the famous Autumn Harvest Uprising was organized there by Hunan native Mao Zedong, who then established a short-lived Hunan Soviet in 1927. The Communists maintained a guerrilla army in the mountains along the Hunan–Jiangxi border until 1934, when they were forced to begin their famous long march to the bases in Shaanxi under the pressure of Nationalist Kuomintang forces. In 1949, the Communists returned as the Nationalists retreated southward.

As the home province of Mao Zedong, Hunan supported the Cultural Revolution that lasted from 1966 to 1976. On the other hand, the province was slow to adopt the reforms implemented by Deng Xiaoping in the years after Mao's era.

Economically, as of the middle of the nineteenth century, Hunan exported rhubarb, musk, honey, tobacco, hemp, and birds with the Lake Dongting area serving as an important center for ramie production and tea production. Other than agricultural products, Hunan has become in recent years an important center for the production of steel, machinery, and electronics.

Hunan's nominal GDP for 2011 was ¥1.90 trillion (US$300 billion) with its per capita GDP around ¥20,226 (US$2961).

For additional readings, refer to Ministry of Commerce (2013); National Bureau of Statistics of China (2013); Roberts (1837); Svanberg (1988); Shih and Shi (2002).

4.4.1 Current State of Investment and Employment Structures of Hunan

All that is necessary for us to do is the computations of the gross domestic outputs of all the economic sectors and industries and the investment and employment situations (Table 4.7) of Hunan

Table 4.7 The GDPs, Total Investments in Fixed Properties, and Employment Situation of Hunan's Three Economic Sectors and Industries (2011)

Sectors (i)	*Total Output of Sector (0.1 Billion Yuan)*	*Total Investment on Fixed Properties (0.1 Billion Yuan)*	*Employment (10,000 People)*
Total	19669.56	11431.48	4005.03
Primary sector	2768.03	*320.30*	1679.94
Secondary sector	9361.99	*4891.74*	932.62
Tertiary sector	7539.54	*6219.44*	1392.47
Industries (*i*)			
Agriculture, forest, animal husbandry, fishery		320.3	1679.94
Mining		432.4	95.12
Manufacturing		3917.84	503.1
Production/supply of electricity, gas, and water		461.88	22.03
Construction	1239.24	79.61	312.37
Communication, transportation, storage, postal service	948.82	1191.83	170.06
Information transfer, computer, and software areas	403.04	75.82	54.29
# Information Transfer			
Wholesale and retail	1662.34	359.71	338.19
Hotels and restaurants	406.87	164.72	190.8
Finance	501.09	19.4	26.19
Real estate/housing	518.04	2256.32	28.47
Leasing and commercial services	307.236255	196.51	91.96
Scientific research, technology service, and geological survey	180.94896	103.19	17.38
Water conservation, environment, and public facility management	75.3954	1150.5	10.19
# Public Facility Management			
Residents and other services	684.213255	43.79	163.31
Education	507.034065	157.02	89.7
Health, safety, and welfare	276.731575	130.93	48.9
Social, sports, and entertainment	252.78049	94.93	60.15
Public administration and social organizations	815	274.77	102.88

Source: The data in italic are from *Statistical Reports of Hunan Province's Domestic Economy and Social Development*, 2011, and rest from *Hunan Province's Statistics Yearbook*, 2012.

Table 4.8 The Structures and Comparative Gain Measures of Hunan's Three Economic Sectors and Industries ($t = 2011$)

Sectors (i)	*Demand Structure $d_i(t)$ (%)*	*Employment Structure $s_i^1(t)$ (%)*	*Comparative Gain of Labor $G_i^1(t)$*	*Investment Structure $s_i^2(t)$ (%)*	*Comparative Gain of Capital $G_i^2(t)$*
Total	100	100	—	100	—
Primary sector	14.07266	41.94575	0.335497	2.801912	5.02252
Secondary sector	47.59634	23.28622	2.043970	42.79183	1.11228
Tertiary sector	38.33100	34.76803	1.102479	54.40625	0.70453
Industries (*i*)					
Agriculture, forest, animal husbandry, fishery		41.94575		2.801912	
Mining		2.375013		3.782537	
Manufacturing		12.56170		34.27238	
Production/supply of electricity, gas, and water		0.550058		4.040422	
Construction	6.300293	7.799442	0.807788	0.69641	9.04681
Communication, transportation, storage, postal service	4.823799	4.246160	1.136038	10.42586	0.46268
Information transfer, computer, and software areas	2.049054	1.355545	1.511609	0.663256	3.08939
# Information Transfer					
Wholesale and retail	8.451333	8.444132	1.000853	3.146662	2.68581
Hotels and restaurants	2.068526	4.764009	0.434199	1.440933	1.43555
Finance	2.547540	0.653928	3.895753	0.169707	15.0114
Real estate/housing	2.633714	0.710856	3.704989	19.73778	0.13344
Leasing and commercial services	1.561988	2.296113	0.680275	1.719025	0.90865
Scientific research, technology service, and geological survey	0.919944	0.433954	2.119910	0.902683	1.01912

continued

Table 4.8 (continued) The Structures and Comparative Gain Measures of Hunan's Three Economic Sectors and Industries ($t = 2011$)

Sectors (i)	*Demand Structure* $d_i(t)$ *(%)*	*Employment Structure* $s_i^1(t)$ *(%)*	*Comparative Gain of Labor* $G_i^1(t)$	*Investment Structure* $s_i^2(t)$ *(%)*	*Comparative Gain of Capital* $G_i^2(t)$
Water conservation, environment, and public facility management	0.383310	0.254430	1.506544	10.06431	0.03809
# Public Facility Management					
Residents and other services	3.478539	4.077622	0.853080	0.383065	9.08081
Education	2.577760	2.239684	1.150948	1.373575	1.87668
Health, safety, and welfare	1.406903	1.220965	1.152288	1.145346	1.22836
Social, sports, and entertainment	1.285135	1.501861	0.855695	0.830426	1.54756
Public administration and social organizations	4.143458	2.568770	1.613013	2.403626	1.72384

Source: Data are from *China Statistics Abstract 2012* (XLS, DOC).

and the relevant comparative gains. By comparing the resultant comparative gains, we can then obtain the current states of investment and employment for the three economic sectors and industries. All the detailed computations are similar to what has been described earlier. The results are given in Table 4.8.

First, in terms of the three major economic sectors, among the six relevant data points of the three sectors of Hunan, all of them are situated in extreme instability. In particular, the comparative gain of labor of the primary sector is way too low and equal to merely 0.335497. That is not even reaching one-sixth, and one-third of that same indicator of the secondary and the tertiary sectors, respectively. In terms of the capital investment, the comparative gains of the primary and the secondary sectors are much too low, and are respectively equal to 5.02252 and 1.11223, while the capital investment in the tertiary sector is in extreme excess (0.70453).

Second, in terms of the employment situation, the relevant data of the 21 industries of the Hunan Province should contain 42 data points. However, because of lack of raw data, we only collected 32 of them. So, we can only base our discussion on these available statistics. Among all the industries, only two are basically in equilibrium, where the comparative gain of labor in "wholesale and retail" and the comparative gain of capital in "scientific research, technology service, and geological survey" are 1.000854 and 1.01912, respectively. All the rest (28/30 ≥ 93%) are situated in extremely imbalanced states. In particular, the comparative gains of labor of the following 10 industries are all over 1. Especially, the comparative gain of labor in "finance" is even equal to 3.895753, which is equal to (3.895753/0.434199≈) 9 times of the lowest value of this indicator that is reached by "hotels and restaurants."

Finance	Real estate/ housing	Scientific research, technology service, and geological survey	Public administration and social organizations	Information transfer, computer, and software areas
Water conservation, environment, and public facility management	Health, safety, and welfare	Education	Communication, transportation, storage, postal service	Wholesale and retail

At the same time, the imbalance in terms of the comparative gains of capital seems to be way too severe. The comparative gains in the following 11 industries are over 1. And the value of this indicator for "water conservation, environment, and public facility management" is too low, equaling only 0.03809. That is merely (0.03809/15.0114 ≈) 0.0025 when compared to the value of this indicator of "finance," (0.03809/9.08081 ≈) 0.0042 when compared to the value of this indicator of "residents and other services," (0.03809/9.04681 ≈) 0.0042 when compared to the value of this indicator of "construction," (0.03809/3.08939 ≈) 0.0123 when compared to the value of this indicator of "information transfer, computer, and software areas," and (0.03809/2.68581 ≈) 0.0142 when compared to the value of this indicator of "wholesale and retail."

Finance	Residents and other services	Construction	Information transfer, computer, and software areas
Wholesale and retail	Education	Public administration and social organizations	Social, sports, and entertainment
Hotels and restaurants	Health, safety, and welfare	Scientific research, technology service, and geological survey	

So, in terms of the magnitudes of the comparative gains of labor, the industries of the Hunan Province with available data can be ranked as follows:

1	2	3	4
Finance	Real estate/housing	Scientific research, technology service, and geological survey	Public administration and social organizations
5	6	7	8
Information transfer, computer, and software areas	Water conservation, environment, and public facility management	Health, safety, and welfare	Education

9	10	11	12
Communication, transportation, storage, postal service	Wholesale and retail	Social, sports, and entertainment	Residents and other services
13	14	15	
Construction	Leasing and commercial services	Hotels and restaurants	

In terms of the magnitudes of the comparative gains of capital, the industries of the Hunan Province with available data can be ranked as follows:

1	2	3	4
Finance	Residents and other services	Construction	Information transfer, computer, and software areas
5	6	7	8
Wholesale and retail	Education	Public administration and social organizations	Social, sports, and entertainment
9	10	11	12
Hotels and restaurants	Health, safety, and welfare	Scientific research, technology service, and geological survey	Leasing and commercial services
13	14	15	
Communication, transportation, storage, postal service	Real estate/ housing	Water conservation, environment, and public facility management	

4.4.2 *Investment and Employment in Hunan's Economic Sectors and Industries*

First, in terms of the comparative gains of labor, the values of this indicator for the following 10 industries:

Finance	Real estate/ housing	Scientific research, technology service, and geological survey	Public administration and social organizations	Information transfer, computer, and software areas
Water conservation, environment, and public facility management	Health, safety, and welfare	Education	Communication, transportation, storage, postal service	Wholesale and retail

are over 1, representing relatively optimal industries for employment. In the other five industries, there has been an excessive supply of labor so that appropriate control on the increase of labor into these industries should be imposed. Specifically, the excessive labor in the following industries should be transferred into the previous 10 industries.

Hotels and restaurants	Leasing and commercial services	Construction	Residents and other services	Social, sports, and entertainment

In other words, no matter whether the job seekers, who look for employment opportunities in the Hunan Province, are from Central China, or other parts of the mainland China, or the rest of the world, they should first consider the secondary and the tertiary economic sectors with the relevant industries ranked as follows:

1	2	3	4	5
Finance	Real estate/ housing	Scientific research, technology service, and geological survey	Public administration and social organizations	Information transfer, computer, and software areas
6	7	8	9	10
Water conservation, environment, and public facility management	Health, safety, and welfare	Education	Communication, transportation, storage, postal service	Wholesale and retail

Without special skills, don't look for employment opportunities in the primary economic sector and other industries, because those areas are experiencing excessive labor supplies.

That is the direction of employment.

Second, in terms of capital, the following 11 industries have enjoyed the comparative gains of more than 1, representing relatively more optimal areas of additional investment.

Finance	Residents and other services	Construction	Information transfer, computer, and software areas
Wholesale and retail	Education	Public administration and social organizations	Social, sports, and entertainment
Hotels and restaurants	Health, safety, and welfare	Scientific research, technology service, and geological survey	

However, for the following four industries, excessive investment of capital has been given. Further capital investment in these industries should be curbed and redirected to the previous 11 industries.

Water conservation, environment, and public facility management	Real estate/housing	Communication, transportation, storage, postal service	Leasing and commercial services

In other words, no matter whether the investors, who look for investment opportunities in the Hunan Province, are from Central China, or other parts of the mainland China, or the rest of the world, they should consider the following industries ranked from the most optimal to the least:

1	2	3	4
Finance	Residents and other services	Construction	Information transfer, computer, and software areas
5	6	7	8
Wholesale and retail	Education	Public administration and social organizations	Social, sports, and entertainment
9	10	11	
Hotels and restaurants	Health, safety, and welfare	Scientific research, technology service, and geological survey	

That is the direction of capital investment.

Third, by combining the two criteria discussed above, we see that the industries in the Hunan Province that are most optimal for both investment and employment are the following, because the comparative gains of labor and capital of each of these industries are over 1.

Finance	Information transfer, computer, and software areas	Public administration and social organizations	Education
Scientific research, technology service, and geological survey	Wholesale and retail	Health, safety, and welfare	

and the worst choice is the industry of "water conservation, environment, and public facility management."

4.5 Inner Mongolia

Inner Mongolia is an autonomous region located in the northern region of China. It shares an international border with Mongolia and Russia with its capital at Hohhot. It is the third largest

Figure 4.5 The geographic location of Inner Mongolia.

subdivision of China spanning about 1,200,000 km^2 (463,000 sq mi) or 12% of China's total land mass. It has a population of about 25 million in 2010, accounting to 1.84% of the total population of the Mainland China, with Han Chinese as the majority and Mongols as a substantial minority. Both Chinese and Mongolian are the accepted official languages. For more details, see Figure 4.5.

Inner Mongolia is officially a provincial-level unit of North China, bordering with the following eight other provinces: Heilongjiang, Jilin, Liaoning, Hebei, Shanxi, Shaanxi, Ningxia, and Gansu. Most of the province is a plateau of around 1200 meters (3940 ft) in altitude and covered by extensive deposits of loess and sand. The northern part of the province sits the Mesozoic era Khingan Mountains covered with forests of such tree species as Manchurian elm, ash, birch, Mongolian oak, pine, and spruce. In the north of Hailar District, the forests are made up of almost exclusively coniferous trees. Natural grassland covers the south, the natural vegetation is sparse in the east, and the west is very arid.

Historically, before the rise of the Mongols in the thirteenth century, the central and western Inner Mongolia were under the control alternatively between Chinese agriculturalists from the south and Xiongnu, Xianbei, Khitan, Jurchen, Tujue, and other Mongol nomads from the north. The eastern Inner Mongolia was mostly controlled alternatively by different Tungusic and Mongol tribes. During the Zhou Dynasty, central and western Inner Mongolia was inhabited by nomadic peoples, while eastern Inner Mongolia was home to the Donghu. During the Warring States period, King Wuling (340–295 BC) of the state of Zhao, which was based in what is now Hebei and Shanxi provinces, expanded into the region. He created the commandery of Yunzhong near modern Hohhot after destroying the Di State of Zhongshan, located in the area of the present Hebei province, the Linhu and Loufan. After Qin Shihuang reunified China in 221 BC, his general Meng Tian drove the Xiongnu away from the region, and 30,000 households were moved there to solidify the region. During the Western Han Dynasty in 127 BC, General Wei Qing was sent to recapture the control of the region from Xiongnu. During the Eastern Han Dynasty (25–220 AD), Xiongnu, who surrendered to the Han Dynasty, started to settle in Hetao, and intermingled with the Han immigrants. During the Western Jin Dynasty, Liu Yuan, a Xiongnu noble, established the Han Zhao Kingdom in the region.

The Sui Dynasty (581–618) and Tang Dynasty (618–907) re-established a unified China and settled Han Chinese into the region. The entire territory of the present Inner Mongolia was then taken over by the Khitan Empire (Liao Dynasty) by a nomadic people originally from what is now the southern part of Manchuria and eastern Inner Mongolia. The Khitans were later replaced by the Jurchens, who were the precursors of the modern Manchus and who established the Jin Dynasty over Manchuria and northern China. In 1206, Genghis Khan unified the Mongol tribes and founded the Mongol Empire. In 1227, he conquered the Tangut Western Xia Empire, and in 1234, the Jurchen Jin Dynasty. Genghis Khan's grandson Khubilai established the Yuan Dynasty. During this time period, Ongud and Khunggirad peoples dominated the area of what is Inner Mongolia today. When the Ming Dynasty defeated the Yuan in 1368, the Ming established the Three Guards composed of the Mongols there. When the Oirat ruler Esen Taishi incidentally captured the Chinese emperor in 1449, Mongols flooded Southern Mongolia. From then on until 1635, Inner Mongolia was the political and cultural center of the Mongols during the Northern Yuan Dynasty. The Manchus gained far-reaching control of the tribes of Inner Mongolian in 1635. Then Manchus invaded Ming China in 1644 and established Qing Dynasty. During the eighteenth century, so many Han Chinese settlers had illegally moved into Inner Mongolian that the jasak petitioned the Qing government to legalize the status of these settlers. During the nineteenth century, when faced with Russian threat, the Manchus began to encourage Han Chinese to settle in both Mongolia and Manchuria.

In the early twentieth century, the newly formed Republic of China promised a new nation of five races: Han, Manchu, Mongol, Tibetan, and Uyghur, while forced the Inner Mongolian princes to recognize the Republic of China. The Communists gradually gained influence in Inner Mongolia during the Japanese period. At the end of World War II, the Chinese Communists gained control of Manchuria and Inner Mongolia and established the Inner Mongolia Autonomous Region in 1947. In 1969, Inner Mongolia was distributed to the surrounding provinces such that Hulunbuir was divided between Heilongjiang and Jilin, Jirim went to Jilin, Juu Uda to Liaoning, and the Alashan and Ejine region was divided between Gansu and Ningxia. This division was reversed in 1979.

Economically, farming takes precedence along the river valleys. In more arid grasslands, herding of goats, sheep, and so on, is a traditional method of living. Forestry and hunting are important in the Greater Khingan ranges in the east. Reindeer herding is carried out in the Evenk Autonomous Banner. And growing grapes and winemaking have become an economic factor in the Wuhai area. In short, Inner Mongolia has developed considerably since the time when Deng Xiaoping instituted economic reform in 1978.

Inner Mongolia has abundant natural resources, such as coal, cashmere, natural gas, and rare earth elements, niobium, zirconium, and beryllium. The economy in Inner Mongolia has been developed mainly around coal, power generation, forestry-related sectors, and related industries. The nominal GDP of the province in 2010 was ¥1.16 trillion (US$172.1 billion), and its per capita GDP reached ¥37,287 (US$5460) in 2009. In 2008, Inner Mongolia's primary, secondary, and tertiary industries were respectively worth ¥90.7 billion, ¥427.1 billion, and ¥258.4 billion.

For additional readings, refer to Ministry of Commerce (2013); National Bureau of Statistics (2013); Borjigin (2004); and Belov (1996).

4.5.1 Current State of Investment and Employment Structures of Inner Mongolia

All that is necessary for us to do is the computations of the gross domestic outputs of all the economic sectors and industries and the investment and employment situations (Table 4.9) of Inner Mongolia and the relevant comparative gains. By comparing the resultant comparative gains, we can then obtain the current states of investment and employment for the three economic sectors and industries. All the detailed computations are similar to what has been described earlier. The results are given in Table 4.10.

First, in terms of the three major economic sectors, among the six relevant data points of the three sectors of Inner Mongolia, only two are basically in equilibrium, where the comparative gains of labor and capital of the tertiary sector are 0.959497 and 0.98933, respectively. All others are situated in extreme instability. In particular, the comparative gain of labor of the primary sector is way too low and equal to merely 0.198337. That is not even reaching one-sixteenth, and one-fifth of that same indicator of the secondary and the tertiary sectors, respectively. In terms of the capital investment, the comparative gains of the primary and the secondary sectors are much too low, and are equal to 1.76039 and 1.18733, respectively.

Second, in terms of the employment situation, the relevant data of the 21 industries of Inner Mongolia should contain 42 data points. However, because of lack of raw data, we only collected 30 of them. Even so, the employment data, both total and those for individual industries, are from *2012 Inner Mongolia's Statistics Yearbook*," where the yearend employment data for individual industries only reflect the situation for the urban population and are not consistent with the overall data. So, these data cannot be used as the basis for our discussion. For details, see the data in Table 4.10 in bold. Therefore, we have no means to address the employment situation.

Among all the effective data, only 15 of them are related to the comparative gains of capital; and one of these data values is in basic equilibrium, where the comparative gain of capital in "Scientific research, technology service, and geological survey" is equal to 1.03097, while the rest 14 (or 14/15 ≥ 93%) are situated in extreme instabilities. In particular, the comparative gains of capital of the following 12 industries are over 1. And the value of the indicator "water conservation, environment, and public facility management," is much too low, only equal to 0.03897, which is only (0.03897/9.93946≈) 0.0039 when compared to that of "public administration and social organizations," (0.03897/8.18450≈) 0.0047 when compared to that of "residents and other services," (0.03897/7.28042≈) 0.0053 when compared to that of "finance," (0.03897/6.43167≈) 0.0060 when compared to that of "construction," and (0.03897/2.96393≈) 0.013 when compared to that of "hotels and restaurants."

Public administration and social organizations	Residents and other services	Finance	Construction
Hotels and restaurants	Wholesale and retail	Information transfer, computer, and software areas	Health, safety, and welfare
Communication, transportation, storage, postal service	Real estate/housing	Education	Scientific research, technology service, and geological survey

Table 4.9 The GDPs, Total Investments in Fixed Properties, and Employment Situation of Inner Mongolia's Three Economic Sectors and Industries (2011)

Sectors (i)	*Total Output of Sector (0.1 Billion Yuan)*	*Total Investment on fixed Properties (0.1 Billion Yuan)*	*Employment (10,000 People)*
Total	14359.89	10899.78	1249.3
Primary sector	1306.30	563.25	573.0
Secondary sector	8037.69	5138.40	221.5
Tertiary sector	5015.90	3848.33	454.8
Industries (*i*)			
Agriculture, forest, animal husbandry, fishery		359.01	26.105
Mining		1009.94	20.018
Manufacturing		2850.09	39.686
Production/supply of electricity, gas, and water		1152.12	10.987
Construction	936.09	110.47	15.919
Communication, transportation, storage, postal service	1040.03	1017.30	16.205
Information transfer, computer, and software areas	147.55	72.96	4.174
# Information Transfer			
Wholesale and retail	1216.60	379.85	7.472
Hotels and restaurants	381.64	97.74	2.536
Finance	447.46	46.65	9.450
Real estate/housing	384.76	263.35	1.999
Leasing and commercial services	174.69	102.43	2.886
Scientific research, technology service, and geological survey	73.61	54.20	2.886
Water conservation, environment, and public facility management	45.62	888.51	7.343
# Public Facility Management			
Residents and other services	202.32	18.76	1.955
Education	251.28	181.23	34.325
Health, safety, and welfare	126.00	67.86	12.671
Social, sports, and entertainment	54.45	77.20	3.218
Public administration and social organizations	469.89	221.49	35.384

Source: The data from *Inner Mongolia's Statistics Yearbook, 2012.*

Table 4.10 The Structures and Comparative Gain Measures of Inner Mongolia's Three Economic Sectors and Industries ($t = 2011$)

Sectors (i)	*Demand Structure $d_i(t)$ (%)*	*Employment Structure $s_i^1(t)$ (%)*	*Comparative Gain of Labor $G_i^1(t)$*	*Investment Structure $s_i^2(t)$ (%)*	*Comparative Gain of Capital $G_i^2(t)$*
Total	100	100	—	100	—
Primary sector	9.096871	45.86568	0.198337	5.167531	1.76039
Secondary sector	55.97322	17.72993	3.156991	47.14219	1.18733
Tertiary sector	34.92991	36.40439	0.959497	35.30646	0.98933
Industries (*i*)					
Agriculture, forest, animal husbandry, fishery		2.089570		3.29387	
Mining		1.602361		9.265716	
Manufacturing		3.176635		26.14815	
Production/supply of electricity, gas, and water		0.879444		10.57009	
Construction	6.5187790	1.274202	**5.115972**	1.013543	6.43167
Communication, transportation, storage, postal service	7.2426143	1.297134	**5.583550**	9.333478	0.77598
Information transfer, computer, and software areas	1.0275154	0.334131	**3.075186**	0.669352	1.53509
# Information Transfer					
Wholesale and retail	8.4722141	0.598119	**14.16476**	3.484947	2.43109
Hotels and restaurants	2.6576819	0.202954	**13.09502**	0.896674	2.96393
Finance	3.1160315	0.756400	**4.119557**	0.428002	7.28042
Real estate/housing	2.6793960	0.160010	**16.74522**	2.416107	1.10897
Leasing and commercial services	1.2165043	0.231001	**5.266221**	0.939713	1.29455

continued

Table 4.10 (continued) The Structures and Comparative Gain Measures of Inner Mongolia's Three Economic Sectors and Industries ($t = 2011$)

Sectors (i)	*Demand Structure $d_i(t)$ (%)*	*Employment Structure $s_i^1(t)$ (%)*	*Comparative Gain of Labor $G_i^1(t)$*	*Investment Structure $s_i^2(t)$ (%)*	*Comparative Gain of Capital $G_i^2(t)$*
Scientific research, technology service, and geological survey	0.5126304	0.231001	**2.219166**	0.497233	1.03097
Water conservation, environment, and public facility management	0.3176718	0.587793	**0.540448**	8.151583	0.03897
# Public Facility Management					
Residents and other services	1.4088952	0.156456	**9.005079**	0.172142	8.18450
Education	1.7498751	2.747571	**0.636881**	1.662704	1.05243
Health, safety, and welfare	0.8774455	1.014272	**0.865099**	0.622589	1.40935
Social, sports, and entertainment	0.3791903	0.257576	**1.472148**	0.708255	0.53539
Public administration and social organizations	3.2722412	2.872329	**1.139229**	0.329217	9.93946

Source: Data are from *China Statistics Abstract 2012* (XLS, DOC).

So, in terms of the magnitudes of the comparative gains of labor, because of lack of effective data, the employment situation of Inner Mongolia cannot be meaningfully addressed.

In terms of the magnitudes of the comparative gains of capital, the industries of Inner Mongolia with available data can be ranked as follows:

1	2	3	4
Public administration and social organizations	Residents and other services	Finance	Construction
5	6	7	8
Hotels and restaurants	Wholesale and retail	Information transfer, computer, and software areas	Health, safety, and welfare

9	10	11	12
Real estate/housing	Education	Scientific research, technology service, and geological survey	Communication, transportation, storage, postal service
13	14		
Social, sports, and entertainment	Water conservation, environment, and public facility management		

4.5.2 Investment and Employment in Inner Mongolia's Economic Sectors and Industries

First, in terms of the comparative gains of labor, because of lack of effective statistics, we cannot address the employment problem in terms of the individual industries. However, in terms of economic sectors, only the comparative gain of labor of the secondary sector is over 1, representing a relatively ideal area for employment. For the primary and the tertiary sectors, there has been an excessive supply of labor so that appropriate control on the increase of labor into these sectors should be imposed. Specifically, the excessive labor in these sectors should be transferred into the secondary sector.

In other words, no matter whether the job seekers, who look for employment opportunities in Inner Mongolia, are from Central China, or other parts of the mainland China, or the rest of the world, they should first consider the secondary economic sector. Without special skills, don't look for employment opportunities in other economic sectors, especially the primary economic sector, because those areas are experiencing excessive labor supplies.

That is the direction of employment.

Second, in terms of capital, the following 11 industries have enjoyed the comparative gains of more than 1, representing relatively more optimal areas of additional investment.

Public administration and social organizations	Residents and other services	Finance	Construction
Hotels and restaurants	Wholesale and retail	Information transfer, computer, and software areas	Health, safety, and welfare
Communication, transportation, storage, postal service	Real estate/ housing	Education	Scientific research, technology service, and geological survey

However, for the following three industries, excessive investment of capital has been given. Further capital investment in these industries should be curbed and redirected to the previous 11 industries.

Water conservation, environment, and public facility management	Social, sports, and entertainment	Communication, transportation, storage, postal service

In other words, no matter whether the investors, who look for investment opportunities in Inner Mongolia, are from Central China, or other parts of the mainland China, or the rest of the world, they should consider the following industries ranked from the most optimal to the least:

1	2	3	4
Public administration and social organizations	Residents and other services	Finance	Construction
5	6	7	8
Hotels and restaurants	Wholesale and retail	Information transfer, computer, and software areas	Health, safety, and welfare
9	10	11	12
Information transfer, computer, and software areas	Real estate/ housing	Education	Scientific research, technology serv ce, and geological survey

That is the direction of capital investment.

Third, by combining the two criteria discussed above, we see that the economic sector in the Inner Mongolia Autonomous Region that is most optimal for both investment and employment is the secondary sector.

4.6 Heilongjiang Province

Heilongjiang is located in the northeastern part of the country. It borders Jilin in the south, Inner Mongolia in the west, and Russia to the north and east. The province is a land of varied topography dominated by mountain ranges. The Greater Khingan Range contains China's largest forest and is an important area for China's forestry industry. The interior of the province is relatively flat and low in altitude, and contains the Muling River, the Naoli River, the Songhua River, the Nen River, and the Mudan River, while the northern border forms part of the Amur valley. The province is predominated by a humid continental climate while its areas in the far north by subarctic climate. Winters there are long and bitter with an average of −31 to −15°C (−24 to 5°F) in the month of January; and summers are short and warm with an average of 18 to 23°C (64 to 73°F) in the month of July. The annual average rainfall is 400–700 mm (16–28 in), concentrated mainly in the summer. In the spring, wind energy is abundant in the Songnen Plain and in the Sanpingjiang Plain. For more details, see Figure 4.6.

Historically, Heilongjiang was far from any literate civilization and was inhabited by such peoples as the Buyeo, the Mohe, and the Khitan. The eastern Heilongjiang was ruled by the Balhae Kingdom between the seventh century and tenth century. The Jurchen Jin Dynasty (1115–1234) arose within the territory of Heilongjiang. As an administrative entity, Heilongjiang was created in 1683 during the Kangxi era of the Qing Dynasty from the northwestern part of the Jilin. In 1858 and 1860, the Qing government was forced to give up all land beyond the Amur and Ussuri Rivers to the Russian Empire, giving Heilongjiang its present northern borders. In 1932, the present-day Heilongjiang was part of the Japanese puppet state of Manchukuo. During the Chinese Civil War,

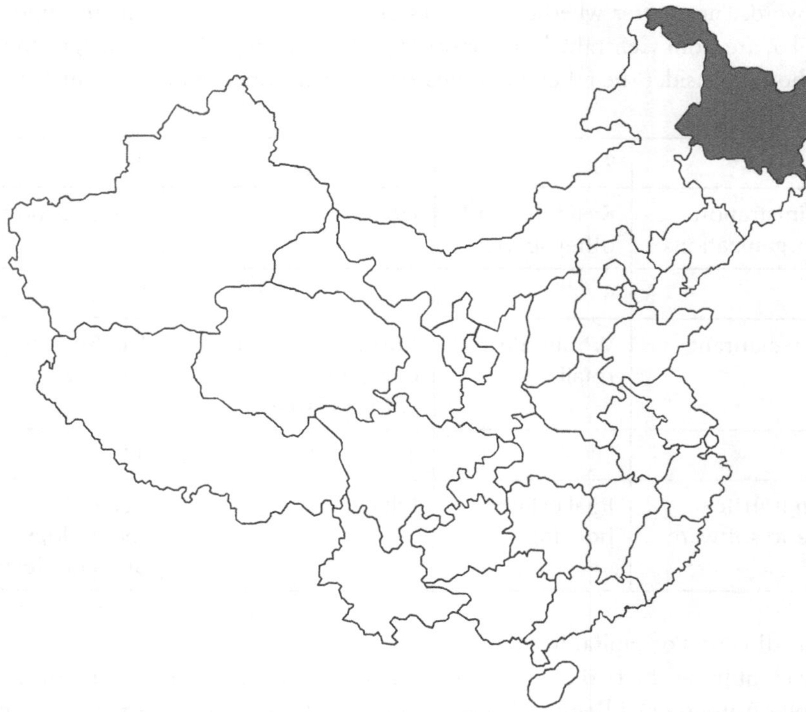

Figure 4.6 The geographic location of Heilongjiang.

Heilongjiang became the first province to be completely controlled by the Chinese communists. In 1954, the present-day Heilongjiang was officially formed.

Economically, Heilongjiang is an important source of lumber in China. The province has the greatest number of milk cows and the highest production of milk among all Chinese provinces. It is part of northeast China, the traditional base of industry for China. Its industry focuses on coal, petroleum, lumber, machinery, and food. Due to its location, Heilongjiang is an important gateway for economic trades with Russia. Its GDP has been rising steadily since 2003, and the incremental value of the private economy grew by 20.1% in 2006 reaching ¥234 billion. Answering to the call of revitalizing the Northeast of the central government, Heilongjiang is now restructuring its six pillar industries: equipment manufacturing, petrochemicals, food processing, energy, pharmaceuticals, and forest and timber processing. In 2011, Heilongjiang's nominal GDP was ¥1250 billion (US$198.5 billion) with per capita GDP ¥21,640 (US$3168). Its primary, secondary, and tertiary industries were worth, respectively, ¥108.9 billion, ¥436.6 billion, and ¥285.5 billion.

For additional readings, refer to Ministry of Commerce (2013); National Bureau of Statistics of China (2013); and Edmonds (1985).

4.6.1 Current State of Investment and Employment Structures of Heilongjiang

All that is necessary for us to do is the computations of the gross domestic outputs of all the economic sectors and industries and the investment and employment situations (Table 4.11) of

Table 4.11 The GDPs, Total Investments in Fixed Properties, and Employment Situation of Heilongjiang's Three Economic Sectors and Industries (2011)

Sectors (i)	*Total Output of Sector (0.1 Billion Yuan)*	*Total Investment on Fixed Properties (0.1 Billion Yuan)*	*Employment (10,000 People)*
Total	12503.8	7475.40	1977.77
Primary sector	1705.6	448.03	
Secondary sector	6317.3	3195.07	
Tertiary sector	4481	3514.85	
Industries (*i*)			
Agriculture, forest, animal husbandry, fishery	1705.6	**655.7**	130.68
Mining		**567.8**	53.47
Manufacturing		**2012.7**	117.39
Production/supply of electricity, gas, and water		**364.9**	19.21
Construction	**727.77**	**246.1**	62.45
Communication, transportation, storage, postal service	**543.81**	**572.7**	47.69
Information transfer, computer, and software areas		**82.4**	16.79
# Information Transfer			
Wholesale and retail	**1060.26**	**211.1**	189.59
Hotels and restaurants	**275.8**	**62.0**	52.36
Finance	**350.82**	**9.3**	17.26
Real estate/housing	**465.61**	**1731.2**	14.83
Leasing and commercial services		**77.3**	30.42
Scientific research, technology service, and geological survey		**30.6**	20.05
Water conservation, environment, and public facility management		**497.7**	14.89
# Public Facility Management			
Residents and other services		**32.2**	69.82
Education		**69.0**	49.73

continued

Table 4.11 (continued) The GDPs, Total Investments in Fixed Properties, and Employment Situation of Heilongjiang's Three Economic Sectors and Industries (2011)

Sectors (i)	*Total Output of Sector (0.1 Billion Yuan)*	*Total Investment on Fixed Properties (0.1 Billion Yuan)*	*Employment (10,000 People)*
Health, safety, and welfare		**59.2**	26.76
Social, sports, and entertainment		**63.8**	9.42
Public administration and social organizations		**130.0**	45.70

Source: The data in bold are *China Statistics Yearbook*, 2012; the data regarding investments on fixed properties and employments are from *Heilongjiang Province's Statistics Yearbook, 2012*. Because *Heilongjiang Province's Statistics Yearbook* does not contain any values of GDPs, these values are from *Statistics Report of Heilongjiang Province's Domestic Economy and Social Development, 2011*.

Heilongjiang and the relevant comparative gains. By comparing the resultant comparative gains, we can then obtain the current states of investment and employment for the three economic sectors and industries. All the detailed computations are similar to what has been described earlier. The results are given in Table 4.12.

First, in terms of the three major economic sectors, among the six relevant data points of the three sectors of Heilongjiang, because of lack of raw data, we obtained only three values for comparative gains of capital and all of them situated in extreme instability. In terms of the capital investment, the comparative gains of the primary and the secondary sectors are much too low, and are respectively equal to 2.27594 and 1.18207, while the capital investment in the tertiary sector is in extreme excess (0.76219).

Second, in terms of the situation with respect to the industries, because of lack of data and inconsistencies in the available data, we cannot make any meaningful discussion.

4.6.2 *Investment and Employment in Heilongjiang's Economic Sectors and Industries*

First, in terms of the comparative gains of labor, because of lack of statistics, we cannot draw any conclusions.

Second, in terms of the comparative gain of capital, no matter whether the investors, who look for investment opportunities in Heilongjiang, are from Central China, or other parts of the mainland China, or the rest of the world, they should consider the primary and the secondary economic sectors.

4.7. Jilin Province

Jilin is a province located in the northeastern part of the country. It borders North Korea and Russia in the east, Heilongjiang in the north, Liaoning in the south, and Inner Mongolia in the west. The province is rich in natural mineral deposits with 136 different types of minerals, and in resources of traditional Chinese medicine with approximately 27,000 kinds of wild plants and 9000 kinds of medicinal herbs. It is also rich in large reserves of oil, gas, coal, iron mine, nickel,

Table 4.12 The Structures and Comparative Gain Measures of Heilongjiang's Three Economic Sectors and Industries (t = 2011)

Sectors (i)	*Demand Structure $d_i(t)$ (%)*	*Employment Structure $s_i^1(t)$ (%)*	*Comparative Gain of Labor $G_i^1(t)$*	*Investment Structure $s_i^2(t)$ (%)*	*Comparative Gain of Capital $G_i^2(t)$*
Total	100	100	—	100	—
Primary sector	13.64065			5.99341	2.27594
Secondary sector	50.52304			42.74113	1.18207
Tertiary sector	35.83711			47.0189	0.76219
Industries (*i*)					
Agriculture, forest, animal husbandry, fishery	13.64065	6.607194	2.064515	8.771437	1.55512
Mining		2.703621		7.59558	
Manufacturing		5.935528		26.92431	
Production/supply of electricity, gas, and water		0.971523		4.881344	
Construction	5.820391	3.157657	1.843252	3.292132	1.76797
Communication, transportation, storage, postal service	4.349158	2.411388	1.803591	7.661129	0.56769
Information transfer, computer, and software areas		0.848794		1.102282	
# Information Transfer					
Wholesale and retail	8.479502	9.586014	0.884570	2.823929	3.00273
Hotels and restaurants	2.205729	2.647446	0.833154	0.829387	2.65947
Finance	2.805707	0.872614	3.215290	0.124408	22.5525
Real estate/housing	3.723748	0.750021	4.964855	23.15863	0.16079
Leasing and commercial services		1.538121		0.406941	

continued

Table 4.12 (continued) The Structures and Comparative Gain Measures of Heilongjiang's Three Economic Sectors and Industries ($t = 2011$)

Sectors (i)	*Demand Structure $d_i(t)$ (%)*	*Employment Structure $s_i^1(t)$ (%)*	*Comparative Gain of Labor $G_i^1(t)$*	*Investment Structure $s_i^2(t)$ (%)*	*Comparative Gain of Capital $G_i^2(t)$*
Scientific research, technology service, and geological survey		1.013723		0.268201	
Water conservation, environment, and public facility management		0.753020		0.199227	
# Public Facility Management					
Residents and other services		3.530416		0.934044	
Education		2.514580		0.665283	
Health, safety, and welfare		1.352983		0.357959	
Social, sports, and entertainment		0.476380		0.126036	
Public administration and social organizations		2.310885		0.611392	

Source: Data are from *China Statistics Abstract 2012* (XLS, DOC).

molybdenum, talc, graphite, gypsum, cement rock, gold, and silver. Its reserves of oil shale are the largest in the country. For more details, see Figure 4.7.

The altitude of Jilin is the highest in the southeast, and falls gradually toward the northwest. The Changbai Mountains run through its southeastern regions. Other mountain ranges include the Jilinhada Mountains, Zhang Guangcai Mountains, and Longgang Mountains. The Yalu and Tumen rivers flow through Jilin is in the extreme southwest; tributaries of the Liao River and the Songhua and Nen rivers run along the southern border. The province has a northerly continental monsoon climate with long, cold winters and short, warm summers.

Historically, various peoples inhabited in Jilin, such as the Xianbei, Koreans, the Mohe, and the Wuji and various Korean kingdoms, including the Buyeo, Goguryeo, and Balhae, were established in this area. The region then fell successively under the control of the Xiongnu, Xianbei state, Khitan Liao Dynasty, the Jurchen Jin Dynasty, and the Mongol Yuan Dynasty. During the Qing Dynasty, the immigration of Han Chinese was strictly controlled. After the Primorsky Krai area was taken by the Russians in 1860, the Qing government began to open the area up to Han Chinese migrants. In 1932, the area was incorporated into Manchukuo, a puppet state set up by Japan. In 1945, when Japan was defeated, the region was taken by the communists. Manchuria

Figure 4.7 The geographic location of Jilin.

was the staging ground from which the communists eventually conquered the rest of China in the Chinese Civil War. In the 1950s, Jilin was expanded to its present borders.

The agricultural production of Jilin centered around rice, maize, and sorghum, where rice is mostly cultivated in the eastern parts; the Changbai Mountains are an important source of lumber. Sheep herding is an important activity in the western parts. Among its natural resources, Jilin has the largest reserves of shale oil and one of the top five mineral reserves in all of China. And compared to other provinces, Jilin has extensive deposits of Kieselguhr, wollastonite, floatstone, and molybdenum. The industry in Jilin concentrates on automobiles, train carriages, and iron alloy. Jilin is one of the most important bases of commodity grains in China and is ranked sixth in timber production. Traditionally, Jilin is known as a major pharmaceutical center, which produces ginseng and deer antlers, used extensively in traditional Chinese medicine. In 2011, the nominal GDP of Jilin totaled ¥1053.1 billion (US$167.1 billion) and has been rising at a double-digit rate since 2003. Its per capita nominal GDP reached ¥26,289 (US$3848) in 2009.

For additional readings, refer to Ministry of Commerce (2013) and National Bureau of Statistics of China (2013).

4.7.1 Current State of Investment and Employment Structures of Jilin

All that is necessary for us to do is the computations of the gross domestic outputs of all the economic sectors and industries and the investment and employment situations (Table 4.13) of Jilin and the relevant comparative gains. By comparing the resultant comparative gains, we can then obtain the current states of investment and employment for the three economic sectors and industries. All the detailed computations are similar to what has been described earlier. The results are given in Table 4.14.

First, in terms of the three major economic sectors, among the six relevant data points of the three sectors of Jilin Province, all of them are situated in extreme instability. In particular, the

Table 4.13 The GDPs, Total Investments in Fixed properties, and Employment Situation of Jilin's Three Economic Sectors and Industries (2011)

Sectors (i)	*Total Output of Sector (0.1 Billion Yuan)*	*Total Investment on Fixed Properties (0.1 Billion Yuan)*	*Employment (10,000 People)*
Total	10568.83	7221.64	1337.8
Primary sector	1277.44	197.07	573.9
Secondary sector	5611.48	4041.63	270.2
Tertiary sector	3679.91	2982.94	493.7
Industries (*i*)			
Agriculture, forest, animal husbandry, fishery	1277.44	195.826	**16.3**
Mining	644.25	394.633	**14.8**
Manufacturing	4027.8	3206.889	**64.9**
Production/supply of electricity, gas, and water	245.9	386.877	**9**
Construction	693.53	38.806	**13.4**
Communication, transportation, storage, postal service	420.98	483.223	**14.3**
Information transfer, computer, and software areas	235.16	38.079	**5.8**
# Information Transfer			
Wholesale and retail	860.47	243.475	**9.4**
Hotels and restaurants	205.69	53.360	**3.1**
Finance	207.65	4.924	**10.5**
Real estate/housing	238.61	219.371	**5.3**
Leasing and commercial services	206.64	33.050	**4.4**
Scientific research, technology service, and geological survey	69.81	28.805	**6.8**
Water conservation, environment, and public facility management	34.38	437.281	**8.5**
# Public Facility Management			
Residents and other services	217.99	25.018	**0.9**

continued

Table 4.13 (continued) The GDPs, Total Investments in Fixed properties, and Employment Situation of Jilin's Three Economic Sectors and Industries (2011)

Sectors (i)	*Total Output of Sector (0.1 Billion Yuan)*	*Total Investment on Fixed Properties (0.1 Billion Yuan)*	*Employment (10,000 People)*
Education	339.15	59.701	**36.8**
Health, safety, and welfare	144.57	40.961	**17.5**
Social, sports, and entertainment	51.85	72.746	**3.7**
Public administration and social organizations	446.96	68.234	**32.5**

Source: The data in bold are from *China Statistics Yearbook*, 2012, the rest from *Jilin Province's Statistics Yearbook*, 2012.

Note: Because the investments on fixed properties reported in *Jilin Province's Statistics Yearbook* (2012) do not agree with those reported in *Statistics Report of Jilin Province's Domestic Economy and Social Development* (2011), which was published on February 29, 2012, the latter source is used here.

Table 4.14 The Structures and Comparative Gain Measures of Jilin's Three Economic Sectors and Industries ($t = 2011$)

Sectors (i)	*Demand Structure $d_i(t)$ (%)*	*Employment Structure $s_i^1(t)$ (%)*	*Comparative Gain of Labor $G_i^1(t)$*	*Investment Structure $s_i^2(t)$ (%)*	*Comparative Gain of Capital $G_i^2(t)$*
Total	100	100	—	100	—
Primary sector	12.08686	42.89943	0.281749	2.728882	4.42924
Secondary sector	53.09463	20.19764	2.628753	55.96554	0.94870
Tertiary sector	34.81852	36.90442	0.943478	41.30558	0.84295
Industries (*i*)					
Agriculture, forest, animal husbandry, fishery	12.08686	1.218437	9.919978	2.711654	4.45738
Mining	6.095755	1.106310	5.509986	5.464586	1.11550
Manufacturing	38.11018	4.851321	7.855629	44.40666	0.85821
Production/supply of electricity, gas, and water	2.326653	0.672756	3.458389	5.357212	0.43430
Construction	6.562032	1.001659	6.551161	0.537361	12.2116

continued

Table 4.14 (continued) The Structures and Comparative Gain Measures of Jilin's Three Economic Sectors and Industries ($t = 2011$)

Sectors (i)	*Demand Structure* $d_i(t)$ *(%)*	*Employment Structure* $s_i^1(t)$ *(%)*	*Comparative Gain of Labor* $G_i^1(t)$	*Investment Structure* $s_i^2(t)$ *(%)*	*Comparative Gain of Capital* $G_i^2(t)$
Communication, transportation, storage, postal service	3.983222	1.068935	3.726346	6.691312	0.59528
Information transfer, computer, and software areas	2.225033	0.433554	5.132078	0.527285	4.21980
# Information Transfer					
Wholesale and retail	8.141582	0.702657	11.58686	3.371457	2.41486
Hotels and restaurants	1.946195	0.231727	8.398646	0.73889	2.63394
Finance	1.964740	0.784882	2.503228	0.068189	28.8129
Real estate/housing	2.257677	0.396179	5.698631	3.037691	0.74322
Leasing and commercial services	1.955183	0.328903	5.944557	0.457655	4.27218
Scientific research, technology service, and geological survey	0.660527	0.508305	1.299471	0.398866	1.65601
Water conservation, environment, and public facility management	0.325296	0.635381	0.511970	6.055152	0.05372
# Public Facility Management					
Residents and other services	2.062575	0.067276	30.65857	0.346425	5.95388
Education	3.208964	2.750826	1.166546	0.82669	3.88170
Health, safety, and welfare	1.367890	1.308137	1.045678	0.567195	2.41167
Social, sports, and entertainment	0.490594	0.276578	1.773801	1.007328	0.48702
Public administration and social organizations	4.229040	2.429398	1.740777	0.94485	4.47588

Source: Data are from *China Statistics Abstract 2012* (XLS, DOC).

comparative gain of labor of the primary sector is way too low and equal to merely 0.281749. That is not even reaching one-ninth, and one-third of that same indicator of the secondary and the tertiary sectors, respectively. In terms of the capital investment, the comparative gains of the primary sector is much too low, and is equal to 4.42924, while the capital investments in the secondary and the tertiary sectors are in relative excess, equaling 0.94870 and 0.84295, respectively.

Second, in terms of the employment situation, the relevant data of the 21 industries of Jilin Province should contain 42 data points. However, because of lack of raw data, we only collected 21 of them. So, we can only base our discussion on these available statistics. All industries are situated in extremely imbalanced states. In particular, the comparative gains of labor of the following 14 industries are all over 1. Especially, the comparative gains of capital in "real estate/housing" are over 10, while the value of this indicator of "water conservation, environment, and public facility management" is too low, equal to merely 0.02634. That is only (0.02634/28.8129≈) 0.0009 when compared to that value of "finance," (0.02634/12.2116≈) 0.0021 when compared to that value of "construction," (0.02634/5.9538≈) 0.0043 when compared to that value of "residents and other services," (0.02634/4.4759 ≈) 0.0058 when compared to that value of "public administration and social organizations," and (0.02634/4.4574≈) 0.0058 when compared to that value of "agriculture, forest, animal husbandry, fishery."

Finance	Construction	Residents and other services	Public administration and social organizations	Agriculture, forest, animal husbandry, fishery
Information transfer, computer, and software areas	Leasing and commercial services	Education	Hotels and restaurants	Wholesale and retail
Health, safety, and welfare	Scientific research, technology service, and geological survey	Manufacturing	Mining	

Therefore, in terms of the magnitudes of the comparative gains of labor, because of lack of data, the problem of employment cannot be addressed.

In terms of the magnitudes of the comparative gains of capital, the industries of Jilin Province with available data can be ranked as follows:

1	2	3	4	5
Finance	Construction	Residents and other services	Public administration and social organizations	Agriculture, forest, animal husbandry, fishery
6	7	8	9	10
Information transfer, computer, and software areas	Leasing and commercial services	Education	Hotels and restaurants	Wholesale and retail

11	12	13	14	15
Health, safety, and welfare	Scientific research, technology service, and geological survey	Mining	Manufacturing	Real estate/ housing
16	17	18	19	
Communication, transportation, storage, postal service	Social, sports, and entertainment	Production/ supply of electricity, gas, and water	Water conservation, environment, and public facility management	

4.7.2 Investment and Employment in Jilin's Economic Sectors and Industries

First, the only direction of employment is in the secondary economic sector.

Second, in terms of capital, the following 13 industries have enjoyed the comparative gains of more than 1, representing relatively more optimal areas of additional investment.

Finance	Construction	Residents and other services	Public administration and social organizations	Agriculture, forest, animal husbandry, fishery
Information transfer, computer, and software areas	Leasing and commercial services	Education	Hotels and restaurants	Wholesale and retail
Health, safety, and welfare	Scientific research, technology service, and geological survey	Mining		

However, for the following six industries, excessive investment of capital has been given. Further capital investment in these industries should be curbed and redirected to the previous 13 industries.

Water conservation, environment, and public facility management	Production/supply of electricity, gas, and water	Social, sports, and entertainment	Communication, transportation, storage, postal service
Real estate/housing	Manufacturing		

In other words, no matter whether the investors, who look for investment opportunities in Jilin Province, are from Central China, or other parts of the mainland China, or the rest of the world, they should consider the following industries ranked from the most optimal to the least:

1	2	3	4	5
Finance	Construction	Residents and other services	Public administration and social organizations	Agriculture, forest, animal husbandry, fishery
6	7	8	9	10
Leasing and commercial services	Information transfer, computer, and software areas	Education	Hotels and restaurants	Wholesale and retail
11	12	13		
Health, safety, and welfare	Scientific research, technology service, and geological survey	Mining		

At the same time, try not to invest additional capital in the following seven industries, because there has been a severe over supply of capital in these industries.

That is the direction of capital investment.

Public facility management	Water conservation, environment, and public facility management	Production/supply of electricity, gas, and water	Social, sports, and entertainment
Communication, transportation, storage, postal service	Real estate/housing	Manufacturing	

4.8 Hubei Province

Hubei is located in the eastern most part of Central China with Wuhan as its capital, which has been the political, cultural, and economic hub and a major transportation thoroughfare of Central China. It borders Henan in the north, Anhui in the east, Jiangxi in the southeast, Hunan in the south, Chongqing in the west, and Shaanxi to the northwest. The well-known Three Gorges Dam is located at a small town called Yichang in the west of the province. For more details, see Figure 4.8.

The most of central and eastern Hubei is taken up by the Jianghan Plain, and the west and the peripheries of the province are more mountainous with such ranges as the Wudang Mountains, the Jing Mountains, the Daba Mountains, and the Wu Mountains. The Dabie Mountains lie in the northeast of the Janghan Plain along the border with Henan and Anhui; the Tongbai Mountains lie in the north along the border with Henan; in the southeast, the Mufu Mountains form the border with Jiangxi. Two major rivers that go through the province are the Yangtze and its left tributary Hanshui, from which the name Jianghan Plain comes. After flowing through most of the province, the two great rivers meet at Wuhan.

The landscape of the Jianghan Plain is also decorated by thousands of lakes with Lake Liangzi and Hong Lake being the largest. The numerous hydroelectric dams created several large reservoirs

Figure 4.8 The geographic location of Hubei.

with Danjiangkou Reservoir, located on the border between Hubei and Henan, on the Hanshui being the largest.

Hubei has four distinct seasons and a humid subtropical climate. Winters are cold with average temperatures ranging from 1 to 6°C (34 to 43°F) in January; summers are hot and humid with average temperatures ranging from 24 to 30°C (75 to 86°F) in July. However, for Wuhan city, the summer temperature can easily go above 40°C (104°F). That makes the mountainous districts of western Hubei attractive for visitors from Wuhan and other lowland cities for their cooler summers.

Historically, by the time of the Spring and Autumn period (770 BC–476 BC), the territory of today's Hubei was part of the state of Chu, a tributary state of the Zhou Dynasty. During the Warring States period (475 BC–221 BC) Chu became the major opposition of the state of Qin in what is now Shaanxi Province. Over time, Chu lost land piece by piece, then its dominance over the Sichuan Basin, and eventually (in 278 BC) its heartland, which corresponds to modern Hubei. In 223 BC, Qin chased down the remnants of the Chu regime as part of Qin's bid for the conquest of all China. When the Han Dynasty succeeded the Qin Dynasty in 206 BC, it established the province (known as zhou at the time) of Jingzhou, which covered the territory in what is now Hubei and Hunan. In the beginning of the third century, near the end of the Han Dynasty, Jingzhou was ruled by warlord Liu Biao, then by Cao Cao, a powerful warlord who conquered nearly all of north China. However, in the battle of Red Cliffs, Liu Bei, and Sun Quan drove Cao Cao out of Jingzhou with Liu Bei taking control of Jingzhou. Liu Bei then went on to conquer the Sichuan Basin, while losing Jingzhou to Sun Quan; consequently, the Kingdom of Wu ruled

Jingzhou for the next few decades. During the Tang Dynasty, the territory of what is now Hubei was placed under several circuits: Jiangnanxi Circuit in the south; Shannandong Circuit in the west, and Huainan Circuit in the east. After Tang disintegrated in the tenth century, Hubei came under the control of several regional regimes: Jingnan in the center, in the east, and the Five Dynasties to the north.

When the Song Dynasty reunified China in 982, it placed most of Hubei into Jinghubei Circuit. Mongols conquered China in 1279 and established the province of Huguang, covering Hubei, Hunan, and parts of Guangdong and Guangxi. In 1334, during the Yuan Dynasty, Hubei was devastated by the outbreak of the Black Death, which spread widely to decimate populations throughout Eurasia in the following three centuries. The Ming Dynasty drove out the Mongols in 1368 with its Huguang province corresponding almost entirely to the combination of modern Hubei and Hunan. The Qing Dynasty, which conquered China in 1644, split Huguang into the modern provinces of Hubei and Hunan in 1664. With the famous modernization reforms of Zhang Zhidong, Hubei, especially Wuhan, became a prosperous commerce and industry center. The south-east area of Wuhan became an important center of mining and metallurgy. In 1911, the Wuchang Uprising took place in modern-day Wuhan, which helped to establish the Republic of China. During World War II the eastern parts of Hubei were occupied by Japan while the western parts remained under Chinese control.

As the fear of a nuclear war increased during the time of Sino-Soviet border conflicts in the late 1960s, the Xianning prefecture of Hubei was chosen as the site of Project 131, an underground military command headquarters. The province, Wuhan in particular, suffered severely during the 1954 Yangtze River floods, which prompted the construction of large-scale dams with the Gezhouba Dam on the Yangtze River near Yichang started in 1970 and completed in 1988, the Three Gorges Dam, further upstream, began in 1993, and a number of smaller dams constructed on the Yangtze's tributaries.

Economically, Hubei is known as the "land of fish and rice." Major agricultural products in Hubei include cotton, rice, wheat, and tea, while industries include automobiles, metallurgy, machinery, power generation, textiles, foodstuffs, and high-tech commodities. The mineral resources in Hubei in significant quantities include borax, copper, garnet, gold amalgam, gypsum, hongshiite, iron, manganese, marlstone, phosphorus, rock salt, rutile, vanadium, and wollastonite.

In terms of nominal GDPs, Hubei's economy is ranked 11th in China in 2011 with ¥1.959 trillion (311 billion USD) and a per capita of ¥21,566 RMB (2863 USD).

For additional readings, refer to Ministry of Commerce (2013) and National Bureau of Statistics of China (2013).

4.8.1 Current State of Investment and Employment Structures of Hubei

All that is necessary for us to do is the computations of the gross domestic outputs of all the economic sectors and industries and the investment and employment situations (Table 4.15) of Hubei and the relevant comparative gains. By comparing the resultant comparative gains, we can then obtain the current states of investment and employment for the three economic sectors and industries. All the detailed computations are similar to what has been described earlier. The results are given in Table 4.16.

First, in terms of the three major economic sectors, among the six relevant data points of the three sectors of Hubei Province, all of them are situated in extreme instability. In particular, the comparative gain of labor of the primary sector is way too low and equal to merely 0.286371. That is not even reaching one-eighth, and one-fourth of that same indicator of the secondary and the tertiary sectors,

Table 4.15 The GDPs, Total Investments in Fixed Properties, and Employment Situation of Hubei's Three Economic Sectors and Industries (2011)

Sectors (i)	*Total Output of Sector (0.1 Billion Yuan)*	*Total Investment on Fixed Properties (0.1 Billion Yuan)*	*Employment (10,000 People)*
Total	19632.26	12935.0	3672
Primary sector	2569.3	440.9	1678.10
Secondary sector	9815.94	5526.4	771.12
Tertiary sector	7247.02	6967.7	1222.78
Industries (*i*)			
Agriculture, forest, animal husbandry, fishery	4252.9	440.9	**9.4**
Mining		265.4	**13.0**
Manufacturing		4759.9	**168.2**
Production/supply of electricity, gas, and water		470.8	**17.8**
Construction	1277.9	30.4	**103.1**
Communication, transportation, storage, postal service	869.48	1152.6	**25.6**
Information transfer, computer, and software areas		82.2	**5.6**
# Information Transfer			
Wholesale and retail	1512.89	401.8	**27.4**
Hotels and restaurants	446.52	216.4	**8.2**
Finance	674.57	30.3	**15.8**
Real estate/housing	634.67	2784.7	**8.1**
Leasing and commercial services		107.0	**4.5**
Scientific research, technology service, and geological survey		38.5	**11.6**
Water conservation, environment, and public facility management		1170.8	**9**
# Public Facility Management			
Residents and other services		47.6	**1.5**

continued

Table 4.15 (continued) The GDPs, Total Investments in Fixed Properties, and Employment Situation of Hubei's Three Economic Sectors and Industries (2011)

Sectors (i)	*Total Output of Sector (0.1 Billion Yuan)*	*Total Investment on Fixed Properties (0.1 Billion Yuan)*	*Employment (10,000 People)*
Education		153.5	**65.9**
Health, safety, and welfare		128.5	**31.3**
Social, sports, and entertainment		119.7	**5.4**
Public administration and social organizations		534.0	**54.7**

Source: The data in bold are from *China Statistics Yearbook* (2012), and the rest from *Hubei Province's Statistics Yearbook* (2012).

Table 4.16 The Structures and Comparative Gain Measures of Hubei's Three Economic Sectors and Industries ($t = 2011$)

Sectors (i)	*Demand Structure $d_i(t)$ (%)*	*Employment Structure $s_i^1(t)$ (%)*	*Comparative Gain of Labor $G_i^1(t)$*	*Investment Structure $s_i^2(t)$ (%)*	*Comparative Gain of Capital $G_i^2(t)$*
Total	100	100	—	100	—
Primary sector	13.08713	45.69989	0.286371	3.408653	3.83939
Secondary sector	49.99903	21	2.380906	42.72433	1.17027
Tertiary sector	36.91383	33.30011	1.108520	53.86702	0.68528
Industries (*i*)					
Agriculture, forest, animal husbandry, fishery	21.66281	0.255991	84.62325	3.408653	6.35524
Mining		0.354031		2.051475	
Manufacturing		4.580610		36.79858	
Production/supply of electricity, gas, and water		0.484749		3.639603	
Construction	6.509184	2.807734	2.318305	0.234682	27.7362
Communication, transportation, storage, postal service	4.428833	0.697168	6.352607	8.910735	0.49702

continued

Table 4.16 (continued) The Structures and Comparative Gain Measures of Hubei's Three Economic Sectors and Industries ($t = 2011$)

Sectors (i)	*Demand Structure $d_i(t)$ (%)*	*Employment Structure $s_i^1(t)$ (%)*	*Comparative Gain of Labor $G_i^1(t)$*	*Investment Structure $s_i^2(t)$ (%)*	*Comparative Gain of Capital $G_i^2(t)$*
Information transfer, computer, and software areas		0.152505		0.635165	
# Information Transfer					
Wholesale and retail	7.706143	0.746187	10.32736	3.106665	2.48052
Hotels and restaurants	2.274420	0.223312	10.18496	1.672601	1.35981
Finance	3.436028	0.430283	7.985504	0.234462	14.6550
Real estate/housing	3.232791	0.220588	14.65532	21.52853	0.15016
Leasing and commercial services		0.122549		0.826866	
Scientific research, technology service, and geological survey		0.315904		0.297266	
Water conservation, environment, and public facility management		0.245098		9.051562	
# Public Facility Management					
Residents and other services		0.040850		0.368129	
Education		1.794662		1.18707	
Health, safety, and welfare		0.852397		0.993617	
Social, sports, and entertainment		0.147059		0.925608	
Public administration and social organizations		1.489651		4.128708	

Source: Data are from *China Statistics Abstract 2012* (XLS, DOC).

respectively. In terms of the capital investment, the comparative gains of the primary and the secondary sectors are much too low, and are equal to 3.83939 and 1.17027, respectively, while the capital investment in the tertiary sector is in extreme excess (0.68528).

Second, in terms of the employment situation, the relevant data of the 21 industries of Hubei Province should contain 42 data points. However, because of lack of raw data, and the data regarding the comparative gains of labor are not reliable, we only have seven data points related to the comparative gains of capital. Even so, all of these data are situated in extremely imbalanced states. In particular, the comparative gains of labor of the following five industries are all over 1. Especially, the comparative gains of capital in "construction" and "finance" are over 10, while the value of this indicator of "real estate and housing" is too low, equal to only 0.15016, which is (0.15016/27.7362≈) 0.0054 when compared to that of "construction," (0.15016/14.6550≈) 0.010 when compared to that of "finance," (0.15016/6.35524≈) 0.023 when compared to that of "agriculture, forest, animal husbandry, fishery," and (0.15016/2.48052≈) 0.060 when compared to that of "wholesale and retail."

Construction	Finance	Agriculture, forest, animal husbandry, fishery	Wholesale and retail	Hotels and restaurants

Therefore, in terms of the comparative gains of labor, we do not have enough effective data to address the problem of employment.

In terms of the magnitudes of the comparative gains of capital, the industries of the Hubei Province with available data can be ranked as follows:

1	2	3	4
Construction	Finance	Agriculture, forest, animal husbandry, fishery	Wholesale and retail
5	6	7	
Hotels and restaurants	Communication, transportation, storage, postal service	Real estate/housing	

4.8.2 Investment and Employment in Hubei's Economic Sectors and Industries

First, although we know that the employment direction is in the secondary and the tertiary economic sectors, we have no knowledge regarding the employment direction in individual industries.

Second, in terms of capital, the following five industries have enjoyed the comparative gains of more than 1, representing relatively more optimal areas of additional investment.

Construction	Finance	Agriculture, forest, animal husbandry, fishery	Wholesale and retail	Hotels and restaurants

However, for the following two industries, excessive investment of capital has been given. Further capital investment in these industries should be curbed and redirected to the previous five industries.

Real estate/housing	Communication, transportation, storage, postal service

In other words, no matter whether the investors, who look for investment opportunities in Hubei Province, are from Central China, or other parts of the mainland China, or the rest of the world, they should consider the following industries ranked from the most optimal to the least:

1	2	3	4	5
Construction	Finance	Agriculture, forest, animal husbandry, fishery	Wholesale and retail	Hotels and restaurants

That is the direction of capital investment.

Third, by combining the two criteria discussed above, we see that the economic sector in the Hubei Province that is most optimal for both investment and employment is the secondary sector, because the comparative gains of labor and capital of each of these industries are over 1.

4.9 Guangxi Autonomous Region

Guangxi, officially Guangxi Autonomous Region, is located in southern China, bordering with Vietnam in the southwest, the Gulf of Tonkin in the south, Yunnan to the west, Guizhou in the north, Hunan in the northeast, and Guangdong in the east and southeast. Because of its location in the mountainous terrain in the far south of China, Guangxi has been on the frontier of Chinese civilization throughout much of the history. For more details, see Figure 4.9.

In Guangxi's mountainous hills, the Nanling Mountains form the northeast border with the Yuecheng and Haiyang Mountains branching from them. In the north are the Duyao and Fenghuang Mountains. Near the center are the Dayao and Daming Mountains. On the southeastern border are the Yunkai Mountains. The highest point of the province reaches 2141 meters (7024 ft) in the Kitten Mountain of the Yuecheng Mountains. And many rivers, most of which form the tributary basin of the West River, cut valleys through the mountains. Along the short coastline on the Gulf of Tonkin, Guangxi has such important seaports as Beihai, Qinzhou, and Fangchenggang. The Detian waterfall along the border with Vietnam separates the two countries. The climate in Guangxi is subtropical with long and hot summers with the average annual temperature ranging from 17°C to 23°C, while the average annual precipitation ranges from 1250 to 1750 mm.

Historically, Guangxi was inhabited by various tribal groups and first became part of China during the Qin Dynasty. In 214 BC, General Zhao Tuo claimed most of southern China for Qin Shihuang. Upon the emperor's death and the ensuing civil war, Zhao established a separate kingdom known as Southern Yue. Then, Southern Yue expanded colonization and sinicization until its collapse in 111 BC during the southward expansion of the Han. The name of the province can be traced to the time of the Eastern Wu, who controlled southeastern

Figure 4.9 The geographic location of Guangxi.

China during the Three Kingdoms period. During the Tang Dynasty, the Zhuang supported the Kingdom of Southern Zhao in Yunnan which successfully resisted imperial armies in both 751 and 754. After the collapse of the Southern Zhao, Liu Yan established the Southern Han (Nanhan) in Xingwangfu (modern Guangdong). The Yuan Dynasty established control over Yunnan during its conquest of Dali in 1253 and eliminated the Southern Song in 1279. The Mongolians then established Guangxi as a proper province. The Qing Dynasty left the region alone until the imposition of direct rule in 1726. However, the nineteenth century suffered from constant unrests. For example, a Yao revolt in 1831 was followed by the Taiping Rebellion in 1850 and the Jintian Uprising in 1851. The execution of St. Auguste Chapdelaine in Guangxi provoked the Second Opium War in 1858 and the legalization of foreign interference. In the 1884 Franco-Chinese War, the Guangxi Army was greatly mobilized and repulsed the French from China in 1885.

Following the Wuchang Uprising, Guangxi withdrew from the Qing Empire on November 6, 1911. General Lu Rongting's Guangxi Clique overran Hunan and Guangdong and helped lead the resistance to YuanShikai's attempt to reestablish an imperial government. After a brief occupation by Chen Jiongming's Cantonese forces, Guangxi fell into disunity and profound banditry for several years until Li Zongren's Guangxi Pacification Army established the New Guangxi Clique. After the death of Sun Yat-sen, Li also repulsed Tang Jiyao's revolt and joined the Northern Expedition to establish the Republican control over other warlords. Within Communist China, Guangxi is known for the Baise Uprising, led by Chen Zhaoli and Deng Xiaoping in 1929. The 1944 Operation Ichi-Go successfully expanded Japanese control along the rail lines through Guangxi into French Indochina, although the lines were harassed constantly by

American bombers and Zhuang guerrillas under Bai Chongxi. Being in the far south, Guangxi was not involved in the Chinese Civil War, and joined the People's Republic in December 1949. In 1958, Guangxi was converted into an autonomous region for the Zhuang at the recommendation of Premier Zhou Enlai, because the Zhuang was the biggest minority group in China and was concentrated in the province.

For the most part of its history, Guangxi was landlocked. In 1952, a small section of Guangdong's coastline was given to Guangxi. While some heavy industry was developed in the 1960s and 1970s, the province remained largely a tourist destination. And, in recent years, there has been a growing amount of industrialization and increasing concentration on cash crops.

Major crops that Guangxi produces include rice, maize, and sweet potatoes. Cash crops include sugar cane, peanuts, tobacco, and kenaf. Eight-five percent of the world's star anise, a major ingredient in the antiviral treatment known as Tamiflu, is grown in Guangxi. The province is one of China's key production centers for nonferrous metals. It holds about one-third of the tin and manganese deposits in China.

In recent years, Guangxi's economy has fallen behind that of its wealthy neighbor, Guangdong. Its nominal GDP in 2011 was about ¥1171.4 billion (US$185.9 billion) and its per capita GDP ¥15,800 (US$2300).

For additional readings, refer to Ministry of Commerce (2013); National Bureau of Statistics of China (2013); and Bonavia (1995).

4.9.1 Current State of Investment and Employment Structures of Guangxi

All that is necessary for us to do is the computations of the gross domestic outputs of all the economic sectors and industries and the investment and employment situations (Table 4.17) of Guangxi and the relevant comparative gains. By comparing the resultant comparative gains, we can then obtain the current states of investment and employment for the three economic sectors and industries. All the detailed computations are similar to what has been described earlier. The results are given in Table 4.18.

First, in terms of the three major economic sectors, among the six relevant data points of the three sectors of the Guangxi Autonomous Region, all of them are situated in extreme instability. In particular, the comparative gain of labor of the primary sector is way too low and equal to merely 0.327637. That is only roughly one-eighth, and one-fourth of that same indicator of the secondary and the tertiary sectors, respectively. In terms of the capital investment, the comparative gains of the primary and the secondary sectors are much too low, and are equal to 6.65734 and 1.24152, respectively, while the capital investment in the tertiary sector is in extreme excess (0.58437).

Second, in terms of the employment situation, the relevant data of the 21 industries of the Guangxi Province should contain 42 data points. However, because of lack of raw data, we only collected 39 of them. So, we can only base our discussion on these available statistics. Among all the industries, only two are basically in equilibrium, where the comparative gain of labor in "social, sports, and entertainment" is equal to 1.015691, and the comparative gain of capital in "leasing and commercial services" is equal to 0.98196. All the rest ($37/39 \geq 95\%$) are situated in extremely imbalanced states. In particular, the comparative gains of labor of the following 13 industries are all over 1. Especially, the comparative gain of labor in "real estate/housing" is even equal to 8.558308, which is equal to ($8.558308/0.227553 \approx$) 38 times of the lowest value of this indicator that is reached by "residents and other services."

Table 4.17 The GDPs, Total Investments in Fixed Properties, and Employment Situation of Guangxi's Three Economic Sectors and Industries (2011)

Sectors (i)	*Total Output of Sector (0.1 Billion Yuan)*	*Total Investment on Fixed Properties (0.1 Billion Yuan)*	*Employment (10,000 People)*
Total	11720.87	9280.3004	2936
Primary sector	2047.23	243.4826	1565.2
Secondary sector	5675.32	3619.4306	561.8
Tertiary sector	3998.33	5417.3872	809.07
Industries (*i*)			
Agriculture, forest, animal husbandry, fishery	2047.23	243.4826	1565.2
Mining	334.12	296.9251	35.7
Manufacturing	4140.52	2819.5226	323.6
Production/supply of electricity, gas, and water	376.73	461.2039	17.1
Construction	823.95	41.779	185.4
Communication, transportation, storage, postal service	588.2	1085.4977	84.2
Information transfer, computer, and software areas	195.1	125.3181	13.77
# Information Transfer			
Wholesale and retail	803.48	254.2199	230.09
Hotels and restaurants	307.88	150.2319	95.1
Finance	445.37	25.6923	14.56
Real estate/housing	465.68	1916.1372	13.63
Leasing and commercial services	160.81	129.6647	29.6
Scientific research, technology service, and geological survey	72.24	31.9583	11.39
Water conservation, environment, and public facility management	25.92	1093.5719	11.82
# Public Facility Management			
Residents and other services	111.79	27.3716	123.06
Education	288.85	211.6958	79.89
Health, safety, and welfare	172.7	103.3969	34.52
Social, sports, and entertainment	51.09	118.5449	12.6
Public administration and social organizations	309.22	144.086	54.84

Source: The data are from *Guangxi Statistics Abstracts 2012*.

Table 4.18 The Structures and Comparative Gain Measures of Guangxi's Three Economic Sectors and Industries ($t = 2011$)

Sectors (i)	*Demand Structure $d_i(t)$ (%)*	*Employment Structure $s_i^1(t)$ (%)*	*Comparative Gain of Labor $G_i^1(t)$*	*Investment Structure $s_i^2(t)$ (%)*	*Comparative Gain of Capital $G_i^2(t)$*
Total	100	100	—	100	—
Primary sector	17.46654	53.31063	0.327637	2.623650	6.65734
Secondary sector	48.42064	19.13488	2.530491	39.00122	1.24152
Tertiary sector	34.11291	27.55688	1.237909	58.37513	0.58437
Industries (*i*)					
Agriculture, forest, animal husbandry, fishery	17.46654	53.31063	0.327637	2.623650	6.65734
Mining	2.850642	1.215940	2.344393	3.199520	0.89096
Manufacturing	35.32605	11.02180	3.205107	30.38180	1.16274
Production/supply of electricity, gas, and water	3.214181	0.582425	5.518618	4.969709	0.64675
Construction	7.029768	6.314714	1.113236	0.450190	15.6151
Communication, transportation, storage, postal service	5.018399	2.867847	1.749883	11.69679	0.42904
Information transfer, computer, and software areas	1.664552	0.469005	3.549111	1.350367	1.23267
# Information Transfer					
Wholesale and retail	6.855123	7.836853	0.874729	2.739350	2.50246
Hotels and restaurants	2.626767	3.239101	0.810956	1.618826	1.62264
Finance	3.799803	0.495913	7.662241	0.276848	13.7252
Real estate/ housing	3.973084	0.464237	8.558308	20.64736	0.19243
Leasing and commercial services	1.371997	1.008174	1.360873	1.397204	0.98196

continued

Table 4.18 (continued) The Structures and Comparative Gain Measures of Guangxi's Three Economic Sectors and Industries ($t = 2011$)

Sectors (i)	*Demand Structure* $d_i(t)$ *(%)*	*Employment Structure* $s_i^1(t)$ *(%)*	*Comparative Gain of Labor* $G_i^1(t)$	*Investment Structure* $s_i^2(t)$ *(%)*	*Comparative Gain of Capital* $G_i^2(t)$
Scientific research, technology service, and geological survey	0.616337	0.387943	1.588730	0.344367	1.78977
Water conservation, environment, and public facility management	0.221144	0.402589	0.549305	11.78380	0.01377
# Public Facility Management					
Residents and other services	0.953769	4.191417	0.227553	0.294943	3.23374
Education	2.464408	2.721049	0.905583	2.281131	1.08034
Health, safety, and welfare	1.473440	1.175749	1.253192	1.114155	1.32247
Social, sports, and entertainment	0.435889	0.429155	1.015691	1.277382	0.34124
Public administration and social organizations	2.638200	1.867847	1.412428	1.552601	1.69921

Source: Data are from *China Statistics Abstract 2012* (XLS, DOC).

Real estate/ housing	Finance	Production/supply of electricity, gas, and water	Information transfer, computer, and software areas	Manufacturing
Mining	Communication, transportation, storage, postal service	Scientific research, technology service, and geological survey	Public administration and social organizations	Leasing and commercial services
Health, safety, and welfare	Construction	Social, sports, and entertainment		

At the same time, the imbalance in terms of the comparative gains of capital seems to be way too severe. The comparative gains in the following 12 industries are over 1. In particular, the comparative gains of capital of "construction" and "finance" are even over 10, while the value of this indicator for "water conservation, environment, and public facility management" is too low, equaling only 0.00988. That is merely (0.00988/15.6151 ≈) 0.00063 when compared to the value of this indicator of "construction," (0.00988/13.7252 ≈) 0.00072 when compared to the value of this indicator of "finance," (0.00988/6.65734 ≈) 0.0015 when compared to the value of this indicator of "agriculture, forest, animal husbandry, fishery," (0.00988/3.23374 ≈) 0.0031 when compared to the value of this indicator of "residents and other services," and (0.00988/2.50246 ≈) 0.0040 when compared to the value of this indicator of "wholesale and retail."

Construction	Finance	Agriculture, forest, animal husbandry, fishery	Residents and other services
Wholesale and retail	Scientific research, technology service, and geological survey	Public administration and social organizations	Hotels and restaurants
Health, safety, and welfare	Information transfer, computer, and software areas	Leasing and commercial services	Education

So, in terms of the magnitudes of the comparative gains of labor, the industries of the Guangxi Province with available data can be ranked as follows:

1	2	3	4	5
Real estate/housing	Finance	Production/supply of electricity, gas, and water	Information transfer, computer, and software areas	Manufacturing
6	7	8	9	10
Mining	Communication, transportation, storage, postal service	Scientific research, technology service, and geological survey	Public administration and social organizations	Leasing and commercial services
11	12	13	14	15
Health, safety, and welfare	Construction	Social, sports, and entertainment	Education	Wholesale and retail
16	17	18	19	
Hotels and restaurants	Water conservation, environment, and public facility management	Agriculture, forest, animal husbandry, fishery	Residents and other services	

In terms of the magnitudes of the comparative gains of capital, the industries of the Guangxi Province with available data can be ranked as follows:

1	2	3	4	5
Construction	Finance	Agriculture, forest, animal husbandry, fishery	Residents and other services	Wholesale and retail
6	7	8	9	10
Scientific research, technology service, and geological survey	Public administration and social organizations	Hotels and restaurants	Health, safety, and welfare	Information transfer, computer, and software areas
11	12	13	14	15
Manufacturing	Education	Leasing and commercial services	Mining	Production/ supply of electricity, gas, and water
16	17	18	19	
Communication, transportation, storage, postal service	Social, sports, and entertainment	Real estate/ housing	Water conservation, environment, and public facility management	

4.9.2 Investment and Employment in Guangxi's Economic Sectors and Industries

First, in terms of the comparative gains of labor, the values of this indicator for the following 13 industries:

Real estate/ housing	Finance	Production/supply of electricity, gas, and water	Information transfer, computer, and software areas	Manufacturing
Mining	Communication, transportation, storage, postal service	Scientific research, technology service, and geological survey	Public administration and social organizations	Leasing and commercial services
Health, safety, and welfare	Construction	Social, sports, and entertainment		

are over 1, representing relatively optimal industries for employment. In the other six industries, there has been excessive supply of labor so that appropriate control on the increase of labor into these industries should be imposed. Specifically, the excessive labor in the following industries should be transferred into the previous 13 industries.

Residents and other services	Agriculture, forest, animal husbandry, fishery	Water conservation, environment, and public facility management
Hotels and restaurants	Wholesale and retail	Education

In other words, no matter whether the job seekers, who look for employment opportunities in the Guangxi Province, are from Central China, or other parts of the mainland China, or the rest of the world, they should first consider the secondary and the tertiary economic sectors with the relevant industries ranked as follows:

1	2	3	4	5
Real estate/ housing	Finance	Production/supply of electricity, gas, and water	Information transfer, computer, and software areas	Manufacturing
6	7	8	9	10
Mining	Communication, transportation, storage, postal service	Scientific research, technology service, and geological survey	Public administration and social organizations	Leasing and commercial services
11	12	13		
Health, safety, and welfare	Construction	Social, sports, and entertainment		

Without special skills, don't look for employment opportunities in the primary economic sector and other industries, because those areas are experiencing excessive labor supplies.

That is the direction of employment.

Second, in terms of capital, the following 12 industries have enjoyed the comparative gains of more than 1, representing relatively more optimal areas of additional investment.

Construction	Finance	Agriculture, forest, animal husbandry, fishery	Residents and other services	Wholesale and retail
Scientific research, technology service, and geological survey	Public administration and social organizations	Hotels and restaurants	Health, safety, and welfare	Information transfer, computer, and software areas
Manufacturing	Education			

However, for the following seven industries, excessive investment of capital has been given. Further capital investment in these industries should be curbed and redirected to the previous 12 industries.

Water conservation, environment, and public facility management	Real estate/ housing	Social, sports, and entertainment	Communication, transportation, storage, postal service
Production/supply of electricity, gas, and water	Mining	Leasing and commercial services	

In other words, no matter whether the investors, who look for investment opportunities in Guangxi Province, are from Central China, or other parts of the mainland China, or the rest of the world, they should consider the following industries ranked from the most optimal to the least:

1	2	3	4
Construction	Finance	Agriculture, forest, animal husbandry, fishery	Residents and other services
5	6	7	8
Wholesale and retail	Scientific research, technology service, and geological survey	Public administration and social organizations	Hotels and restaurants
9	10	11	12
Health, safety, and welfare	Information transfer, computer, and software areas	Manufacturing	Education

At the same time, try not to invest additional capital in the following eight industries, because there has been a severe over supply of capital in these industries.

That is the direction of capital investment.

Public facility management	Water conservation, environment, and public facility management	Real estate/ housing	Social, sports, and entertainment
Communication, transportation, storage, postal service	Production/supply of electricity, gas, and water	Mining	Leasing and commercial services

Third, by combining the two criteria discussed above, we see that the industries in the Guangxi Autonomous Region that are most optimal for both investment and employment are the following, because the comparative gains of labor and capital of each of these industries are over 1.

Finance	Construction	Information transfer, computer, and software areas	Manufacturing
Scientific research, technology service, and geological survey	Public administration and social organizations	Health, safety, and welfare	

The secondary choices are the industries of "education" and "leasing and commercial services." And the worst choice is the industry of "water conservation, environment, and public facility management."

4.10 Anhui Province

Anhui is located across the basins of the Yangtze River and the Huai River and part of the East China region. It borders Jiangsu in the east, Zhejiang in the southeast, Jiangxi in the south, Hubei in the southwest, Henan in the northwest, and Shandong for a small section in the north. The province is quite topographically diverse. The north of it is part of the North China Plain while the north-central regions are part of the Huai He River watershed. Both of these areas are flat and densely populated. Moving further south the land becomes more uneven with the Dabie Mountains occupying much of the southwest and hills and ranges cutting through the southeast, between which the Yangtze River finds its way through south Anhui. The Lotus Peak, part of Huangshan in southeastern Anhui, locates the highest peak in Anhui that has an altitude of 1873 m. Major rivers of the province include the Huai He in the north and the Yangtze River in the south. The largest lake is Lake Chaohu in the center. There are also many lakes in the southeastern part of the province near the Yangtze River. For more details, see Figure 4.10.

In terms of the topography, Anhui's climate changes from north to south, where the north is more temperate with distinct seasons. The average temperature in January is around −1 to 2°C in the north of the Huai He, and 0 to 3°C in the south of the Huai He. The average temperature in July reaches 27°C or higher. In June and July plum rains occur and consequent floods.

Figure 4.10 The geographic location of Anhui.

Historically, the province of Anhui was formed in the seventeenth century. Before that time, northern Anhui was culturally a part of the North China Plain; central Anhui made up the most of the fertile and densely populated Huai He River watershed. Southern Anhui was culturally closer to Hubei and southern Jiangsu. And, the hills of southeastern Anhui had its unique and distinct cultural sphere of its own.

During the time period of Warring States, Shouchun (modern Shou County) in the central area became a refugee capital for the state of Chu when its heartlands in modern Hubei was taken by the state of Qin in the west in 278 BC. During the Qin Dynasty and the Han Dynasty, Anhui was administered under several different commanderies. Near the end of the Han Dynasty, Shouchun was under the control of the warlord Cao Cao who founded the Kingdom of Wei in the era of the Three Kingdoms. In the fourth century, nomadic tribes from central Asia flooded north China, causing several centuries of political division of China into north and south. Because of its particular location at the juncture of north and south, the land comprising modern Anhui changed hands frequently and was often bisected through the middle politically. The well-known Battle of Feishui between the Former Qin of the north and the Eastern Jin Dynasty of the south took place in 383 AD in modern Anhui. During the Sui Dynasty (581–618) and the Tang Dynasty (618–907), Anhui was ruled under several different jurisdictions. During the division of China between the Jurchen Jin Dynasty in the north and the Southern Song Dynasty in the south, Anhui was bisected along the Huai He River. This division ended when the Mongols reunified China in 1279. During the Ming Dynasty since 1368, both Jiangsu and Anhui kept their special status as a territory that was directly governed by the central government. During the Qing Dynasty in 1666, Jiangsu and Anhui were split apart as separate provinces. Since then, Anhui has kept almost the same borders as today.

Economically, the agriculture in Anhui varies from one climate zone to another that the province crosses. Wheat and sweet potatoes are grown north of the Huai He river, while rice and wheat are planted south of the Huai He. In terms of natural resources, Anhui has iron in Ma'anshan, coal in Huainan, and copper in Tongling along with relevant industries related to these natural resources. Consequently, most of the wealth in Anhui is concentrated in the industrial regions close to the Yangtze River, such as Hefei, Wuhu, and Ma'anshan. As a mid-size economy in terms of economic output, the nominal GDP of the province for 2011 was about ¥1.51 trillion (US$239.8 billion) with a per capita GDP of ¥16,391 (US$2400).

For additional readings, refer to Ministry of Commerce (2013); National Bureau of Statistics of China (2013); and Thechinaperspective.com (2013).

4.10.1 Current State of Investment and Employment Structures of Anhui

All that is necessary for us to do is the computations of the gross domestic outputs of all the economic sectors and industries and the investment and employment situations (Table 4.19) of Anhui and the relevant comparative gains. By comparing the resultant comparative gains, we can then obtain the current states of investment and employment for the three economic sectors and industries. All the detailed computations are similar to what has been described earlier. The results are given in Table 4.20.

First, in terms of the three major economic sectors, among the six relevant data points of the three sectors of Anhui Province, all of them are situated in extreme instability. In particular, the comparative gain of labor of the primary sector is way too low and equal to merely 0.339471. That is roughly equal to one-sixth, and one-third of that same indicator of the secondary and the tertiary sectors, respectively. In terms of the capital investment, the comparative gain of the primary

Table 4.19 The GDPs, Total Investments in Fixed Properties, and Employment Situation of Anhui's Three Economic Sectors and Industries (2011)

Sectors (i)	*Total Output of Sector (0.1 Billion Yuan)*	*Total Investment on Fixed Properties (0.1 Billion Yuan)*	*Employment (10,000 People)*
Total	15300.65	12147.7794	4120.9
Primary sector	2015.31	176.103	1598.9
Secondary sector	8309.38	6086.7005	1038.5
Tertiary sector	4975.95	5884.9759	1483.5
Industries (*i*)			
Agriculture, forest, animal husbandry, fishery		176.103	1598.9
Mining		387.4157	69.8
Manufacturing		5091.5803	547.4
Production/supply of electricity, gas, and water		332.6035	10.7
Construction	1247.38	275.101	410.6
Communication, transportation, storage, postal service	589.82	559.2573	171.7
Information transfer, computer, and software areas	193.61	51.9675	28.3
# Information Transfer			
Wholesale and retail	193.61	240.3612	481.5
Hotels and restaurants	252.62	165.6556	176.5
Finance	503.85	57.3198	21.7
Real estate/housing	634.92	3230.0774	26.3
Leasing and commercial services	284.67	131.1066	37.3
Scientific research, technology service, and geological survey	117.34	47.0426	8.5
Water conservation, environment, and public facility management	97.83	821.6068	31.1
# Public Facility Management			
Residents and other services	136.72	25.6597	312.2
Education	382.15	169.6338	68.7
Health, safety, and welfare	205.21	93.9603	32.4
Social, sports, and entertainment	81.24	106.2765	30
Public administration and social organizations	445.38	185.0508	57.3

Source: The data are from *China Statistics Yearbook* (2012).

Table 4.20 The Structures and Comparative Gain Measures of Anhui's Three Economic Sectors and Industries (t = 2011)

Sectors (i)	*Demand Structure $d_i(t)$ (%)*	*Employment Structure $s_i^1(t)$ (%)*	*Comparative Gain of Labor $G_i^1(t)$*	*Investment Structure $s_i^2(t)$ (%)*	*Comparative Gain of Capital $G_i^2(t)$*
Total	100	100	—	100	—
Primary sector	13.17140	38.79978	0.339471	1.449672	9.08578
Secondary sector	54.30737	25.20081	2.154985	50.10546	1.08386
Tertiary sector	32.52117	35.99942	0.903380	48.44487	0.67130
Industries (*i*)					
Agriculture, forest, animal husbandry, fishery		38.79978		1.449672	
Mining		1.693805		3.189189	
Manufacturing		13.28351		41.91367	
Production/supply of electricity, gas, and water		0.259652		2.737978	
Construction	8.152464	9.963843	0.818205	2.264620	3.59993
Communication, transportation, storage, postal service	3.854869	4.166566	0.925191	4.603782	0.83733
Information transfer, computer, and software areas	1.265371	0.686743	1.842568	0.427794	2.95790
# Information Transfer					
Wholesale and retail	1.265371	11.68434	0.108296	1.978643	0.63951
Hotels and restaurants	1.651041	4.283045	0.385483	1.363670	1.21073
Finance	3.292997	0.526584	6.253508	0.471854	6.97885
Real estate/housing	4.149628	0.638210	6.501977	26.58986	0.15606
Leasing and commercial services	1.860509	0.905142	2.055489	1.079264	1.72387
Scientific research, technology service, and geological survey	0.766896	0.206266	3.718000	0.387253	1.98035

continued

Table 4.20 (continued) The Structures and Comparative Gain Measures of Anhui's Three Economic Sectors and Industries ($t = 2011$)

Sectors (i)	*Demand Structure $d_i(t)$ (%)*	*Employment Structure $s_i^1(t)$ (%)*	*Comparative Gain of Labor $G_i^1(t)$*	*Investment Structure $s_i^2(t)$ (%)*	*Comparative Gain of Capital $G_i^2(t)$*
Water conservation, environment, and public facility management	0.639385	0.754690	0.847215	6.763432	0.09454
# Public Facility Management					
Residents and other services	0.893557	7.576015	0.117945	0.211230	4.23026
Education	2.497606	1.667112	1.498164	1.396418	1.78858
Health, safety, and welfare	1.341185	0.786236	1.705830	0.773477	1.73397
Social, sports, and entertainment	0.530958	0.727996	0.729341	0.874864	0.60690
Public administration and social organizations	2.910857	1.390473	2.093429	1.523330	1.91085

Source: Data are from *China Statistics Abstract 2012* (XLS, DOC).

sector is much too low and equal to 9.08578, while the capital investment in the tertiary sector is in extreme excess (0.67130).

Second, in terms of the employment situation, the relevant data of the 21 industries of Anhui Province should contain 42 data points. However, because of lack of raw data, we only collected 30 of them. So, we can only base our discussion on these available statistics. All the industries are situated in extremely imbalanced states. In particular, the comparative gains of labor of the following eight industries are all over 1. Especially, the comparative gain of labor in "real estate/housing" is even equal to 6.501977, which is more than (6.501977/0.108296≈) 65 times of the lowest value of this indicator that is reached by "residents and other services."

Real estate/housing	Finance	Scientific research, technology service, and geological survey	Public administration and social organizations
Leasing and commercial services	Information transfer, computer, and software areas	Health, safety, and welfare	Education

At the same time, the imbalance in terms of the comparative gains of capital seems to be way too severe. The comparative gains in the following ten industries are over 1. In particular, the value of this indicator for "water conservation, environment, and public facility management" is too low, equaling only 0.09454. That is merely (0.09454/6.97885≈) 0.013 when compared to the value of this indicator of "finance," (0.09454/4.23026≈) 0.021 when compared to the value of this indicator of "residents and other services," (0.09454/3.59993≈) 0.025 when compared to the value of this indicator of "construction," and (0.09454/2.95790≈) 0.03 when compared to the value of this indicator of "Information transfer, computer, and software areas."

Finance	Residents and other services	Construction	Information transfer, computer, and software areas	Scientific research, technology service, and geological survey
Public administration and social organizations	Education	Health, safety, and welfare	Leas ng and commercial services	Hotels and restaurants

So, in terms of the magnitudes of the comparative gains of labor, the industries of the Anhui Province with available data can be ranked as follows:

1	2	3	4
Real estate/housing	Finance	Scientific research, technology service, and geological survey	Public administratior and social organizations
5	6	7	8
Leasing and commercial services	Information transfer, computer, and software areas	Health, safety, and welfare	Education
9	10	11	12
Communication, transportation, storage, postal service	Water conservation, environment, and public facility management	Construction	Social, sports, and entertainment
13	14	15	
Hotels and restaurants	Residents and other services	Wholesale and retail	

In terms of the magnitudes of the comparative gains of capital, the industries of the Anhui Province with available data can be ranked as follows:

1	2	3	4
Finance	Residents and other services	Construction	Information transfer, computer, and software areas
5	6	7	8
Scientific research, technology service, and geological survey	Public administration and social organizations	Education	Health, safety, and welfare

9	10	11	12
Leasing and commercial services	Hotels and restaurants	Communication, transportation, storage, postal service	Wholesale and retail
13	14	15	
Social, sports, and entertainment	Real estate/housing	Water conservation, environment, and public facility management	

4.10.2 Investment and Employment in Anhui's Economic Sectors and Industries

First, in terms of the comparative gains of labor, the values of this indicator for the following eight industries

Real estate/housing	Finance	Scientific research, technology service, and geological survey	Public administration and social organizations
Leasing and commercial services	Information transfer, computer, and software areas	Health, safety, and welfare	Education

are over 1, representing relatively optimal industries for employment. In all the other seven industries, there has been excessive supply of labor so that appropriate control on the increase of labor into these industries should be imposed. Specifically, the excessive labor in the following industries should be transferred into the previous eight industries.

Wholesale and retail	Residents and other services	Hotels and restaurants	Social, sports, and entertainment
Construction	Water conservation, environment, and public facility management	Communication, transportation, storage, postal service	

In other words, no matter whether the job seekers, who look for employment opportunities in the Anhui Province, are from Central China, or other parts of the mainland China, or the rest of the world, they should first consider the secondary economic sector with the relevant industries ranked as follows:

Real estate/housing	Finance	Scientific research, technology service, and geological survey	Public administration and social organizations
Leasing and commercial services	Information transfer, computer, and software areas	Health, safety, and welfare	Education

Without special skills, don't look for employment opportunities in the primary and the tertiary economic sectors and other industries, because those areas are experiencing excessive labor supplies.

That is the direction of employment.

Second, in terms of capital, the following 10 industries have enjoyed the comparative gains of more than 1, representing relatively more optimal areas of additional investment.

Finance	Residents and other services	Construction	Information transfer, computer, and software areas	Scientific research, technology service, and geological survey
Public administration and social organizations	Education	Health, safety, and welfare	Leasing and commercial services	Hotels and restaurants

However, for the following five industries, excessive investment of capital has been given. Further capital investment in these industries should be curbed and redirected to the previous 10 industries.

Communication, transportation, storage, postal service	Wholesale and retail	Social, sports, and entertainment	Real estate/ housing	Water conservation, environment, and public facility management

In other words, no matter whether the investors, who look for investment opportunities in the Anhui Province, are from Central China, or other parts of the mainland China, or the rest of the world, they should consider the following industries ranked from the most optimal to the least:

1	2	3	4	5
Finance	Residents and other services	Construction	Information transfer, computer, and software areas	Water conservation, environment, and public facility management
6	7	8	9	10
Public administration and social organizations	Education	Health, safety, and welfare	Leasing and commercial services	Hotels and restaurants

That is the direction of capital investment.

Third, by combining the two criteria discussed above, we see that the industries in the Anhui Province that are most optimal for both investment and employment are the following, because the comparative gains of labor and capital of each of these industries are over 1, and the worst choice is the industry of "water conservation, environment, and public facility management."

1	2	3	4
Finance	Scientific research, technology service, and geological survey	Information transfer, computer, and software areas	Public administration and social organizations
5	6	7	
Leasing and commercial services	Education	Health, safety, and welfare	

4.11 Jiangxi Province

Jiangxi is located in the southeast of the country, spanning from the banks of the Yangtze River in the north into hillier areas in the south and east. It shares a border with Anhui in the north, Zhejiang in the northeast, Fujian in the east, Guangdong in the south, Hunan in the west, and Hubei in the northwest. For more details, see Figure 4.11.

Geographically, Jiangxi is surrounded by mountains on three sides with the Mufu, Jiulin, and Luoxiao Mountains in the west, Huaiyu and Wuyi Mountains in the east, and the Jiulian and Dayu Mountains in the south. The southern half of the province is interspersed with ranges and valleys. The northern half is flatter and lower in altitude. Jiangxi reaches its highest point of the altitude of 2157 meters (7077 ft) at Mount Huanggang in the Wuyi Mountains along the border with Fujian.

Figure 4.11 The geographic location of Jiangxi.

The Gan River flows through the entire length of the province from south to north, and enters in the north Lake Poyang, the largest freshwater lake of China. Important reservoirs of the province include the Xiushui Tuolin Reservoir in the northwest on the Xiushui River, and the Wan'an Reservoir in the upper section of the Gan River.

The climate in Jiangxi is humid subtropical with short, cool, damp winters, and very hot, humid summers. The average temperature in January ranges from 3°C to 9°C (37 to 48°F) and the average temperature in July ranges from 27°C to 30°C (81 to 86°F). The annual precipitation of the province is within the range from 1200 to 1900 mm (47–75 in.) with much of the precipitation falling as heavy rains in late spring and summer.

Historically, Jiangxi is centered on the Gan River valley, through which the main north–south transport route of south China has been constructed. The corridor along the Gan River is one convenient route through the mountainous and rugged terrain of the south-eastern mountains, and has been the primary route of trade and communication between the North China Plain and the Yangtze River valley in the north and the territory of modern Guangdong Province in the south. As a result, Jiangxi has represented a strategically important geographic location throughout much of China's history.

During the Shang Dynasty (sixteenth to eleventh centuries BC), Jiangxi was outside the sphere of influence of early Chinese civilization. During the Spring and Autumn period, the northern part of modern Jiangxi formed the western frontier of the state of Wu. When Wu was conquered by the state of Yue based in modern northern Zhejiang in 473 BC, the state of Chu based in modern Hubei took over northern Jiangxi and then subjugated Yue in 333 BC. When Qin conquered Chu in 223 BC, majority of Jiangxi area was put under Jiujiang Commandery. Yuzhang Commandery was established in Jiangxi in the Han Dynasty by 202 BC. In 291 AD, during the Western Jin Dynasty, Jiangxi became its own *Zhou* called Jiangzhou. During the Northern and Southern Dynasties, Jiangxi was under the control of the southern dynasties. During the Tang Dynasty, circuits were established as a new top-level administrative division with Jiangxi as part of the Jiangnan Circuit at first. In 733, this circuit was divided into western and eastern halves. Jiangxi was found in the western half. During the period of Five Dynasties and Ten Kingdoms, Jiangxi first belonged to Wu, then to Southern Tang. During the Yuan Dynasty, the circuit was divided into 13 different circuits, and Jiangxi Province was established for the first time and included the majority of modern Guangdong. Jiangxi acquired its modern borders during the Ming Dynasty. After the fall of the Qing Dynasty, Jiangxi became one of the earliest bases for the Communists. The Nanchang Uprising took place in Jiangxi on August 1, 1927. Later the Communist leadership occupied the mountains of southern and western Jiangxi. In 1931, the government of Chinese Soviet Republic was established in Ruijin. In 1935, after complete encirclement by the Nationalist forces, the Communists broke through and began the Long March to Yan'an.

Economically, rice has been the dominant crop in Jiangxi. The cash crops that are commonly grown include cotton and rapeseed. The province is rich in such mineral deposits as copper, tungsten, gold, silver, uranium, thorium, tantalum, niobium, among others. The nominal GDP of Jiangxi in 2011 was ¥1158.3 billion (US$183.8 billion) and a per capita GDP of ¥17,248 (US$2525).

For additional readings, refer to National Bureau of Statistics of China (2013); Thechinaperspective.com (2012b).

4.11.1 Current State of Investment and Employment Structures of Jiangxi

All that is necessary for us to do is the computations of the gross domestic outputs of all the economic sectors and industries and the investment and employment situations (Table 4.21) of

Table 4.21 (continued) The GDPs, Total Investments in Fixed Properties, and Employment Situation of Jiangxi's Three Economic Sectors and Industries (2011)

Sectors (i)	*Total Output of Sector (0.1 Billion Yuan)*	*Total Investment on Fixed Properties (0.1 Billion Yuan)*	*Employment (10,000 People)*
Total	11702.82	9087.5985	2532.6
Primary sector	1391.07	235.0877	870.5
Secondary sector	6390.55	5207.5587	763.3
Tertiary sector	3921.2	3644.9521	898.8
Industries (*i*)			
Agriculture, forest, animal husbandry, fishery		235.0877	870.55
Mining		228.6780	
Manufacturing		4644.4059	
Production/supply of electricity, gas, and water		276.9078	11.41
Construction	978.69	57.5670	217.81
Communication, transportation, storage, postal service	507.44	456.8303	82.60
Information transfer, computer, and software areas		44.5428	34.73
# Information Transfer			
Wholesale and retail	1102.26	283.5115	422.03
Hotels and restaurants	270.29	205.3814	128.66
Finance	357.44	23.6083	11.34
Real estate/housing	402.51	1349.4136	11.63
Leasing and commercial services		87.3643	16.91
Scientific research, technology service, and geological survey		23.6662	6.43
Water conservation, environment, and public facility management		744.0841	5.66
# Public Facility Management			
Residents and other services		52.9374	60.14
Education		106.2972	47.96
Health, safety, and welfare		74.2648	18.10
Social, sports, and entertainment		106.4990	6.51
Public administration and social organizations		86.5515	46.10

Source: The data are from *Jiangxi Province's Statistics Yearbook* (2012).

Jiangxi and the relevant comparative gains. By comparing the resultant comparative gains, we can then obtain the current states of investment and employment for the three economic sectors and industries. All the detailed computations are similar to what has been described earlier. The results are given in Table 4.22.

First, in terms of the three major economic sectors, among the six relevant data points of the three sectors of the Jiangxi Province, only one is basically in equilibrium, where the comparative gain of capital of the secondary economic sector is 0.95293. The remaining five are situated in

Table 4.22 The Structures and Comparative Gain Measures of Jiangxi's Three Economic Sectors and Industries ($t = 2011$)

Sectors (i)	*Demand Structure $d_i(t)$ (%)*	*Employment Structure $s_i^1(t)$ (%)*	*Comparative Gain of Labor $G_i^1(t)$*	*Investment Structure $s_i^2(t)$ (%)*	*Comparative Gain of Capital $G_i^2(t)$*
Total	100	100	—	100	—
Primary sector	11.886622	34.3718	0.345825	2.586907	4.59492
Secondary sector	54.606924	30.1390	1.811837	57.30401	0.95293
Tertiary sector	33.506454	35.4892	0.944130	40.10908	0.83538
Industries (*i*)					
Agriculture, forest, animal husbandry, fishery		34.3738		2.586907	
Mining				2.516374	
Manufacturing				51.10708	
Production/supply of electricity, gas, and water		0.45053		3.047095	
Construction	8.3628561	8.60025	0.972397	0.633468	13.2017
Communication, transportation, storage, postal service	4.3360489	3.26147	1.329477	5.026964	0.8526
Information transfer, computer, and software areas		1.37132		0.490149	
# Information Transfer					
Wholesale and retail	9.4187555	16.6639	0.565219	3.119763	3.0191

continued

Table 4.22 The Structures and Comparative Gain Measures of Jiangxi's Three Economic Sectors and Industries ($t = 2011$)

Sectors (i)	*Demand Structure $d_i(t)$ (%)*	*Employment Structure $s_i^1(t)$ (%)*	*Comparative Gain of Labor $G_i^1(t)$*	*Investment Structure $s_i^2(t)$ (%)*	*Comparative Gain of Capital $G_i^2(t)$*
Hotels and restaurants	2.3096143	5.08015	0.454635	2.260018	1.0219
Finance	3.0543066	0.44776	6.821285	0.259786	11.7570
Real estate/ housing	3.4394274	0.45921	7.489849	14.84896	0.2316
Leasing and commercial services		0.66769		0.961357	
Scientific research, technology service, and geological survey		0.25389		0.260423	
Water conservation, environment, and public facility management		0.22349		8.187907	
# Public Facility Management					
Residents and other services		2.37463		0.582521	
Education		1.89371		1.169695	
Health, safety, and welfare		0.71468		0.81721	
Social, sports, and entertainment		0.25705		1.171916	
Public administration and social organizations		1.82026		0.952413	

Source: Data are from *China Statistics Abstract 2012* (XLS, DOC).

extreme instability. In particular, the comparative gain of labor of the primary sector is way too low and equal to merely 0.345825. That is not even reaching one-fifth of that same indicator of the secondary sector. In terms of the capital investment, the comparative gain of the primary sector is much too low, and is equal to 4.59492, while the capital investment in the tertiary sector is in extreme excess (0.83538).

Second, in terms of the employment situation, the relevant data of the 21 industries of the Jiangxi Province should contain 42 data points. However, because of lack of raw data, we only collected 12 of them. So, we can only base our discussion on these available statistics. Among all these data points, only two are basically in equilibrium, where the comparative gain of labor in "construction" and the comparative gain of capital in "hotels and restaurants" are equal to 0.972397 and 1.02194, respectively. All the rest (10/12 ≥ 83%) are situated in extremely imbalanced states. In particular, the comparative gains of labor of the following three industries are all over 1. Especially, the comparative gain of labor in "real estate/housing" is even equal to 7.489849, which is equal to (7.489849/0.454635≈) 16 times of the lowest value of this indicator that is reached by "hotels and restaurants."

Real estate/housing	Finance	Communication, transportation, storage, postal service

At the same time, the imbalance in terms of the comparative gains of capital seems to be way too severe. The comparative gains in the following four industries are over 1. In particular, the comparative gain of capital of "real estate/housing" is too low, equaling only 0.2316. That is merely (0.2316/13.2017≈) 0.017 when compared to the value of this indicator of "construction," and (0.2316/11.7570≈) 0.019 when compared to the value of this indicator of "finance."

Construction	Finance	Wholesale and retail	Hotels and restaurants

So, in terms of the magnitudes of the comparative gains of labor, the industries of the Jiangxi Province with available data can be ranked as follows:

1	2	3	4
Real estate/housing	Finance	Communication, transportation, storage, postal service	Construction
5	6		
Wholesale and retail	Hotels and restaurants		

In terms of the magnitudes of the comparative gains of capital, the industries of the Jiangxi Province with available data can be ranked as follows:

1	2	3	4
Construction	Finance	Wholesale and retail	Hotels and restaurants
5	6		
Communication, transportation, storage, postal service	Real estate/housing		

4.11.2 Investment and Employment in Jiangxi's Economic Sectors and Industries

First, in terms of the comparative gains of labor, the values of this indicator for the following three industries

Real estate/housing	Finance	Communication, transportation, storage, postal service

are over 1, representing relatively optimal industries for employment. In all the other three industries, there has been excessive supply of labor so that appropriate control on the increase of labor into these industries should be imposed. Specifically, the excessive labor in the following industries should be transferred into the previous three industries.

Hotels and restaurants	Wholesale and retail	Construction

In other words, no matter whether the job seekers, who look for employment opportunities in the Jiangsu Province, are from Central China, or other parts of the mainland China, or the rest of the world, they should first consider the secondary economic sector with the relevant industries ranked as follows:

Real estate/housing	Finance	Communication, transportation, storage, postal service

Without special skills, don't look for employment opportunities in the primary and tertiary economic sectors and other industries, because those areas are experiencing excessive labor supplies.

That is the direction of employment.

Second, in terms of capital, the following four industries have enjoyed the comparative gains of more than 1, representing relatively more optimal areas of additional investment.

Construction	Finance	Wholesale and retail	Hotels and restaurants

However, for the following two industries, excessive investment of capital has been given. Further capital investment in these industries should be curbed and redirected to the previous four industries.

Real estate/housing	Communication, transportation, storage, postal service

In other words, no matter whether the investors, who look for investment opportunities in the Jiangxi Province, are from Central China, or other parts of the mainland China, or the rest of the world, they should first consider the following industries ranked from the most optimal to the least:

Construction	Finance	Wholesale and retail	Hotels and restaurants

That is the direction of capital investment.

Third, by combining the two criteria discussed above, we see that the industry in the Jiangxi Province that is most optimal for both investment and employment is "finance," because the comparative gains of labor and capital of this industry are over 1.

Chapter 5

West China

The Most Difficult Area

From Chapter 3, it follows that if China is divided into three parts: the east, center, and west, then the western China represents the worst area for investment and employment. Although this area is the worst in terms of the opportunities for investment and employment, the regions in this part of China can still be ranked from the most optimal to the least optimal based on their respective characteristics in terms of investment and employment. On the basis of the theoretical foundation, models, and methods, provided in Chapter 2, this chapter calculates and analyze the statistics available for the year of 2011, and then provides the consequent guidelines for investment and employment opportunities in this region.

For the geographic location in reference to the west of China, refer to Figure 5.1.

5.1 Regional Industries and Economic of the Most Difficult Area

5.1.1 Current State of the Regional Economic Structure of West China

From Table 2.1, the 2011 GDP of West China and the relevant investment and employment are shown in Table 5.1.

Based on the data presented in Table 5.1, we can calculate and obtain the data in Table 5.2 by using the relevant formulas discussed in Chapter 1.

Table 5.2 indicates that the economic structure of this region is severely imbalanced, no matter whether one looks at the supply and demand of labor or the supply and demand of capital.

In fact, from the fifth column (the comparative gains of labor) and the seventh column (the comparative gains of capital) of the table, it follows that among the 20 relevant data of the 10 provinces, municipalities, and autonomous regions of this area, only eight of them are basically in equilibrium, including the comparative gains of labor of Guizhou, Qinghai, and the comparative gains of capital of Chongqing, Guizhou, Sichuan, Yunnan, Shaanxi, and Xingjiang, equaling respectively 1.015, 1.008, and 0.971, 0.976, 0.957, 1.041, 0.962, and 1.034. And all others, amounting to (12/20≥) 60%, are in various states of imbalance. Here, the most severe

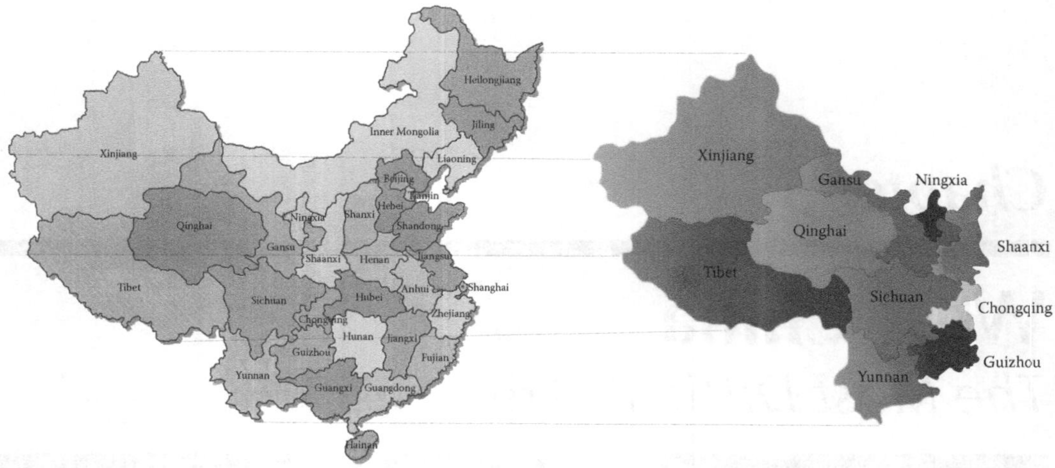

Figure 5.1 The geographic location of West China.

imbalance of supply and demand appears in such areas as Tibet and Qinghai. In fact, Tibet's comparative gain of labor is the lowest, only equaling to 0.555, while Qinghai's comparative gain of capital is the lowest, equal to merely 0.844. At the same time, the comparative gain of labor of Shaanxi and the comparative gain of capital of Yunnan are the highest, equaling 1.233 and 1.041, respectively.

Table 5.1 The GDPs, Total Investments in Fixed Properties, and Employment Situation of West China (for the year of 2011)

Area	*Province/ Municipality/ Autonomous Region (i)*	*Regional GDP (0.1 Billion Yuan)*	*Value of Fixed Properties (0.1 Billion Yuan)*	*Employment (10,000 People)*
West China	22. Chongqing	10011.37	7473.4	1580.5
	23. Sichuan	21026.68	14222.2	2647.7
	24. Guizhou	5701.84	4235.9	789.9
	25. Yunnan	8893.12	6191.0	1753.6
	26. Tibet	605.83	516.3	153.4
	27. Shaanxi	12512.30	9431.1	1427.2
	28. Gansu	5020.37	3965.8	747
	29. Qinghai	1670.44	1435.6	233.1
	30. Ningxia	2102.21	1644.7	321.3
	31. Xinjiang	6610.05	4632.1	771.5
	Total	74154.21	53747.8	10425.2

Source: Data is from *China Statistics Yearbook* 2012 (XLS; DOC).

Table 5.2 The Regional Economic Structures and Measures of Comparative Gains of the Provinces, Municipalities, and Autonomous Regions of West China (for the year of 2011)

Area	*Province/ Municipality/ Autonomous Region (i)*	*Demand Structure $d_i(t)$ (%)*	*Employment Structure $s_i^1(t)$ (%)*	*Comparative Gain of Labor $G_i^1(t)$*	*Investment Structure $s_i^2(t)$ (%)*	*Comparative Gain of Capital $G_i^2(t)$*
West China	22. Chongqing	13.501	15.160	0.891	13.905	0.971
	23. Sichuan	28.355	25.397	1.116	26.461	0.957
	24. Guizhou	7.689	7.577	1.015	7.881	0.976
	25. Yunnan	11.993	16.821	0.713	11.519	1.041
	26. Tibet	0.817	1.471	0.555	0.961	0.850
	27. Shaanxi	16.873	13.690	1.233	17.547	0.962
	28. Gansu	6.770	7.165	0.945	7.379	0.917
	29. Qinghai	2.253	2.236	1.008	2.671	0.844
	30. Ningxia	2.835	3.082	0.920	3.060	0.920
	31. Xinjiang	8.914	7.400	1.205	8.618	1.034
	Total	100	0.1651	—	100	—

So, in terms of the magnitudes of comparative gains of labor, the 10 provinces, municipalities, and autonomous regions of this area can be ordered as follows:

1	2	3	4	5
Shaanxi	Xingjiang	Sichuan	Guizhou	Qinghai
6	7	8	9	10
Gansu	Ningxia	Chongqing	Yunnan	Tibet

In terms of the magnitudes of the comparative gains of capital, these 10 provinces, municipalities, and autonomous regions of this area can be ranked as follows:

1	2	3	4	5
Yunnan	Xingjiang	Guizhou	Chongqing	Shaanxi
6	7	8	9	10
Sichuan	Ningxia	Gansu Mongolia	Tibet	Qinghai

5.1.2 Investment and Employment Opportunities in West China

First, in terms of labor, the comparative gains of the 10 provinces, municipalities, and autonomous regions reach the maximum with Shaanxi and the minimum with Tibet. The comparative gains of labor of the five provinces, autonomous regions, and municipalities as Shaanxi, Xingjiang,

Sichuan, Guizhou, and Qinghai are all over 1, representing the most ideal places for employment. The remaining five provinces, autonomous regions, and municipalities experience excessive labor so that limitations should be imposed to control the import of labors. In particular, for the four provinces, municipalities, and autonomous regions Tibet, Yunnan, Chongqing, and Ningxia, they should try different methods to transport their excessive labors to the previous five provinces and municipalities.

In other words, no matter whether the job seekers, who look for employment opportunities in the western area of China, are from western China, or other parts of the mainland China, or the rest of the world, the order of the provinces, autonomous regions, and municipalities in terms of their employment opportunities is given as follows:

1	2	3	4	5
Shaanxi	Xingjiang	Sichuan	Guizhou	Qinghai

Without special skills, don't look for employment opportunities in the other five provinces, autonomous regions, and municipalities, because these places are experiencing excessive labor supplies.

Second, in terms of capital, among the 10 provinces, autonomous regions, and municipalities, the comparative gains of both Yunan and Xingjiang are barely over 1, while for all other provinces, autonomous regions, and municipalities, their comparative gains of capital are all smaller than 1 with Qinghai and Tibet being the worst. Therefore, it makes almost no economical sense to invest in the western area of China. Other than Yunan and Xingjiang that can absorb a small amount of investment, all potentially successive investment capitals should be directed to other regions not located within the western area.

Third, by combining the two criteria discussed above, it follows that the most optimal provinces and regions in western China are Xingjiang, Shaanxi, and Sichuan.

In the following sections, we will analyze computationally the investment and employment problems of these 11 provinces and municipalities of eastern China by making use of the recent statistics (2011) and produce our best conclusions for the reader to make his/her appropriate decisions.

Note: Since we have no access to relevant statistics of regions internal to individual provinces, municipalities, and autonomous regions, we have not been able to produce any computational conclusions on investment and employment regarding these finer divisions of China. Instead, we can only focus on the investigation of the corresponding investment and employment problems in terms of economic sectors of the individual provinces, municipalities, and autonomous regions. The same situation holds true for the rest of this book.

5.2 Xinjiang Autonomous Region

Xinjiang, officially known as Xinjiang Uyghur Autonomous Region, and the largest Chinese administrative division that spans over 1.6 million km^2, is a sparsely populated area, located in the northwest of the country. It borders with Russia, Mongolia, Kazakhstan, Kyrgyzstan, Tajikistan, Afghanistan, Pakistan, and India. The province has abundant oil reserves and is China's largest natural gas-producing region. For more details, see Figure 5.2.

Geographically, Xinjiang accounts for more than one-sixth of China's total territory and a quarter of its boundary length. The province is split by the Tian Shan mountain range into two

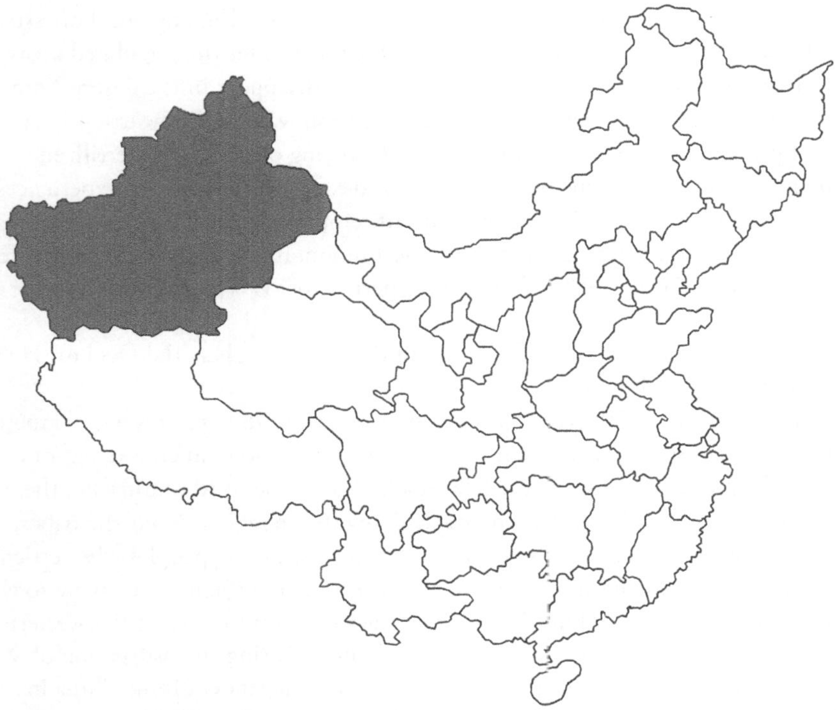

Figure 5.2 The geographic location of Xinjiang.

large basins: the Dzungarian Basin in the north and the Tarim Basin in the south. A V-shaped wedge between these two basins is the basin of the Ili River, which flows into Kazakhstan's Lake Balkhash. Farther north is a smaller wedge, known as the Emin Valley. Other major mountain ranges of the province include the Pamir Mountains and Karakoram in the southwest, the Kunlun Mountains in the south (along the border with Tibet), and the Altai Mountains in the northeast (shared with Mongolia). The highest point of the area is the mountain K2, reaching 8611 m above the sea level in the Karakoram Mountains on the border with Pakistan. Much of the Tarim Basin is occupied by the Taklamakan Desert. North of the basin is the Turpan Depression, which contains the lowest point in Xinjiang and in the entire mainland China, which is 155 m below the sea level. The Dzungarian Basin is cooler and receives relatively more precipitation than the Tarim Basin. It contains a large Gurbantünggüt Desert, also known as Dzoosotoyn Elisen, in its center. The Tian Shan mountain range marks the Xinjiang–Kyrgyzstan border at the Torugart Pass.

Most part of Xinjiang is geologically young, formed from the collision between the Indian plate and the Eurasian plate. That is how the Tian Shan, Kunlun Shan, and Pamir mountain ranges are formed. Tha also implies that Xinjiang is a major earthquake zone. Within its borders Xinjiang has such a land area that is the remotest from the sea, about 2647 km (1645 miles) from the nearest coastline.

Due to the hot summer and low precipitation, rivers in Xinjiang either disappear in the desert or terminate in salt lakes instead of flowing toward an ocean with the exception that in the northernmost part of the province, the Irtysh River, which is originated in the Altai Mountains, flows (via Kazakhstan and Russia) toward the Arctic Ocean. As of this writing, a significant part of the water in this river is now artificially diverted via the Irtysh–Karamay–Ürümqi Canal to the drier areas of southern Dzungarian Basin. Most other rivers in Xinjiang are short streams fed by the snows of the several ranges of the Tian Shan. Once they enter the populated areas, the waters are extensively used for irrigation so that the rivers often disappear in the desert. Some of the salt lakes in Xinjiang, having lost much of their inflows of fresh water, are now extensively used for the production of mineral salts to be used in the manufacturing of potassium fertilizers.

The climate in Xinjiang is generally semi-arid or desert. The entire area experiences great seasonal differences in temperature. In summer, the air temperature in the Turpan Depression easily exceeds 40°C (104°F). In the far north at the highest mountain elevations, the winter temperature regularly drops below −20°C (−4°F). Continuous permafrost is typically found in the Tian Shan Mountains.

Xinjiang has been home to many different ethnic groups, such as the Uyghur, Han, Kazakh, Tajiks, Hui, Kyrgyz, and Mongol.

Historically, in the past 2500 years, a succession of people and empires has competed for control over this territory. In 60 BC, the Han Dynasty drove the Xiongnu empire out of the region in order to secure the profitable routes to the Silk Road. During the third century BC, there were some light-skinned and reddish or blond-haired people living in the region. Nomadic tribes, such as the Yuezhi, were part of the large migration of Indo-European speaking peoples who settled in eastern Central Asia. In the second century BC, Han China sent Zhang Qian as an envoy to the states in the Xinjiang region. In 60 BC, Han China established the protectorate of the western regions at Wulei to oversee the entire region as far west as the Pamir. During the usurpation of Wang Mang in China, the dependent states of the protectorate became independent from China in AD 13. Over the next 100 years, Han China conducted several expeditions into the region, which expanded Han sovereignty into the Tarim Basin and Central Asia. After the Han Dynasty, the protectorate continued to be maintained by Cao Wei (until 265) and the Western Jin Dynasty (from 265 onward). When the Northern Wei Empire reunified northern China, its protectorate controlled what is now the southeastern region of Xinjiang, while other local states, such as Shule, Yutian, Guizi, and Qiemo, controlled the western region with the central region around Turpan controlled by Gaochang, remnants of the Northern Liang that once ruled part of what is now Gansu province.

During the Tang Dynasty, a series of expeditions were conducted against the Western Turkic Khaganate, and their vassals, the oasis states of southern Xinjiang. The expansion into Central Asia continued in 657 headed by Su Dingfang, who formally strengthened the rule of Tang in southern Xinjiang and ended the Indo-European influence in Xinjiang. During the Anshi Rebellion, Tibet invaded the Tang on a wide front from Xinjiang to Yunnan, occupied the Tang capital of Chang'an in 763 for 16 days, and took control of southern Xinjiang by the end of the century. At the same time, the Uyghur Khaganate took control of northern Xinjiang and much of the Central Asia, including Mongolia. In the tenth and eleventh century, the Kara-Khanid Khanate, a confederation of Turkic tribes, such as the Karluks, Chigils, and Yaghmas, took control of western Xinjiang. When the Uyghur khanate in Mongolia had been smashed by the Kirghiz in 840, branches of the Uyghurs established themselves in Qocha (Karakhoja) and Beshbalik near the present-day cities of Turfan and Urumchi. This Uyghur state existed in eastern Xinjiang until the thirteenth century. In 1132, defeated by the Jurchens, remnants of the Khitan Empire from Manchuria entered Xinjiang and established an exile Chinese empire, the Western Liao. When Genghis Khan unified Mongolia, the Uyghur state in the Turpan-Urumchi area offered its allegiance to the Mongols in 1209 by contributing taxes and troops to the Mongol imperial effort of expanding westward, which conquered the Western Liao in 1218. Then the Yuan Dynasty competed with the Chagatai Khanate for the ruling over the area. In the seventeenth century, the Mongolian Dzungars ruled much of the region. The Qing Dynasty gained control of the eastern Xinjiang after a long struggle with the Dzungars since the seventeenth century. After 1759, state farms were established, and the Chinese population in Xinjiang grew rapidly. In 1884, the Qing dynasty established Xinjiang as a province and formally applied to it the political system of the rest of China.

In the Kashgar region on November 12, 1933, the short-lived East Turkistan Republic was declared. That in theory encompassed Kashgar, Khotan, and Aqsu prefectures in southwestern Xinjiang. The Chinese Muslim Kuomintang 36th Division destroyed the East Turkestan Republic in 1934. In 1937, the entire province was brought under the control of northeast Chinese warlord Sheng Shicai for the next decade with support from the Soviet Union. A second East Turkestan Republic existed from 1944 to 1949 with Soviet support in northern Xinjiang. This Republic came to an end when the People's Liberation Army entered Xinjiang in 1949. Since 1949, the education system has been expanded greatly in Xinjiang with over 6000 primary schools up from around 1000; over 2000 middle schools up from 9, and over 20 institutions of higher learning up from 1. The illiteracy rate among young and middle-age people has decreased to less than 2%. And currently, Xinjiang maintains nearly 100 libraries and 30 museums, compared to none of each in 1949. Also, Xinjiang has over 100 newspapers in nearly 50 languages up from 4 newspapers in 1952. According to official statistics, the ratios of doctors, medical workers, medical clinics, and hospital beds to people surpass the national average, and immunization rates have reached 85%.

Economically, Xinjiang is a major area of irrigated agriculture, with wheat being the main staple crop; some maize and millets are grown in the south. The province is nationally known for its fruits production, including grapes, melons, pears, and walnuts. The main livestock of the region has traditionally been sheep. Although there is limited amount of fresh water, there is a significant amount of fishing in Lake Ulungur and Lake Bosten and in the Irtysh River. Since the 1970s, a large number of fish ponds have been constructed, producing over 85% of the market supply of fish. Since the late nineteenth century, the region has been noted for the production of salt, soda, borax, gold, jade, and coal. Currently, the oil and gas extraction industry in Aksu and Karamay is booming. And, the oil and petrochemical sector accounts for 60% of the province's economy. In terms of foreign trade, the province's exports amounted to US$19.3 billion and the imports reached US$2.9 billion in 2008. Most of the overall import/export volume in Xinjiang was directed to and from Kazakhstan. The Horgos Free Trade Zone, the first border-free trade zone of China, is located at Horgos, a border city between Xinjiang and Kazakhstan. In 2006, the Jeminay Border Trade Zone, the second border trade market to Kazakhstan, was opened. The nominal GDP of Xinjiang was about ¥657.4 billion (US$104.3 billion) in 2011; and its per capita GDP for 2009 was ¥19,798 (US$898).

For additional readings, refer to Beckwith (2009); Findley (2005); Hill (2009); Kim (2004); Millward (2007).

5.2.1 Current State of Investment and Employment Structures of Xinjiang

All that is necessary for us to do is the computations of the gross domestic outputs of all the economic sectors and industries and the investment and employment situations (Table 5.3) of Xingjiang Autonomous Region and the relevant comparative gains. By comparing the resultant comparative gains, we can then obtain the current states of investment and employment for the three economic sectors and industries. All the detailed computations are similar to what has been described earlier. The results are given in Table 5.4. Table 5.4 indicates the following facts.

First, in terms of the three major economic sectors, among the six relevant data points of the three sectors of Xingjiang Autonomous Region, only one is basically in equilibrium, where

Table 5.3 The GDPs, Total Investments in Fixed Properties, and Employment Situation of Xinjiang's Three Economic Sectors and Industries (2011)

Sectors (i)	*Total Output of Sector (0.1 Billion Yuan)*	*Total Investment on Fixed Properties (0.1 Billion Yuan)*	*Employment (10,000 People)*
Total	6610.05	4712.7699	953.34
Primary sector	1139.03	237.7851	463.91
Secondary sector	3225.9	2465.3475	149.04
Tertiary sector	2245.12	2009.6373	340.39
Industries (*i*)			
Agriculture, forest, animal husbandry, fishery		237.7851	**57.9**
Mining		651.8237	**18.7**
Manufacturing		1156.8245	**30.1**
Production/supply of electricity, gas, and water		570.0481	**7.6**
Construction	525.7	86.6512	**20.6**
Communication, transportation, storage, postal service	256.72	394.092	**12.3**
Information transfer, computer, and software areas	99.79	48.1474	**2.5**
# Information Transfer			
Wholesale and retail	371.9	46.8815	**8.5**
Hotels and restaurants	77.87	21.2911	**2.9**
Finance	288.77	5.1154	**8.1**
Real estate/housing	176.22	993.5788	**2.8**
Leasing and commercial services	51.69	17.2723	**5.3**
Scientific research, technology service, and geological survey	58.32	18.6848	**2**
Water conservation, environment, and public facility management	33.57	203.2966	**5.3**
# Public Facility Management			
Residents and other services	57.51	4.8747	**1.6**
Education	219.02	66.2563	**33.9**
Health, safety, and welfare	107.12	35.1039	**14.8**

Table 5.3 (continued) The GDPs, Total Investments in Fixed Properties, and Employment Situation of Xinjiang's Three Economic Sectors and Industries (2011)

Sectors (i)	*Total Output of Sector (0.1 Billion Yuan)*	*Total Investment on Fixed Properties (0.1 Billion Yuan)*	*Employment (10,000 People)*
Social, sports, and entertainment	24.05	18.1151	**3.2**
Public administration and social organizations	422.57	136.9274	**38.2**

Source: The data in bold are from *China Statistics Yearbook* (2012), while the rest are from *Xinjiang China Statistics Yearbook* (2012).

Table 5.4 The Structures and Comparative Gain Measures of Xinjiang's Three Economic Sectors and Industries (t = 2011)

Sectors (i)	*Demand Structure $d_i(t)$ (%)*	*Employment Structure $s_i^1(t)$ (%)*	*Comparative Gain of Labor $G_i^1(t)$*	*Investment Structure $s_i^2(t)$ (%)*	*Comparative Gain of Capital $G_i^2(t)$*
Total	100	100	—	100	—
Primary sector	17.23179	48.66155	0.354115	5.045549	3.41525
Secondary sector	48.80296	15.63346	3.121700	52.31207	0.93292
Tertiary sector	33.96525	35.70500	0.951274	42.64238	0.79651
Industries (*i*)					
Agriculture, forest, animal husbandry, fishery		6.073384		5.045549	
Mining		1.961525		13.83101	
Manufacturing		3.157321		24.54659	
Production/supply of electricity, gas, and water		0.797197		12.09582	
Construction	7.953041	2.160824	**3.680559**	1.838647	4.32549
Communication, transportation, storage, postal service	3.883783	1.290201	**3.010216**	8.362216	0.46444
Information transfer, computer, and software areas	1.509671	0.262236	**5.756919**	1.021637	1.47770

continued

Table 5.4 (continued) The Structures and Comparative Gain Measures of Xinjiang's Three Economic Sectors and Industries ($t = 2011$)

Sectors (i)	*Demand Structure $d_i(t)$ (%)*	*Employment Structure $s_i^1(t)$ (%)*	*Comparative Gain of Labor $G_i^1(t)$*	*Investment Structure $s_i^2(t)$ (%)*	*Comparative Gain of Capital $G_i^2(t)$*
# Information Transfer					
Wholesale and retail	5.626281	0.891602	**6.310305**	0.994776	5.65583
Hotels and restaurants	1.178055	0.304194	**3.872712**	0.451775	2.60762
Finance	4.368651	0.849644	**5.141740**	0.108543	40.2480
Real estate/ housing	2.665941	0.293704	**9.076956**	21.08269	0.12645
Leasing and commercial services	0.781991	0.555940	**1.406610**	0.3665	2.13367
Scientific research, technology service, and geological survey	0.882293	0.209789	**4.205625**	0.396472	2.22536
Water conservation, environment, and public facility management	0.507863	0.555940	**0.913521**	4.313739	0.11773
# Public Facility Management					
Residents and other services	0.870039	0.167831	**5.184017**	0.103436	8.41137
Education	3.313439	3.555919	**0.931810**	1.405889	2.35683
Health, safety, and welfare	1.620563	1.552437	**1.043883**	0.744868	2.17564
Social, sports, and entertainment	0.363840	0.335662	**1.083947**	0.384383	0.94655
Public administration and social organizations	6.392841	4.006965	**1.595432**	2.905455	2.20029

the comparative gain of labor of the tertiary economic sector is equal to 0.951274. And all of the rest, that is, (5/6 ≈) 83%, are situated in extreme instability, where the comparative gain of labor of the primary sector is way too low and equal to merely 0.354115. That is roughly equal to one-ninth and one-third of that same indicator of the secondary and the tertiary sectors, respectively. In terms of the capital investment, the comparative gain of the primary sector is much too low, and is equal to 3.41525, while the capital investment in the tertiary sector is in extreme excess (0.79651).

Second, in terms of the employment situation, the relevant data of the 21 industries of the Xingjiang Autonomous Region should contain 42 data points. However, because of lack of raw data, we only collected 30 of them. At the same time, the employment data, be they the overall or those of individual industries, come from *Xingjiang Statistics Yearbook*. They only contain information regarding the urban population, and are not consistent with the overall statistics. So, these data, in bold in Table 5.4, cannot be used in our discussion. Therefore, we cannot address the situation of employment. Instead, we can only look at the situation of investment.

There are only 15 effective data on the comparative gains of capital, while all of them are situated in extremely imbalanced states. In particular, the comparative gains of labor of the following 11 industries are all over 1. Especially, the comparative gain of capital in "finance" is equal to 40.2480, while the value of this indicator of "water conservation, environment, and public facility management" is way too low, equaling only 0.11773. That value is only (0.11773/40.2480 ≈) 0.0029 when compared to that of "finance," (0.11773/8.41137 ≈) 0.014 when compared to that of "residents and other services," (0.11773/5.65583 ≈) 0.021 when compared to that of "wholesale and retail", and (0.11773/4.32549 ≈) 0.027 when compared to that of "construction."

Finance	Residents and other services	Wholesale and retail	Construction
Hotels and restaurants	Education	Scientific research, technology service, and geological survey	Public administration and social organizations
Health, safety, and welfare	Leasing and commercial services	Information transfer, computer, and software areas	

Therefore, in terms of the magnitudes of the comparative gains of capital, the industries of the Xingjiang Autonomous Region with available data can be ranked as follows:

1	2	3	4
Finance	Residents and other services	Wholesale and retail	Construction
5	6	7	8
Hotels and restaurants	Education	Scientific research, technology service, and geological survey	Public administration and social organizations
9	10	11	12
Health, safety, and welfare	Leasing and commercial services	Information transfer, computer, and software areas	Social, sports, and entertainment
13	14	15	
Communication, transportation, storage, postal service	Real estate/housing	Water conservation, environment, and public facility management	

5.2.2 Investment and Employment in Xinjiang's Economic Sectors and Industries

First, in terms of the comparative gains of labor, no matter whether the job seekers, who look for employment opportunities in Xingjiang, are from west China, or other parts of the mainland China, or the rest of the world, the economic sector with the most employment opportunities is the secondary sector.

Without special skills, don't look for employment opportunities in the primary and the tertiary sectors, because these provinces and regions are experiencing excessive labor supplies.

This is the direction of employment.

Second, in terms of the comparative gains of capital, capital investments should be directed into the primary economic sector instead of the secondary and the tertiary sectors, where excessive investment has been done. The following industries

Finance	Residents and other services	Wholesale and retail	Construction
Hotels and restaurants	Education	Scientific research, technology service, and geological survey	Public administration and social organizations
Health, safety, and welfare	Leasing and commercial services	Information transfer, computer, and software areas	

have enjoyed the comparative gains of more than 1, representing relatively more optimal areas of additional investment. However, for the following four industries, excessive investment of capital has been given. Further capital investment in these industries should be curbed and redirected to the previous 11 industries.

Water conservation, environment, and public facility management	Real estate/ housing	Communication, transportation, storage, postal service	Social, sports, and entertainment

In other words, no matter whether the investors, who look for investment opportunities in the Xingjiang Autonomous Region, are from western China, or other parts of the mainland China, or the rest of the world, they should consider the following industries ranked from the most optimal to the least:

That is the direction of capital investment.

1	2	3	4
Finance	Residents and other services	Wholesale and retail	Construction
5	6	7	8
Hotels and restaurants	Education	Scientific research, technology service, and geological survey	Public administration and social organizations
9	10	11	
Health, safety, and welfare	Leasing and commercial services	Information transfer, computer, and software areas	

5.3 Shaanxi Province

Shaanxi is officially part of the northwest China region. It includes portions of the Loess Plateau on the middle reaches of the Yellow River in addition to the Qin Mountains across the southern part of this province. The province borders, going clockwise, with Shanxi in the northeast, Henan in the east, Hubei in the southeast, Chongqing in the south, Sichuan in the southwest, Gansu in the west, Ningxia in the northwest, and Inner Mongolia in the north. For more details, see Figure 5.3.

Geographically, Shaanxi can be described as being part of the Ordos Desert in the north along the border with Inner Mongolia, the Loess Plateau in the central part of the province, the Qin Mountains running east to west in the south central part, and subtropical climate south of the Qin Mountains. In between the Loess Plateau and the Qin Mountains lies the Wei River Valley, a cradle of early Chinese civilization.

Because of its large span in latitude, Shaanxi has a variety of climates. For example, the northern parts, including the Loess Plateau, have either a cold arid or cold semi-arid climates with cold and very dry winters, dry springs and autumns, and hot summers. The area, commonly known as the Guanzhong Plain, is mostly semi-arid, although there are a few areas with a humid subtropical climate, with cool to cold winters, and hot, humid summers that often experience early-season heatwaves. The southern part is much more humid and lies in the humid subtropical zone with more temperate winters and long, hot, humid summers. The

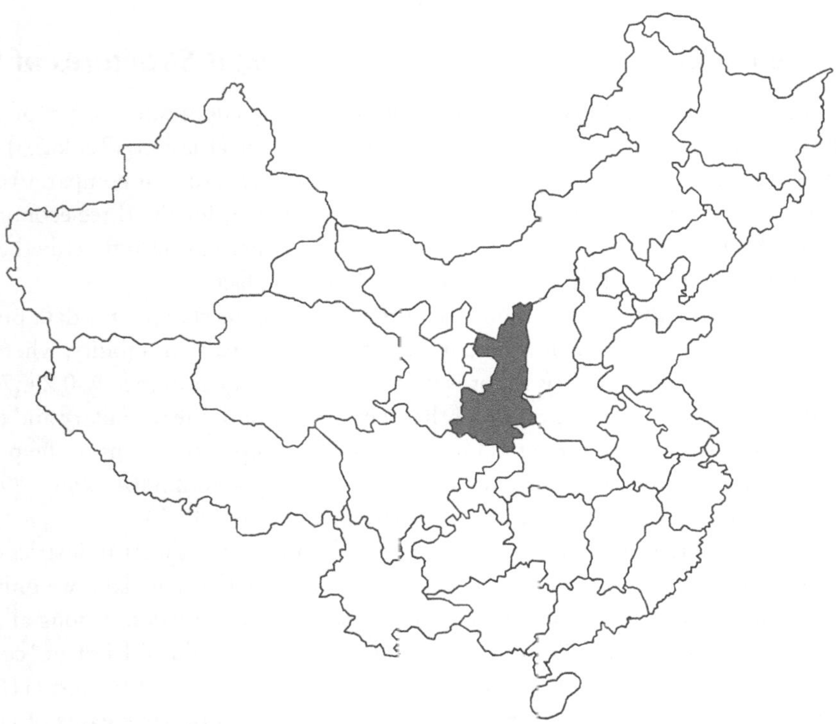

Figure 5.3 The geographic location of Shaanxi.

temperatures in January range from −11 to 3.5°C (12 to 38.3°F); and the temperatures in July range from 21 to 28°C (70 to 82°F).

Historically, Shaanxi is one of the cradles of Chinese civilization with 13 feudal dynasties having their capitals in the province for a time span of more than 1100 years from the Zhou Dynasty to the Tang Dynasty. Xi'an, the province's principal city and current capital, is the eastern terminus of the Silk Road, which leads to Europe, the Arabian Peninsula, and Africa. In the Han Dynasty, the Northern Silk Road was expanded to advance exploration and military purposes to the west. This Northern Silk Road connected Xi'an, the ancient Chinese capital, to the west over the Wushao Ling Pass to Wuwei and emerging in Kashgar before linking to ancient Parthia. Many Chinese from Shaanxi immigrated to Xinjiang during the Qing Dynasty. During modern times, the end of the short-lived Jiangxi Soviet signaled the beginning of the Long March of the communist army and the Chinese communists to the Shaanxi Soviet located at Yan'an.

Economically, as of the mid-nineteenth century, Shaanxi was known for its exports of animal skins, wine, liquor, and musk. It also imported animal skins and watches from Europe. Currently, the fossil fuel and high technology sectors compose the two major industries in the province. As the home of several leading universities and research institutes in western China, Shaanxi plays a major role in China's burgeoning aircraft and aerospace industry, producing more than 50% of the R&D and equipment for China's domestic commercial air industry. The nominal GDP in 2011 was ¥1.239 billion (US$196.7 billion) and the GDP per capita was ¥21,729 (US$3179).

For additional reading, refer to Edmund (1837); Millward (1998); Ministry of Commerce (2013); National Bureau of Statistics of China (2013); Qian (1995).

5.3.1 Current State of Investment and Employment Structures of Shaanxi

All that is necessary for us to do is the computations of the gross domestic outputs of all the economic sectors and industries and the investment and employment situations (Table 5.5) of Shaanxi Province and the relevant comparative gains. By comparing the resultant comparative gains, we can then obtain the current states of investment and employment for the three economic sectors and industries. All the detailed computations are similar to what has been described earlier. The results are given in Table 5.6. Table 5.6 indicates the following facts:

First, in terms of the three major economic sectors, among the six relevant data points of the three sectors of Shaanxi Province, all of them are situated in extreme instability, where the comparative gain of labor of the primary sector is way too low and equal to merely 0.243749. That is not even reaching one-eighth and one-fifth of that same indicator of the secondary and the tertiary sectors, respectively. In terms of the capital investment, the comparative gains of the primary and the secondary sectors are much too low, and are respectively equal to 2.84018 and 1.49227, while the capital investment in the tertiary sector is in extreme excess (0.58587).

Second, in terms of the employment situation, the relevant data of the 21 industries of Shaanxi Province should contain 42 data points. However, because of lack of raw data, we only collected 12 of them. So, we can only base our discussion on these available statistics. Among all the industries, only one is basically in equilibrium, where the comparative gain of labor in "communication, transportation, storage, and postal service" is equal to 1.023212. All the rest ($11/12 \geq 91\%$) are situated in extremely imbalanced states. In particular, the comparative gains of labor of the following three industries are all over 1. Especially, the comparative gain of labor in "finance" is

Table 5.5 The GDPs, Total Investments in Fixed Properties, and Employment Situation of Shaanxi's Three Economic Sectors and Industries (2011)

Sectors (i)	*Total Output of Sector (0.1 Billion Yuan)*	*Total investment on Fixed Properties (0.1 Billion Yuan)*	*Employment (10,000 People)*
Total	12512.3	10023.5264	2059
Primary sector	1220.9	344.3637	824.2436
Secondary sector	6935.59	3723.2231	585.335421
Tertiary sector	4355.81	5955.93957	649.532171
Industries (*i*)			
Agriculture, forest, animal husbandry, fishery		344.3637	824.2436
Mining		1230.022	39.0785283
Manufacturing		1867.3484	301.118967
Production/supply of electricity, gas, and water		308.3759	10.5292373
Construction	1077.67	317.4768	234.608689
Communication, transportation, storage, postal service	552.54	1050.7907	88.8622406
Information transfer, computer, and software areas		62.5043	20.9462712
# Information Transfer			
Wholesale and retail	1036.35	259.9364	191.569221
Hotels and restaurants	266.92	126.656	97.4438177
Finance	432.11	2.8988	14.3238
Real estate/housing	398.03	665.5898	13.8656787
Leasing and commercial services		110.508	13.4249659
Scientific research, technology service, and geological survey		56.0286	17.6843893
Water conservation, environment, and public facility management		1030.4529	9.03525398

continued

Table 5.5 (continued) The GDPs, Total Investments in Fixed Properties, and Employment Situation of Shaanxi's Three Economic Sectors and Industries (2011)

Sectors (i)	*Total Output of Sector (0.1 Billion Yuan)*	*Total investment on Fixed Properties (0.1 Billion Yuan)*	*Employment (10,000 People)*
# Public Facility Management			
Residents and other services		25.4434	23.7868665
Education		155.351	64.8269218
Health, safety, and welfare		119.3129	23.9707042
Social, sports, and entertainment		54.6902	7.27124806
Public administration and social organizations	12512.3	10023.5264	2059

Source: Data are from *China Statistics Yearbook* (2012).

Table 5.6 The Structures and Comparative Gain Measures of Shaanxi's Three Economic Sectors and Industries ($t = 2011$)

Sectors (i)	*Demand Structure $d_i(t)$ (%)*	*Employment Structure $s_i^1(t)$ (%)*	*Comparative Gain of Labor $G_i^1(t)$*	*Investment Structure $s_i^2(t)$ (%)*	*Comparative Gain of Capital $G_i^2(t)$*
Total	100	100	—	100	—
Primary sector	9.757599	40.03126	0.243749	3.43555	2.84018
Secondary sector	55.43018	28.42814	1.949835	37.1448	1.49227
Tertiary sector	34.81222	31.54600	1.103538	59.4196	0.58587
Industries (*i*)					
Agriculture, forest, animal husbandry, fishery		40.03126		3.43555	
Mining		1.897937		12.2714	
Manufacturing		14.62452		18.6297	
Production/supply of electricity, gas, and water		0.511376		3.07652	
Construction	8.612885	11.39430	0.755894	3.16732	2.71930
Communication, transportation, storage, postal service	4.415975	4.315796	1.023212	10.4832	0.42124

Table 5.6 (continued) The Structures and Comparative Gain Measures of Shaanxi's Three Economic Sectors and Industries (t = 2011)

Sectors (i)	*Demand Structure $d_i(t)$ (%)*	*Employment Structure $s_i^1(t)$ (%)*	*Comparative Gain of Labor $G_i^1(t)$*	*Investment Structure $s_i^2(t)$ (%)*	*Comparative Gain of Capital $G_i^2(t)$*
Information transfer, computer, and software areas		1.017303		0.62358	
# Information Transfer					
Wholesale and retail	8.282650	9.303993	0.890225	2.59326	3.19391
Hotels and restaurants	2.133261	4.732580	0.450761	1.26359	1.68826
Finance	3.453486	0.695668	4.964269	0.02892	119.415
Real estate/housing	3.181110	0.673467	4.723485	6.64028	0.47906
Leasing and commercial services		0.652014		1.10249	
Scientific research, technology service, and geological survey		0.858882		0.55897	
Water conservation, environment, and public facility management		0.438818		10.2803	
# Public Facility Management					
Residents and other services		1.155263		0.25384	
Education		3.148466		1.54986	
Health, safety, and welfare		1.164192		1.19033	
Social, sports, and entertainment		0.353145		0.54562	
Public administration and social organizations		2.964507		5.01595	

even equal to 4.964269, which is equal to 94.964269/0.450761 times of the lowest value of this indicator that is reached by "residents and other services."

Finance	Real estate/housing	Communication, transportation, storage, postal service

At the same time, the imbalance in terms of the comparative gains of capital seems to be way too severe. The comparative gains of capital in the following four industries are all over 1. In particular, the comparative gain of capital of "finance" is even equal to 119.415, while the values of this indicator of "real estate and housing" and "communication, transportation, storage, postal service" are too low, equaling to merely 0.47906 and 0.42124, which are only (0.47906/119.415 ≈) 0.004 and (0.42124/119.415 ≈) 0.0037, respectively, when compared to that of "finance."

Finance	Wholesale and retail	Construction	Hotels and restaurants

So, in terms of the magnitudes of the comparative gains of labor, the industries of Shaanxi Province with available data can be ranked as follows:

1	2	3
Finance	Real estate/housing	Communication, transportation, storage, postal service
4	5	6
Wholesale and retail	Construction	Hotels and restaurants

And, in terms of the magnitudes of the comparative gains of capital, the industries of Shaanxi Province with available data can be ranked as follows:

1	2	3
Finance	Wholesale and retail	Construction
4	5	6
Hotels and restaurants	Communication, transportation, storage, postal service	Real estate/housing

5.3.2 Investment and Employment in Shaanxi's Economic Sectors and Industries

First, in terms of the comparative gains of labor, the values of this indicator of the following industries

Finance	Real estate/housing	Communication, transportation, storage, postal service

are over 1, representing relatively optimal industries for employment. In all the other three industries, there has been excessive supply of labor so that appropriate control on the increase of labor into these industries should be imposed. Specifically, the excessive labor in the following industries should be transferred into the previous three industries.

Hotels and restaurants	Construction	Wholesale and retail

In other words, no matter whether the job seekers, who look for employment opportunities in the Shaanxi Province, are from western China, or other parts of the mainland China, or the rest of the world, they should first consider the secondary and tertiary economic sectors with the relevant industries ranked as follows:

Finance	Real estate/housing	Communication, transportation, storage, postal service

Without special skills, don't look for employment opportunities in the primary economic sector and other industries, because those areas are experiencing excessive labor supplies.

That is the direction of employment.

Second, in terms of the comparative gains of capital, the following four industries have enjoyed the comparative gains of more than 1, representing relatively more optimal areas of additional investment.

Finance	Wholesale and retail	Construction	Hotels and restaurants

However, for the following two industries, excessive investment of capital has been given. Further capital investment in these industries should be curbed and redirected to the previous four industries.

Communication, transportation, storage, postal service	Real estate/housing

In other words, no matter whether the investors, who look for investment opportunities in the Shaanxi Province, are from western China, or other parts of the mainland China, or the rest of the world, they should consider the following industries ranked from the most optimal to the least:

Finance	Wholesale and retail	Construction	Hotels and restaurants

That is the direction of capital investment.

Third, by combining the two criteria discussed above, it follows that the industry in Shaanxi Province that is most optimal for both investment and employment is "finance," because the comparative gains of labor and capital of this industry is over 1. And, the worst choice is the industry of "hotels and restaurants."

5.4 Sichuan Province

Sichuan is located in the southwest of the country with its capital in Chengdu, a key economic center of western China. It borders with Qinghai in the northwest, Gansu in the north, Shaanxi in the northeast, Chongqing in the east, Guizhou in the southeast, Yunnan in the south, and the Tibet Autonomous Region in the west. For more details, see Figure 5.4.

The province, within its borders, consists of two very geographically distinct parts. The eastern part is mostly within the fertile Sichuan basin, shared by Sichuan and Chongqing Municipality. The western part consists of the numerous mountain ranges forming the easternmost part of the Qinghai-Tibet Plateau, which are commonly known as the Hengduan Mountains. Lesser mountain ranges surround the Sichuan Basin in the north, the east, and the south. Plate tectonics formed the Longmen Shan fault, which runs under the north-easterly mountain location of the 2008 Sichuan earthquake. The Yangtze River and its tributaries flow through the mountains of western Sichuan and the Sichuan Basin. Therefore, the province is located in the upstream of such great cities found along the Yangtze River further to the east as Chongqing, Wuhan, Nanjing, and Shanghai. One of the major tributaries of the Yangtze within the province is the Min River of central Sichuan. The four main rivers of Sichuan, as the name Sichuan means literally, are the Jaling Jiang River, Tuo Jiang River, Yalong Jiang River, and Jinsha Jiang River.

Because of the significant differences in the terrain, the province experiences highly variable climate. Even so, it is generally influenced by strong monsoons with rainfall heavily concentrated in summer. The Sichuan Basin, within which Chengdu is located, in the eastern half of the province experiences a humid subtropical climate with long, hot, humid summers and short, mild, dry and cloudy winters. The western mountainous areas have a cooler but sunnier climate with cold

Figure 5.4 The geographic location of Sichuan.

winters and mild summers. The region is geologically active with landslides and earthquakes. The average elevation of the province ranges from 2000 to 3500 m; and the average temperature from 0 to 15°C.

Historically, the region represents the cradle of unique local civilizations that coincided with the later years of the Shang Dynasty and Zhou Dynasty in north China. The existence of the early Kingdom of Shu was recorded in history as an ally of the Zhou who defeated the Shang. The area was on the trade route from the Yellow River Valley to foreign countries of the southwest, especially India. The leaders of the Qin Dynasty, which was based in present day Gansu and Shaanxi, realized that the province's military importance as well as its commercial and agricultural significance. Qin armies finished their conquest of the kingdoms of Shu and Ba by 316 BC. Under the control of the Qin Dynasty, Li Bing engineered the Dujiangyan Irrigation System to control the Min River, a tributary of the Yangtze River. This large-scale hydraulic system was composed of movable weirs which could be adjusted for high or low water flow, according to the season, to either provide irrigation or prevent floods. The increased agricultural output and taxes made the area a source of provisions and men for Qin's unification of China. Following the decline of the Han Dynasty in the second century, the Sichuan Basin, surrounded by mountains and easily defensible, became a place for upstart generals to found kingdoms that challenged the authority of Yangtze valley emperors of China. In 221, during the era of the Three Kingdoms, following the fall of the Eastern Han Dynasty, Liu Bei founded the southwest kingdom of Shu-Han (221–263) in parts of Sichuan, Guizhou, and Yunnan with Chengdu as its capital. In 263, the Jin Dynasty from north China conquered the Kingdom of Shu-Han as its first step on the path to unify China.

The province came under the firm control of the central government during the Sui Dynasty, and regained its previous political and cultural prominence during the subsequent Tang Dynasty. During the Song Dynasty (960–1279), Sichuan saw a cultural revival. In the twelfth and thirteenth centuries, the Southern Song Dynasty established coordinated defenses against the Yuan Dynasty in Sichuan and Xiangyang, which were finally broken through in 1273. During the Ming Dynasty, major architectural works were created in Sichuan. In the middle of the seventeenth century, the peasant rebel leader Zhang Xianzhong (1606–1646) from Yan'an, Shaanxi Province, conquered Sichuan. In response to the resistance from the locals, he massacred a large native population. As a result, the population of Sichuan fell sharply, which required a massive resettlement of people from neighboring Huguang province (modern Hubei and Hunan). During the Qing Dynasty, Sichuan was merged with Shaanxi and Shanxi to become "Shenzhuan" during 1680–1731 and 1735–1748. The current borders of Sichuan were established in the early eighteenth century. In the early twentieth century, the newly founded Republic of China established the Chuanbian Special Administrative District by acknowledging the unique culture of the region. The Special District later became the province of Xikang by incorporating the areas inhabited by Yi, Tibetan, and Qiang ethnic minorities to its west. When Beijing, Shanghai, Nanjing, and Wuhan were all occupied by the Japanese during the Second Sino-Japanese War, the capital of the Republic of China was temporarily relocated to Chongqing, then a major city in Sichuan. The Second Sino-Japanese War was soon followed by the Chinese Civil War. When the cities of East China fell to the Communists one after another, the Kuomintang government again tried to make Sichuan its stronghold on the mainland. Chiang Kai-Shek himself flew to Chongqing from Taiwan in November 1949 to lead the defense. But the same month Chongqing fell to the Communists, followed by Chengdu on December 10, 1949.

The People's Republic of China split Sichuan into four areas and separated out Chongqing municipality. Sichuan was reconstituted in 1952, with Chongqing added in 1954, while the former Xikang province was split between Tibet in the west and Sichuan in the east. In 1978, when Deng Xiaoping took power, he employed Sichuan as one of the first provinces to undergo limited experimentation with market economy. In 1997, the city of Chongqing and the surrounding counties of Fuling and Wanxian were split off from Sichuan into the new Chongqing Municipality, which was formed to spearhead China's effort to economically develop its western provinces.

Economically, Sichuan is one of the major agricultural production bases of China. Its cash crops include citrus fruits, sugarcane, sweet potatoes, peaches, and grapeseeds. The province is rich in mineral resources totaling more than 132 kinds, including vanadium, titanium, and lithium. The Panxi region alone possesses 13.3% of the reserves of iron, 93% of titanium, 69% of vanadium, and 83% of the cobalt of the whole country. It also possesses China's largest proven natural gas reserves. At the same time, Sichuan is one of the major industrial centers of China. Other than heavy industries, such as coal, energy, iron, and steel, the province has also developed a light industrial sector that comprises building materials, wood processing, food and silk processing. The wine production of the province in 2000 accounted for 21.9% of the total production of China. The auto industry is an important key sector of the machinery industry in Sichuan, with most of the auto manufacturing companies located in Chengdu, Mianyang, Nanchong, and Luzhou. Other important industries in the province include aerospace and defense (military) industries. For example, the Xichang Satellite Launch Center is located near the city of Xichang. The particular landscape and rich historical relics in Sichuan have also made the province a major center for tourism. The nominal GDP of the province for 2011 was ¥2.15 trillion (US$340 billion), which is equivalent to ¥17,380 (US$2545) per capita.

For additional reading, refer to Ministry of Commerce (2013); National Bureau of Statistics of China (2013); Sage (2006); Boland-Crewe and Lea (2004); Parsons (1957); Dai (2009); and Roberts (1837).

5.4.1 Current State of Investment and Employment Structures of Sichuan

All that is necessary for us to do is the computations of the gross domestic outputs of all the economic sectors and industries and the investment and employment situations (Table 5.7) of Sichuan Province and the relevant comparative gains. By comparing the resultant comparative gains, we can then obtain the current states of investment and employment for the three economic sectors and industries. All the detailed computations are similar to what has been described earlier. The results are given in Table 5.8. Table 5.8 indicates the following facts:

First, in terms of the three major economic sectors, among the six relevant data points of the three sectors of Sichuan Province, we are only able to collect 4. Only one of the four data points is basically in equilibrium, where the comparative gain of labor of the tertiary sector is 1.042447, and the other three are situated in extreme instability, where the comparative gain of labor of the primary sector is way too low and equal to merely 0.332304. That is not even reaching one-sixth and one-third of that same indicator of the secondary and the tertiary sectors, respectively. In terms of the capital investment, we do not have any data related to the primary and the secondary sectors, while the capital investment in the tertiary sector is in extreme excess (0.56489).

Second, in terms of the employment situation, the relevant data of the 21 industries of Sichuan Province should contain 42 data points. However, because of lack of raw data, we only collected 12 of them. So, we can only base our discussion on these available statistics. Among all the industries, only one is basically in equilibrium, where the comparative gain of labor in "communication, transportation, storage, postal service" is equal to 0.953901. All the rest ($11/12 \geq 91\%$) are situated

Table 5.7 The GDPs, Total Investments in Fixed Properties, and Employment Situation of Sichuan's Three Economic Sectors and Industries (2011)

Sectors (i)	*Total Output of Sector (0.1 Billion Yuan)*	*Total Investment on Fixed Properties (0.1 Billion Yuan)*	*Employment (10,000 People)*
Total	21026.68	15124.09	4785.47
Primary sector	2983.51		2043.364
Secondary sector	11029.13		1210.781
Tertiary sector	7014.04	**8931**	1531.3283
Industries (*i*)			
Agriculture, forest, animal husbandry, fishery		280.29	2043.364
Mining		573.43	111.7498
Manufacturing		3947.29	554.7925
Production/supply of electricity, gas, and water		1211.21	27.2749
Construction	1538.08	51.66	516.9638
Communication, transportation, storage, postal service	638.76	2141.41	152.4011
Information transfer, computer, and software areas		79.87	33.2062
# Information Transfer			
Wholesale and retail	1186.58	225.23	424.1979
Hotels and restaurants	562.63	155.31	210.8148
Finance	868.15	31.78	23.7772
Real estate/housing	620.62	4013.79	29.5479
Leasing and commercial services		134.58	52.8901

continued

Table 5.7 (continued) The GDPs, Total Investments in Fixed Properties, and Employment Situation of Sichuan's Three Economic Sectors and Industries (2011)

Sectors (i)	*Total Output of Sector (0.1 Billion Yuan)*	*Total Investment on Fixed Properties (0.1 Billion Yuan)*	*Employment (10,000 People)*
Scientific research, technology service, and geological survey		47.06	18.6207
Water conservation, environment, and public facility management		1377.71	19.8168
# Public Facility Management			
Residents and other services		150.27	265.63
Education		208.6	100.5912
Health, safety, and welfare		123.99	60.2888
Social, sports, and entertainment		154.07	33.202
Public administration and social organizations		216.54	106.3436

Source: Data are from *China Statistics Yearbook* (2012) (XLS, DOC).

in extremely imbalanced states. In particular, the comparative gains of labor of the following two industries are all over 1. Especially, the comparative gain of labor in "finance" is even equal to 8.30975, which is more than (8.30975/0.607401 ≈) 14 times of the lowest value of this indicator that is reached by "hotels and restaurants."

Finance	Real estate/housing

At the same time, the imbalance in terms of the comparative gains of capital seems to be way too severe. The comparative gains of capital in the following four industries are all over 1. In particular, the comparative gain of capital of "construction" is equal to 21.4152, while the value of this indicator of "real estate and housing" is too low, equaling to merely 0.111217, which is only (0.111217/21.4152 ≈) 0.0052 when compared to that of "construction," (0.111217/19.6490 ≈) 0.0056 when compared to that of "finance," (0.111217/3.78939 ≈) 0.029 when compared to that of "wholesale and retail," and (0.111217/2.60569 ≈) 0.019 when compared to that of "hotels and restaurants."

Construction	Finance	Wholesale and retail	Hotels and restaurants

Table 5.8 The Structures and Comparative Gain Measures of Sichuan's Three Economic Sectors and Industries ($t = 2011$)

Sectors (i)	*Demand Structure* $d_i(t)$ *(%)*	*Employment Structure* $s_i^1(t)$ *(%)*	*Comparative Gain of Labor* $G_i^1(t)$	*Investment Structure* $s_i^2(t)$ *(%)*	*Comparative Gain of Capital* $G_i^2(t)$
Total	100	100	—	100	—
Primary sector	14.18916	42.69934	0.332304		
Secondary sector	52.45303	25.30119	2.073144		
Tertiary sector	33.35781	31.99954	1.042447	59.05149	0.56489
Industries (*i*)					
Agriculture, forest, animal husbandry, fishery		42.69934		1.853269	
Mining		2.335190		3.791501	
Manufacturing		11.59327		26.09936	
Production/supply of electricity, gas, and water		0.569952		8.008482	
Construction	7.314897	10.80278	0.677131	0.341574	21.4152
Communication, transportation, storage, postal service	3.037855	3.184663	0.953901	14.15893	0.21455
Information transfer, computer, and software areas		0.693896		0.528098	
# Information Transfer					
Wholesale and retail	5.643211	8.864239	0.636623	1.489214	3.78939
Hotels and restaurants	2.675791	4.405310	0.607401	1.026905	2.60569
Finance	4.128802	0.496862	8.309750	0.210128	19.6490
Real estate/ housing	2.951583	0.617450	4.780277	26.53905	0.111217

continued

Table 5.8 (continued) The Structures and Comparative Gain Measures of Sichuan's Three Economic Sectors and Industries ($t = 2011$)

Sectors (i)	*Demand Structure $d_i(t)$ (%)*	*Employment Structure $s_i^1(t)$ (%)*	*Comparative Gain of Labor $G_i^1(t)$*	*Investment Structure $s_i^2(t)$ (%)*	*Comparative Gain of Capital $G_i^2(t)$*
Leasing and commercial services		1.105223		0.889839	
Scientific research, technology service, and geological survey		0.389109		0.311159	
Water conservation, environment, and public facility management		0.414104		9.109375	
# Public Facility Management					
Residents and other services		5.550761		0.993580	
Education		2.102013		1.379257	
Health, safety, and welfare		1.259830		0.819818	
Social, sports, and entertainment		0.693809		1.018706	
Public administration and social organizations		2.222219		1.431756	

So, in terms of the magnitudes of the comparative gains of labor, the industries of the Sichuan Province with available data can be ranked as follows:

1	2	3
Finance	Real estate/housing	Communication, transportation, storage, postal service
4	5	6
Construction	Wholesale and retail	Hotels and restaurants

And, in terms of the magnitudes of the comparative gains of capital, the industries of the Sichuan Province with available data can be ranked as follows:

1	2	3
Construction	Finance	Wholesale and retail
4	5	6
Hotels and restaurants	Communication, transportation, storage, postal service	Real estate/housing

5.4.2 Investment and Employment in Sichuan's Economic Sectors and Industries

First, in terms of the comparative gains of labor, the values of this indicator of the following two industries

Finance	Real estate/housing

are over 1, representing relatively optimal industries for employment. In all the other four industries, there has been excessive supply of labor so that appropriate control on the increase of labor into these industries should be imposed. Specifically, the excessive labor in the following industries should be transferred into the previous two industries.

Hotels and restaurants	Whole sale and retail	Construction	Communication, transportation, storage, postal service

In other words, no matter whether the job seekers, who look for employment opportunities in the Sichuan Province, are from central China, or other parts of the mainland China, or the rest of the world, they should first consider the secondary and the tertiary economic sectors with the relevant industries ranked as follows:

Finance	Real estate/housing

Without special skills, don't look for employment opportunities in the primary economic sector and other industries, because those areas are experiencing excessive labor supplies.

That is the direction of employment.

Second, in terms of capital, the following four industries have enjoyed the comparative gains of more than 1, representing relatively more optimal areas of additional investment.

Construction	Finance	Wholesale and retail	Hotels and restaurants

However, for the following two industries, excessive investment of capital has been given. Further capital investment in these industries should be curbed and redirected to the previous four industries.

Real estate/housing	Communication, transportation, storage, postal service

In other words, no matter whether the investors, who look for investment opportunities in the Sichuan Province, are from western China, or other parts of the mainland China, or the rest of the world, they should consider the following industries ranked from the most optimal to the least:

1	2	3	4
Construction	Finance	Wholesale and retail	Hotels and restaurants

That is the direction of capital investment.

Third, by combining the two criteria discussed above, it follows that the industry in Sichuan Province that is most optimal for both investment and employment is "finance," because the comparative gains of labor and capital of each of these industries are over 1. The worst choice is the industry of "Communication, transportation, storage, postal service".

5.5 Guizhou Province

Guizhou is located in the southwestern part of the country, with Guiyang as its capital. It adjoins Sichuan Province and Chongqing Municipality in the north, Yunnan Province in the west, Guangxi Province in the south, and Hunan Province in the east. Although Guizhou is a mountainous province, it is hilly in the west with the eastern and southern portions relatively flat. The western part is part of the Yunnan-Guizhou Plateau. The climate in Guizhou is subtropical and humid with few seasonal changes. The annual average temperature of the region ranges from 10 to 20°C with January temperatures in the range of 1–10°C and July temperatures in the range of 17–28°C. For more details, see Figure 5.5.

The mountains of Guizhou that border with Guangxi and Hunan have been identified as one of the eight plant diversity hotspots in China. The main ecosystem types include an evergreen broad-leaved forest, coniferous and broad-leaved mixed forest, and a montane elfin forest. The plant species that are endemic in this region include *Abies ziyuanensis*, *Cathaya argyrophylla*, and *Keteleeria pubescens*. The animals that are only known from Guizhou include Leishan moustache toad, Kuankuoshui salamander, Shuicheng salamander, Guizhou salamander, and Zhijin warty newt. The Caohai Lake area is a National Nature Reserve and an important Bird Area identified by BirdLife International.

Historically, from around 1046 BCE to the time of the Qin Dynasty, northwest Guizhou was part of the State of Shu. During the period of Warring States, the State of Chu conquered the area. During the Han Dynasty, the Lingnan province was established over the region. During the Three Kingdoms period, parts of Guizhou were governed by the State of Shu Han based in Sichuan, then by Cao Wei (220–265), and later by the Jin Dynasty (265–420). During the eighth and ninth centuries, the soldiers of the Tang Dynasty moved into Guizhou. Although the Song Dynasty did not govern the lands south of the Dadu River, the Song's commercial, urban, and Confucian culture greatly influenced the culture of the area. During the Ming Dynasty, Guizhou was formally made a province in 1413. After the Second Opium War during the Qing Dynasty, criminal triads set up shops in Guangxi and Guizhou to sell British opium. The Taiping Rebels took control of Guizhou for a brief period time, and then they were suppressed by the Qing forces. During 1934–1935, while the province was formally ruled by the Guomindang warlord Wang Jialie, the communist Zunyi Conference in Guizhou established

Figure 5.5 The geographic location of Guizhou.

Mao Zedong as the leader of the Communist Party. After 1949, the People's Republic of China promoted the relocation of heavy industry into inland provinces, such as Guizhou, to better protect them from potential Soviet and American attacks.

Economically, Guizhou is a relatively poor and economically undeveloped province, although it is rich in natural, cultural, and environmental resources. Its natural industry includes timber and forestry. It is also the third largest producer of tobacco in China with the established brand Guizhou Tobacco. Other industries of the province include energy (electricity generation) and mining, especially in coal, limestone, arsenic, gypsum, and oil shale. The nominal GDP of Guizhou for 2012 was ¥680.22 billion (US$107.758 million), and its per capita GDP ¥19,566 (US$3100 USD).

For additional readings, refer to Hutchings (2003); Jin, Ma and Bender (2006); Maygew, Miller and English (2002); Ministry of Commerce (2013); National Bureau of Statistics of China (2013); and Roberts (1837).

5.5.1 Current State of Investment and Employment Structures of Guizhou

All that is necessary for us to do is the computations of the gross domestic outputs of all the economic sectors and industries and the investment and employment situations (Table 5.9) of Guizhou Province and the relevant comparative gains. By comparing the resultant comparative gains, we can then obtain the current states of investment and employment for the three economic sectors and industries. All the detailed computations are similar to what has been described earlier. The results are given in Table 5.10. Table 5.10 indicates the following facts:

Table 5.9 The GDPs, Total Investments in Fixed Properties, and Employment Situation of Guizhou's Three Economic Sectors and Industries (2011)

Sectors (i)	*Total Output of Sector (0.1 Billion Yuan)*	*Total Investment on Fixed Properties (0.1 Billion Yuan)*	*Employment (10,000 People)*
Total	5701.84	5101.55	1792.8
Primary sector	726.22	163.20	1194.39
Secondary sector	2194.33	1852.10	215.86
Tertiary sector	2781.29	3086.25	382.55
Industries (*i*)			
Agriculture, forest, animal husbandry, fishery		129.18	**1.9**
Mining		522.83	**13.5**
Manufacturing		952.08	**39.9**
Production/supply of electricity, gas, and water		346.34	**8.3**
Construction	365.13	30.86	**28.3**
Communication, transportation, storage, postal service	590.91	781.87	**9.7**
Information transfer, computer, and software areas		46.19	**2.7**
# Information Transfer			
Wholesale and retail	448.77	49.77	**10.6**
Hotels and restaurants	224.4	52.66	**2.9**
Finance	297.27	4.62	**7.2**
Real estate/housing	160.3	1045.42	**4.3**
Leasing and commercial services		13.38	**3.1**
Scientific research, technology service, and geological survey		10.61	**4.6**
Water conservation, environment, and public facility management		680.77	**3.8**

Table 5.9 (continued) The GDPs, Total Investments in Fixed Properties, and Employment Situation of Guizhou's Three Economic Sectors and Industries (2011)

Sectors (i)	*Total Output of Sector (0.1 Billion Yuan)*	*Total Investment on Fixed Properties (0.1 Billion Yuan)*	*Employment (10,000 People)*
# Public Facility Management			
Residents and other services		10.55	**1.3**
Education		85.85	**42.5**
Health, safety, and welfare		39.17	**13.1**
Social, sports, and entertainment		31.94	**1.9**
Public administration and social organizations		58.27	**41.4**

Source: The data in bold are from *China Statistics Yearbook* (2012), and the rest are from *Guizhou province's Statistics Yearbook* (2012).

First, in terms of the three major economic sectors, the six relevant data points of the three sectors are all situated in extreme instability. Specifically, the comparative gain of labor of the primary sector is way too low and is equal to merely 0.191178. That is not even reaching one-seventeenth and one-twelfth of that same indicator of the secondary and the tertiary sectors, respectively. In terms of the capital investment, the comparative gains of the primary and the secondary sectors are much too low, and are respectively equal to 3.98139 and 1.06005, while the capital investment in the tertiary sector is in extreme excess (0.80631).

Second, in terms of the employment situation, the relevant data of the 21 industries of Guizhou Province should contain 42 data points. However, because of lack of raw data, we only collected 12 of them. Even so, the employment data in Table 5.9 in bold represent those of urban residents. Since these data do not agree with the overall statistics, they cannot be used in our discussion. Hence, we have only six effective data on the comparative gains of capital. And, all of these data are situated in extremely imbalanced states.

As for the situation of investment, the comparative gains of capital in the following four industries are all over 1. In particular, the comparative gains of capital of "finance" and "construction" are over 10, equaling respectively 57.5718 and 10.5879, which are (57.5718/0.13719 ≈) 420 times and (10.5879/0.13719 ≈) 77 times of the lowest value of this indicator reached by "real estate and housing."

Finance	Construction	Wholesale and retail	Hotels and restaurants

Therefore, because of lack of effective data, we will not address the problem of employment for the Guizhou Province. And in terms of the magnitudes of the comparative gains of capital, the industries of the Guizhou Province with available data can be ranked as follows:

1	2	3
Finance	Construction	Wholesale and retail
4	5	6
Hotels and restaurants	Communication, transportation, storage, postal service	Real estate/housing

Table 5.10 The Structures and Comparative Gain Measures of Guizhou's Three Economic Sectors and Industries ($t = 2011$)

Sectors (i)	*Demand Structure* $d_i(t)$ *(%)*	*Employment Structure* $s_i^1(t)$ *(%)*	*Comparative Gain of Labor* $G_i^1(t)$	*Investment Structure* $s_i^2(t)$ *(%)*	*Comparative Gain of Capital* $G_i^2(t)$
Total	100	100	—	100	—
Primary sector	12.7365	66.62149	0.191178	3.199028	3.98139
Secondary sector	38.48459	12.04038	3.196293	36.30465	1.06005
Tertiary sector	48.77882	21.33813	2.285993	60.49632	0.80631
Industries (*i*)					
Agriculture, forest, animal husbandry, fishery		0.105979		2.532111	
Mining		0.753012		10.24839	
Manufacturing		2.225569		18.66258	
Production/supply of electricity, gas, and water		0.462963		6.788894	
Construction	6.403722	1.578536	**4.056747**	0.604816	10.5879
Communication, transportation, storage, postal service	10.36350	0.541053	**19.15431**	15.32604	0.67620
Information transfer, computer, and software areas		0.150602		0.905484	
# Information Transfer					
Wholesale and retail	7.870617	0.591254	**13.31174**	0.975586	8.06758
Hotels and restaurants	3.935572	0.161758	**24.32998**	1.032294	3.81245
Finance	5.213580	0.401606	**12.98181**	0.090558	57.5718
Real estate/housing	2.811373	0.239848	**11.72146**	20.49212	0.13719

Table 5.10 (continued) The Structures and Comparative Gain Measures of Guizhou's Three Economic Sectors and Industries ($t = 2011$)

Sectors (i)	*Demand Structure $d_i(t)$ (%)*	*Employment Structure $s_i^1(t)$ (%)*	*Comparative Gain of Labor $G_i^1(t)$*	*Investment Structure $s_i^2(t)$ (%)*	*Comparative Gain of Capital $G_i^2(t)$*
Leasing and commercial services		0.172914		0.262345	
Scientific research, technology service, and geological survey		0.256582		0.208026	
Water conservation, environment, and public facility management		0.211959		13.34444	
# Public Facility Management					
Residents and other services		0.072512		0.20685	
Education		2.370593		1.682892	
Health, safety, and welfare		0.730701		0.767731	
Social, sports, and entertainment		0.105979		0.626106	
Public administration and social organizations		2.309237		1.142153	

5.5.2 Investment and Employment in Guizhou's Economic Sectors and Industries

First, in terms of the comparative gains of labor, no matter whether the job seekers, who look for employment opportunities in the Guizhou Province, are from western China, or other parts of the mainland China, or the rest of the world, they should first consider the secondary and the tertiary economic sectors. As for specific industries, because of lack of data, we cannot address further.

Without special skills, don't look for employment opportunities in the primary economic sector, because that area is experiencing excessive labor supplies.

That is the direction of employment.

Second, in terms of capital, the primary and the secondary economic sectors, and following four industries have enjoyed the comparative gains of more than 1, representing relatively more optimal areas of additional investment.

Finance	Construction	Wholesale and retail	Hotels and restaurants

However, for the tertiary sector and specifically following two industries, excessive investment of capital has been given. Further capital investment in these industries should be curbed and redirected to the previous sectors and industries.

Real estate/housing	Communication, transportation, storage, postal service

In other words, no matter whether the investors, who look for investment opportunities in Guizhou Province, are from western China, or other parts of the mainland China, or the rest of the world, they should consider the following industries ranked from the most optimal to the least:

1	2	3	4
Finance	Construction	Wholesale and retail	Hotels and restaurants

That is the direction of capital investment.

5.6 Qinghai Province

Qinghai is located in the northwest of the country and mostly on the Qinghai-Tibet Plateau, and borders with Gansu in the northeast, the Xinjiang in the northwest, Sichuan in the southeast, and the Tibet Autonomous Region in the southwest. It has long been a melting pot for a variety of ethnic groups, including the Han, Tibetans, Hui, Tu, Mongols, and Salars. It is the fourth largest province-level administrative division of China by area, but has the third smallest population. As a province, Qinghai was established in 1928 in the period of the Republic of China. For more details, see Figure 5.6.

Geographically, Qinghai is situated on the northeastern part of the Tibetan Plateau. The Yellow River originates in southern part of the province, while the Yangtze and Mekong have their sources in the southwestern part. The province is separated by the Riyue Mountain into pastoral and agricultural zones in the west and east. On average, Qinghai has an elevation of over 3000 m (9800 ft) above the sea level. The major mountain ranges include the Tanggula Mountains and Kunlun Mountains with the highest point at Bukadaban Feng at 6860 meters (22,510 ft). Because of its high altitude, the province has cold winters, mild summers, and a large diurnal temperature variation. Its average annual temperature is around −5 to 8°C (23 to 46°F) with January temperatures between −18 and −7°C (0 and 19°F) and July temperatures between 15 and 21°C (59 and 70°F). During the months from February to April, the area is prone to heavy winds and sandstorms. Major rainfall occurs mostly in the summer, while in the winter and spring, the general precipitation is very low so that much of the province is kept semi-arid or arid. Qinghai Lake is a salt water lake, the largest in China and the second largest in the world. In the northwest part

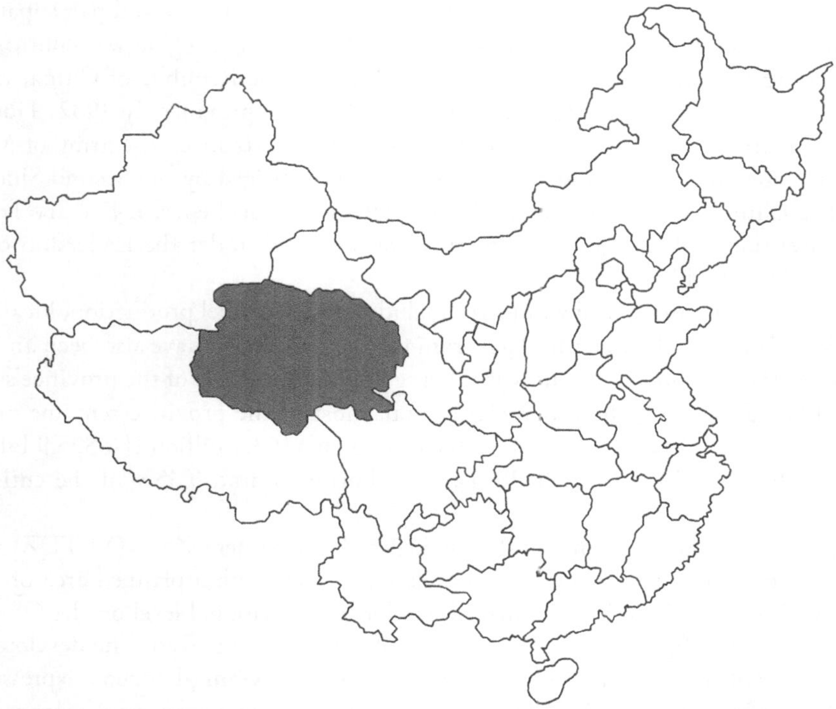

Figure 5.6 The geographic location of Qinghai.

of the province lies the Qaidam Basin, about one-third of which is desert, but rich in natural resources. The headwaters of the Yellow River, Yangtze River, and Mekong River are found in the Sanjiangyuan National Nature Reserve, established to protect the headwaters of these three rivers. The reserve consists of 18 subareas, each of which contains three zones managed with differing degrees of strictness.

Historically, in the Bronze Age, Qinghai was home to the Qiang people who made their livings in agriculture and husbandry. The eastern part of the area was ruled by the Han Dynasty about 2000 years ago. It was a battleground during the Tang and the following dynasties between the Han Chinese and successive Tibetan tribes. In the middle of the third century BC, Xianbei, a nomadic people, migrated to pasture lands around Qinghai Lake and established Tuyuhun Kingdom. Since the seventh century, this kingdom was attacked by both the Tibetan Empire and Tang Dynasty as they sought control over trade routes. After the military conflicts severely weakened the kingdom, it was incorporated into the Tibetan Empire. During the administrative rule of the Yuan Dynasty over Tibet, the region comprising the headwaters of the Yellow and Yangtze Rivers was apportioned to different administrative divisions than Tibet proper. Most of Qinghai was once also under the control of early Ming Dynasty for a short time period, and then gradually lost to the Khoshut Mongols, who controlled a big part of the area that is now Qinghai from 1640 to 1724 when it was conquered by the Qing Dynasty. During the 1720s, the Xining Prefecture was established with its borders roughly the same as those of modern Qinghai. Xining, the capital of present Qinghai, was initially built in this period as the administrative center. Although the region was governed by a viceroy of the Qing Emperor, the local ethnic groups enjoyed their autonomy with many chiefs retaining their traditional authority and participating in the local administration. Following the end of the Qing Dynasty, the region was controlled by the Muslim warlord Ma Qi until the Northern Expedition of the Republic of China, which consolidated the central control in 1928 and created the Qinghai province. In 1932, Tibet invaded Qinghai in an attempt to occupy the southern part of the province. The army of Ma Bufang defeated the Tibetans. When most of eastern China was ravaged by the Second Sino-Japanese War and the Chinese Civil War, Qinghai was relatively untouched. On January 1, 1950, the People's Government of the Qinghai Province was declared under the leadership of the new People's Republic of China.

Economically, Qinghai's heavy industry includes iron and steel production, located near its capital city of Xining. Oil and natural gas from the Qaidam Basin have also been an important contributor to the economy with salt works being operated at many of the province's numerous salt lakes. Outside Xining, the provincial capital, most of the province remains underdeveloped. The nominal GDP of Qinghai for 2011 was about ¥163.4 billion (US$25.9 billion) with per capita GDP about ¥19,407 (US$2841), contributing around 0.35% of the entire China's economy.

In 2000, the Xining Economic & Technological Development Zone (XETDZ) was established in the east of the city as state-level development zone with a planned area of 4.4 square kilometers (1.7 square miles). It is the first of its kind at the national level on the Qinghai-Tibet Plateau, established to fulfill the nation's strategy of developing the west. The development zone enjoys a convenient transportation system, connected by the Xining-Lanzhou expressway, 4 km away from the railway station, 15 km from Xining Airport. It focuses on the development of various industries, such as the chemicals based on salt lake resources, nonferrous metals, petroleum and natural gas processing, special medicine, foods and bio-chemicals using the local plateau animals and plants, new products involving ecological and environmental protection, high technology, new materials as well as information technology, and services such as logistics, banking, real estate, tourism, hotel, catering, agency and international trade.

For additional readings, refer with (National Bureau of Statistics of China, 2013; Smith, 2009; Betta, 2004; Goldstein, 1994; Cooke, 2008; Schram, 2006; Hutchings, 2003; Goodman, 2004).

5.6.1 Current State of Investment and Employment Structures of Qinghai

All that is necessary for us to do is the computations of the gross domestic outputs of all the economic sectors and industries and the investment and employment situations (Table 5.11) of Qinghai Province and the relevant comparative gains. By comparing the resultant comparative gains, we can then obtain the current states of investment and employment for the three economic sectors and industries. All the detailed computations are similar to what has been described earlier. The results are given in Table 5.12. Table 5.12 indicates the following facts:

First, in terms of the three major economic sectors, the six relevant data points of the three sectors of Qinghai Province are all situated in extreme instability, where the comparative gain of labor of the primary sector is way too low and equal to merely 0.235676. That is not even reaching one-tenth of that same indicator of the secondary sector. In terms of the capital investment, the comparative gains of the primary and the secondary sectors are much too low, and are respectively equal to 1.46619 and 1.25501, while the capital investment in the tertiary sector is in extreme excess (0.68582).

Table 5.11 The GDPs, Total Investments in Fixed Properties, and Employment Situation of Qinghai's Three Economic Sectors and Industries (2011)

Sectors (i)	*Total Output of Sector (0.1 Billion Yuan)*	*Total Investment on Fixed Properties (0.1 Billion Yuan)*	*Employment (10,000 People)*
Total	1670.44	1434.33	309.2
Primary sector	155.08	90.82	121.8
Secondary sector	975.18	667.20	73.9
Tertiary sector	540.18	676.31	113.5
Industries (*i*)			
Agriculture, forest, animal husbandry, fishery		86.24	**1.8**
Mining		88.08	**2.6**
Manufacturing		360.57	**12.1**
Production/supply of electricity, gas, and water		183.42	**1.5**
Construction	163.453608	35.09	**6.6**
Communication, transportation, storage, postal service	67.53	133.49	**3.4**
Information transfer, computer, and software areas	49.36	0.99	**0.9**
# Information Transfer			
Wholesale and retail	93.7	13.58	**2.3**
Hotels and restaurants	18.93	8.23	**0.8**
Finance	62.56	0.29	**2**
Real estate/housing	29.05	132.89	**1**
Leasing and commercial services	20.85	7.58	**0.8**
Scientific research, technology service, and geological survey	17.8	4.91	**2.7**
Water conservation, environment, and public facility management	4.26	61.60	**1**

continued

Table 5.11 (continued) The GDPs, Total Investments in Fixed Properties, and Employment Situation of Qinghai's Three Economic Sectors and Industries (2011)

Sectors (i)	*Total Output of Sector (0.1 Billion Yuan)*	*Total Investment on Fixed Properties (0.1 Billion Yuan)*	*Employment (10,000 People)*
# Public Facility Management			
Residents and other services	5.55	3.63	**0.3**
Education	34.62	48.26	**7.6**
Health, safety, and welfare	21.37	19.80	**3.4**
Social, sports, and entertainment	14.19	15.53	**0.7**
Public administration and social organizations	100.41	46.79	**9.2**

Source: The data in bold are from *China Statistics Yearbook* (2012), and the rest are from *Qinghai Province's Statistics Yearbook* (2012).

Second, in terms of the employment situation, the relevant data of the 21 industries of Qinghai Province should contain 42 data points. However, because of lack of raw data, we only collected 30 of them. Even so, the employment data in Table 5.11 in bold are on urban residents, and do not agree with the overall statistics. So, these data cannot be used in our discussion. So, we have only 15 effective data points on the comparative gains of capital. And, all the data are situated in extremely imbalanced states.

At the same time, the imbalance in terms of the comparative gains of capital seems to be way too severe. The comparative gains of capital in the following nine industries are all over 1. In particular, the comparative gains of capital of both "finance" and "information transfer, computer, and software areas" are over 10, while the value of this indicator of "water conservation, environment, and public facility management" is too low, equaling to merely 0.05938, which is only (0.05938/182.588 ≈) 0.00033 when compared to that of "finance," (0.05938/42.9936 ≈) 0.0064 when compared to that of "information transfer, computer, and software areas," (0.05938/5.92418 ≈) 0.01 when compared to that of "wholesale and retail," and (0.05938/3.99993 ≈) 0.015 when compared to that of "construction."

Finance	Information transfer, computer, and software areas	Wholesale and retail	Construction	Scientific research, technology service, and geological survey
Leasing and commercial services	Hotels and restaurants	Public administration and social organizations	Residents and other services	

Table 5.12 The Structures and Comparative Gain Measures of Qinghai's Three Economic Sectors and Industries ($t = 2011$)

Sectors (i)	*Demand Structure* $d_i(t)$ *(%)*	*Employment Structure* $s_i^1(t)$ *(%)*	*Comparative Gain of Labor* $G_i^1(t)$	*Investment Structure* $s_i^2(t)$ *(%)*	*Comparative Gain of Capital* $G_i^2(t)$
Total	100	100	—	100	—
Primary sector	9.283761	39.39198	0.235676	6.331876	1.46619
Secondary sector	58.37872	23.90039	2.442585	46.51649	1.25501
Tertiary sector	32.33752	36.70763	0.880948	47.15163	0.68582
Industries (*i*)					
Agriculture, forest, animal husbandry, fishery		0.582147		6.012682	
Mining		0.840880		6.140888	
Manufacturing		3.913325		25.13863	
Production/supply of electricity, gas, and water		0.485123		12.78758	
Construction	9.785042	2.134541	**4.584144**	2.446306	4.000
Communication, transportation, storage, postal service	4.042639	1.099612	**3.676423**	9.307042	0.434
Information transfer, computer, and software areas	2.954904	0.291074	**10.15174**	0.068729	42.994
# Information Transfer					
Wholesale and retail	5.609288	0.743855	**7.540835**	0.946846	5.924
Hotels and restaurants	1.133232	0.258732	**4.379941**	0.573585	1.976
Finance	3.745113	0.646831	**5.789945**	0.020511	182.588
Real estate/housing	1.739059	0.323415	**5.377170**	9.264821	0.188

continued

Table 5.12 (continued) The Structures and Comparative Gain Measures of Qinghai's Three Economic Sectors and Industries ($t = 2011$)

Sectors (i)	*Demand Structure* $d_i(t)$ *(%)*	*Employment Structure* $s_i^1(t)$ *(%)*	*Comparative Gain of Labor* $G_i^1(t)$	*Investment Structure* $s_i^2(t)$ *(%)*	*Comparative Gain of Capital* $G_i^2(t)$
Leasing and commercial services	1.248171	0.258732	**4.824183**	0.528372	2.362
Scientific research, technology service, and geological survey	1.065585	0.873221	**1.220292**	0.34209	3.115
Water conservation, environment, and public facility management	0.255022	0.323415	**0.788528**	4.294604	0.05938
# Public Facility Management					
Residents and other services	0.332247	0.097025	**3.424360**	0.253226	1.31206
Education	2.072503	2.457956	**0.843182**	3.364672	0.61596
Health, safety, and welfare	1.279301	1.099612	**1.163411**	1.380317	0.92682
Social, sports, and entertainment	5	0.226391	**3.752252**	1.082896	0.78445
Public administration and social organizations	6.010978	2.975420	**2.020211**	3.262227	1.84260

Therefore, because of lack of data, we will not address the employment problem for the Qinghai Province. And, in terms of the magnitudes of the comparative gains of capital, the industries of the Qinghai Province with available data can be ranked as follows:

1	2	3	4	5
Finance	Information transfer, computer, and software areas	Wholesale and retail	Construction	Scientific research, technology service, and geological survey
6	7	8	9	10
Leasing and commercial services	Hotels and restaurants	Public administration and social organizations	Residents and other services	Health, safety, and welfare

11	12	13	14	15
Social, sports, and entertainment	Education	Communication, transportation, storage, postal service	Real estate/ housing	Water conservation, environment, and public facility management

5.6.2 *Investment and Employment in Qinghai's Economic Sectors and Industries*

First, in terms of the comparative gains of labor, no matter whether the job seekers, who look for employment opportunities in the Qinghai Province, are from western China, or other parts of the mainland China, or the rest of the world, they should first consider the secondary economic sector. For specific industries, without effective data, we cannot draw any conclusion.

Without special skills, don't look for employment opportunities in the primary and the tertiary economic sectors, because those areas are experiencing excessive labor supplies.

That is the direction of employment.

Second, in terms of capital, the following nine industries have enjoyed the comparative gains of more than 1, representing relatively more optimal areas of additional investment.

Finance	Information transfer, computer, and software areas	Wholesale and retail	Construction	Scientific research, technology service, and geological survey
Leasing and commercial services	Hotels and restaurants	Public administration and social organizations	Residents and other services	

However, for the following six industries, excessive investment of capital has been given. Further capital investment in these industries should be curbed and redirected to the previous nine industries.

Health, safety, and welfare	Social, sports, and entertainment	Education
Communication, transportation, storage, postal service	Real estate/housing	Water conservation, environment, and public facility management

In other words, no matter whether the investors, who look for investment opportunities in the Qinghai Province, are from western China, or other parts of the mainland China, or the rest of the world, they should consider the following industries ranked from the most optimal to the least:

That is the direction of capital investment.

1	2	3	4	5
Finance	Information transfer, computer, and software areas	Wholesale and retail	Construction	Scientific research, technology service, and geological survey
6	7	8	9	
Leasing and commercial services	Hotels and restaurants	Public administration and social organizations	Residents and other services	

5.7 Gansu Province

Gansu is located in the northwest of the country between the Tibetan and Huangtu plateaus. It borders with Mongolia, Inner Mongolia, and Ningxia in the north, Xinjiang and Qinghai in the west, Sichuan in the south, and Shaanxi in the east, while the Yellow River cuts through the southern part of the province. For more details, see Figure 5.7.

Geographically, Gansu covers an area of 454,000 square kilometers (175,000 square miles) with the vast majority of the land more than 1000 m (3300 ft) above the sea level. The geographical center of China is located within the province. Several major deserts, such as the Gobi Desert, the Badain Jaran Desert, and Tengger Desert, stretch into Gansu. Although the Yellow River only cuts through the southern part of the province, it gets most of its water from the region. Gansu is mountainous in the south and flat in the north. The mountains in the south are part of the Qilian Mountains, while the far western Altyn-Tagh contains the province's highest point, at 5830 m (19,130 ft). A natural land passage, known as Hexi Corridor, stretching some 1000 km (620 miles) from Lanzhou to the Jade Gate, is situated within the province. The corridor is bounded in the north by the Gobi Desert and Qilian Mountains in the south. The climate of the province is generally semi-arid to arid and continental with warm to hot summers and cold to very cold winters. Most of the precipitation is seen in the summer months. However, due to its extreme altitude, some areas of Gansu exhibit the subarctic climate, where winter temperatures can drop to as low as −40°C.

In prehistoric times, Gansu was host to a number of Neolithic cultures. The Dadiwan culture flourished in the eastern side of Gansu during about 6000 BC to 3000 BC. The Majiayao

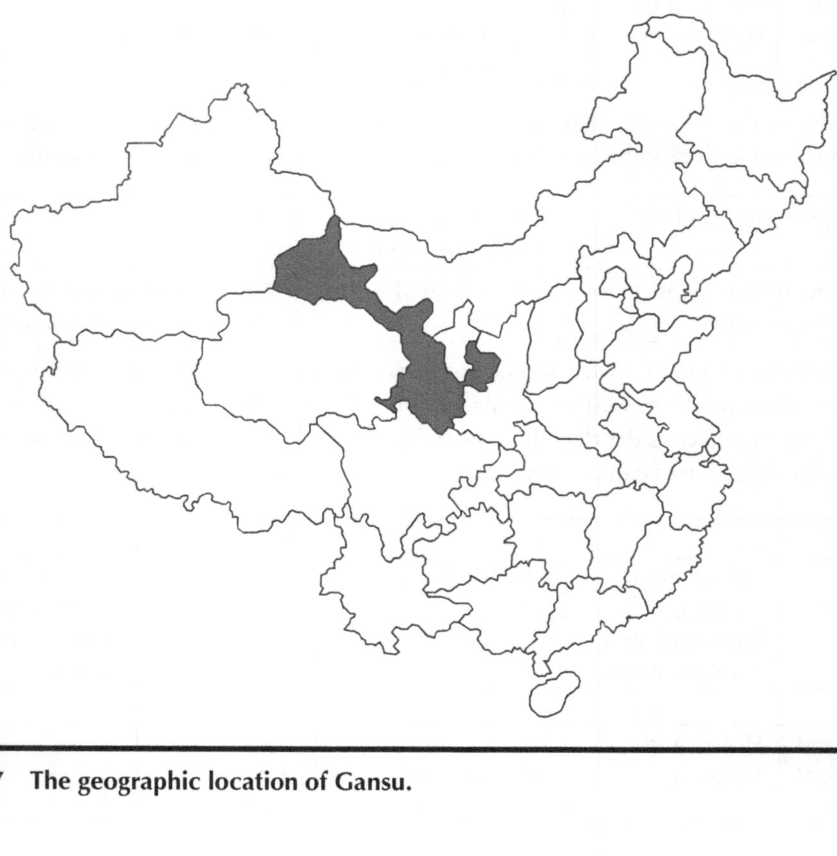

Figure 5.7 The geographic location of Gansu.

culture and part of the Qijia culture also took root in Gansu from about 3100 BC to 2700 BC and 2400 BC to 1900 BC, respectively. The Yuezhi once lived in the western part of the province until they were forced to emigrate by the Xiongnu around 177 BC. The State of Qin grew out the Tianshui area of the southeastern part of Gansu. In imperial times, Gansu was an important strategic outpost and communications link for Chinese empires, because the Hexi corridor runs along the "neck" of the province. The Han dynasty extended the Great Wall across this corridor and constructed the strategic Yumenguan (Jade Gate Pass) and Yangguan fort towns along the wall. The Ming dynasty built the Jiayuguan outpost in Gansu. Because of the Qingshui treaty, signed in 823 between the Tibetan Empire and the Tang Dynasty, China lost the control over a portion of the province for a long while. After the fall of the Uyghur Empire, a Uyghur state was established in parts of Gansu that lasted from 848 to 1036 AD. Since it is situated along the Silk Road, Gansu was economically important, and was a cultural transmission path. However, frequent earthquakes, droughts, and famines have often slowed the economic development of the province until recently the abundant mineral resources have helped to establish the region as a vital industrial center.

Economically, the agricultural production of Gansu includes cotton, linseed oil, maize, melons (such as the honeydew melon), millet, and wheat. The province is known as a source for wild medicinal herbs widely used in Chinese medicine. Even so, most of Gansu's economy is based on mining and the extraction of minerals, especially rare earth elements. The region has significant deposits of antimony, chromium, coal, cobalt, copper, fluorite, gypsum, iridium, iron, lead, limestone, mercury, mirabilite, nickel, crude oil, platinum, troilite, tungsten, zinc, among others. There are significant oil fields at Yumen and Changqing. Industries other than mining include electricity generation, petrochemicals, oil exploration machinery, and building materials. And tourism has been a bright spot in contributing to Gansu's overall economy, where a wide variety of choices are available for national and international tourists. The nominal GDP of the province for 2011 was about ¥502.0 billion (US$79.69 billion) and per capita GDP of ¥12,836 (US$1879).

For additional readings, refer to Ministry of Commerce (2013); National Bureau of Statistics of China (2013); and Gernet (1996).

5.7.1 Current State of Investment and Employment Structures of Gansu

All that is necessary for us to do is the computations of the gross domestic outputs of all the economic sectors and industries and the investment and employment situations (Table 5.13) of Ganxu Province and the relevant comparative gains. By comparing the resultant comparative gains, we can then obtain the current states of investment and employment for the three economic sectors and industries. All the detailed computations are similar to what has been described earlier. The results are given in Table 5.14. Table 5.14 indicates the following facts:

First, in terms of the three major economic sectors, among the six relevant data points of the three sectors of Gansu Province, only one of them is basically in equilibrium, where the comparative gain of capital of the secondary sector is 0.97389, and all the rest are situated in extreme instability. The comparative gain of labor of the primary sector is way too low and equal to merely 0.220697. That is not even reaching one-fourteenth and one-seventh of that same indicator of the secondary and the tertiary sectors, respectively. In terms of the capital investment, the comparative gain of the primary sector is much too low, and is equal to 2.95202, while the capital investment in the tertiary sector is in extreme excess (0.83606).

Table 5.13 The GDPs, Total Investments in Fixed Properties, and Employment Situation of Gansu's Three Economic Sectors and Industries (2011)

Sectors (i)	*Total Output of Sector (0.1 Billion Yuan)*	*Total Investment on Fixed Properties (0.1 Billion Yuan)*	*Employment (10,000 People)*
Total	5020.37	4180.24	1500.26
Primary sector	678.75	191.45	919.06
Secondary sector	2377.83	2032.99	231.49
Tertiary sector	1963.79	1955.80	349.71
Industries (*i*)			
Agriculture, forest, animal husbandry, fishery		191.49	**5.13**
Mining		223.26	**9.29**
Manufacturing		787.15	**31.53**
Production/supply of electricity, gas, and water		508.61	**6.67**
Construction	453.88	513.96	**17.2**
Communication, transportation, storage, postal service	280.33	237.79	**9.93**
Information transfer, computer, and software areas	136.99	36.10	**1.49**
# Information Transfer			
Wholesale and retail	351.97	103.65	**5.39**
Hotels and restaurants	123.61	43.38	**1.76**
Finance	145.05	8.89	**6.38**
Real estate/housing	134.25	278.07	**1.54**
Leasing and commercial services		18.21	**1.34**
Scientific research, technology service, and geological survey	65.9	26.32	**4.78**
Water conservation, environment, and public facility management	27.75	281.65	**3.63**

Table 5.13 (continued) The GDPs, Total Investments in Fixed Properties, and Employment Situation of Gansu's Three Economic Sectors and Industries (2011)

Sectors (i)	*Total Output of Sector (0.1 Billion Yuan)*	*Total Investment on Fixed Properties (0.1 Billion Yuan)*	*Employment (10,000 People)*
# Public Facility Management			
Residents and other services		12.01	**0.36**
Education	180.5	70.57	**34.26**
Health, safety, and welfare	87.57	52.92	**10.36**
Social, sports, and entertainment	27.83122273	39.09	**2.34**
Public administration and social organizations	278.4171305	384.27	**34.92**

Source: The data are from *Gansu Province's Yearbook of Development 2012* (XLS, DOC), where the numbers of industrial employments come from "Section 4.4: Numbers of Employed in Various Industries at the End of the Year."

Second, in terms of the employment situation, the relevant data of the 21 industries of Gansu Province should contain 42 data points. However, because of lack of raw data, we only collected 26 of them. Even so, the employment data in Table 5.13 in bold are on urban residents, and do not agree with the overall statistics. So, these data cannot be used in our discussion. So, we have only 13 effective data points on the comparative gains of capital. And, only one of the data is basically in equilibrium, where the comparative gain of capital of "communication, transportation, storage, postal service" is equal to 0.98160, and all the rest are situated in extremely imbalanced states.

At the same time, the imbalance in terms of the comparative gains of capital seems to be way too severe. The comparative gains of capital in the following seven industries are all over 1. In particular, the comparative gain of capital of "finance" is over 10, while the value of this indicator of "water conservation, environment, and public facility management" is too low, equaling to merely 0.08203, which is only (0.08203/13.5905 ≈) 0.006 when compared to that of "finance," (0.08203/3.1597 ≈) 0.026 when compared to that of "information transfer, computer, and software areas," (0.08203/2.37251 ≈) 0.029 when compared to that of "wholesale and retail" and (0.08203/2.12961 ≈) 0.034 when compared to that of "hotels and restaurants."

Finance	Information transfer, computer, and software areas	Wholesale and retail	Hotels and restaurants
Education	Scientific research, technology service, and geological survey	Health, safety, and welfare	

Table 5.14 The Structures and Comparative Gain Measures of Gansu's Three Economic Sectors and Industries ($t = 2011$)

Sectors (i)	*Demand Structure $d_i(t)$ (%)*	*Employment Structure $s_i^1(t)$ (%)*	*Comparative Gain of Labor $G_i^1(t)$*	*Investment Structure $s_i^2(t)$ (%)*	*Comparative Gain of Capital $G_i^2(t)$*
Total	100	100	—	100	—
Primary sector	13.51992	61.26005	0.220697	4.579881	2.95202
Secondary sector	47.36364	15.42999	3.069583	48.63333	0.97389
Tertiary sector	39.11644	23.30996	1.678100	46.78679	0.83606
Industries (*i*)					
Agriculture, forest, animal husbandry, fishery		0.341941		4.579847	
Mining		0.619226		5.340904	
Manufacturing		2.101636		18.83036	
Production/supply of electricity, gas, and water		0.444590		12.16703	
Construction	9.040768	1.146468	**7.885757**	12.29493	0.73532
Communication, transportation, storage, postal service	5.583851	0.661885	**8.436283**	5.688494	0.98160
Information transfer, computer, and software areas	2.728683	0.099316	**27.47473**	0.863592	3.15969
# Information Transfer					
Wholesale and retail	7.010838	0.359271	**19.51406**	2.479551	2.82746
Hotels and restaurants	2.462169	0.117313	**20.98803**	1.037790	2.37251
Finance	2.889229	0.425260	**6.794036**	0.212591	13.5906
Real estate/housing	2.674106	0.102649	**26.05100**	6.651924	0.40200
Leasing and commercial services		0.089318		0.435590	
Scientific research, technology service, and geological survey	1.312652	0.318611	**4.119916**	0.629682	2.08463

Table 5.14 (continued) The Structures and Comparative Gain Measures of Gansu's Three Economic Sectors and Industries ($t = 2011$)

Sectors (i)	*Demand Structure* $d_i(t)$ *(%)*	*Employment Structure* $s_i^1(t)$ *(%)*	*Comparative Gain of Labor* $G_i^1(t)$	*Investment Structure* $s_i^2(t)$ *(%)*	*Comparative Gain of Capital* $G_i^2(t)$
Water conservation, environment, and public facility management	0.552748	0.241958	**2.284479**	6.737694	0.08204
# Public Facility Management					
Residents and other services		0.023996		0.287244	
Education	3.595353	2.283604	**1.574420**	1.688262	2.12963
Health, safety, and welfare	1.744294	0.690547	**2.525960**	1.266001	1.37780
Social, sports, and entertainment	0.554366	0.155973	**3.554244**	0.935099	0.59284
Public administration and social organizations	5.545749	2.327597	**2.382608**	9.192628	0.60328

Therefore, because of lack of data, we will not address the employment problem for the Gansu Province. And, in terms of the magnitudes of the comparative gains of capital, the industries of Gansu Province with available data can be ranked as follows:

1	2	3	4	5
Finance	Information transfer, computer, and software areas	Wholesale and retail	Hotels and restaurants	Education
6	7	8	9	10
Scientific research, technology service, and geological survey	Health, safety, and welfare	Communication, transportation, storage, postal service	Construction	Public administration and social organizations
11	12	13		
Social, sports, and entertainment	Real estate/ housing	Water conservation, environment, and public facility management		

5.7.2 Investment and Employment in Gansu's Economic Sectors and Industries

First, in terms of comparative gains of labor, no matter whether the job seekers, who look for employment opportunities in Gansu Province, are from western China, or other parts of the mainland China, or the rest of the world, they should first consider the secondary and the tertiary economic sectors, where we do not have any conclusion about specific industries.

Without special skills, don't look for employment opportunities in the primary economic sector, because that area is experiencing excessive labor supplies.

That is the direction of employment.

Second, in terms of capital, the following 7 industries have enjoyed the comparative gains of more than 1, representing relatively more optimal areas of additional investment.

Finance	Information transfer, computer, and software areas	Wholesale and retail	Hotels and restaurants
Education	Scientific research, technology service, and geological survey	Health, safety, and welfare	

However, for the following six industries, excessive investment of capital has been given. Further capital investment in these industries should be curbed and redirected to the previous seven industries.

Communication, transportation, storage, postal service	Construction	Public administration and social organizations
Social, sports, and entertainment	Real estate/housing	Water conservation, environment, and public facility management

In other words, no matter whether the investors, who look for investment opportunities in Gansu Province, are from western China, or other parts of the mainland China, or the rest of the world, they should consider the following industries ranked from the most optimal to the least:

That is the direction of capital investment.

1	2	3	4
Finance	Information transfer, computer, and software areas	Wholesale and retail	Hotels and restaurants
5	6	7	
Education	Scientific research, technology service, and geological survey	Health, safety, and welfare	

5.8 Ningxia Autonomous Region

Ningxia is located in the northwest part of the country. It is bounded by Shaanxi in the east, Gansu in the south and west, and Inner Mongolia in the north. This sparsely settled, mostly desert region is situated partially on the Loess Plateau and in the vast plain of the Yellow River. It features the Great Wall of China along its northeastern boundary. An extensive system of canals has been constructed over the years, making increased cultivation through the extensive land reclamation and irrigation projects. For more details, see Figure 5.8.

Geographically, although the Yellow River flows through Ningxia, the province is a relatively dry, desert-like region. As supported by the Yellow River, the northern section of the region is the best agricultural land. The province is 1200 km (750 miles) from the sea and has a continental climate with average summer temperatures in the range from 17 to 24°C (63 to 75°F) in July and average winter temperatures in the range from −7 to −15°C (19 to 5°F) in January. Extreme temperatures in the summer can reach as high as 39°C (102°F) and in winter −30°C (−22°F). The diurnal temperature variation can reach above 17°C (31°F), especially in spring. The annual rainfall averages around 190–700 mm (7.5–27.6 in.) with more rainfalls appearing in the south of the region.

Historically, Ningxia was incorporated into the Qin Dynasty as early as the third century BC. Throughout the Han Dynasty and the Tang Dynasty, several large cities were established in the region. By the eleventh century, the Tangut tribe had established the Western Xia Dynasty on the outskirts of the then Song Dynasty. Since the early thirteenth century, the area was under the Mongol domination. After the Mongols left, some Turkic-speaking Muslims began moving

Figure 5.8 The geographic location of Ningxia.

into Ningxia from the west. In 1928, Ningxia was detached from Gansu and became a province. Between 1914 and 1928, Muslim Kuomintang General Ma Hongkui was the military Governor of Ningxia and had absolute authority in the province. In 1958, Ningxia formally became an autonomous region.

Economically, Ningxia is the principal region of China, where wolfberries are grown. In 1992, the Yinchuan Economic and Technological Development Zone was established, covering 32 km^2 (12 sq mi) with an annual economic output of ¥23.7 billion (US$3.5 billion). The major industries of the development zone include machinery and equipment manufacturing, new materials, fine chemicals, and the animation industry. Desheng Industrial Park is a base for about 400 enterprises, including industrial chains from Muslim food and commodities to trade and logistics, new materials and bio-pharmaceuticals. Desheng achieved a total output value of ¥4.85 billion in 2008. Additionally, Ningxia wines are a promising area of development, for which the eastern base of the Helan Mountains has been approved to serve as an area suitable for wine production. To aid the effort, the major oil company China Petroleum and Chemical Corporation has founded a grape plantation near the Helan Mountains; and the household appliance company Midea has begun participating in Ningxia's wine industry.

The nominal GDP of Ningxia in 2011 was ¥200.0 billion (US$32.7 billion) and a per capita GDP of ¥21,470 (US$3143). It contributes 0.44% of the national economy.

For additional reading, refer to National Bureau of Statistics of China (2013); and Close and McCormick (1922).

5.8.1 Current State of Investment and Employment Structures of Ningxia

All that is necessary for us to do is the computations of the gross domestic outputs of all the economic sectors and industries and the investment and employment situations (Table 5.15) of Ningxia Autonomous Region and the relevant comparative gains. By comparing the resultant comparative gains, we can then obtain the current states of investment and employment for the three economic sectors and industries. All the detailed computations are similar to what has been described earlier. The results are given in Table 5.16. Table 5.16 indicates the following facts:

First, in terms of the three major economic sectors, among the six relevant data points of the three sectors of Ningxia Autonomous Region, only one is basically in equilibrium, where the comparative gain of capital is equal to 1.00125, and all the rest five data points are situated in extreme instability. Specifically, the comparative gain of labor of the primary sector is way too low and equal to merely 0.179019. That is not even reaching one-seventeenth and one-sixth of that same indicator of the secondary and the tertiary sectors, respectively. In terms of the capital investment, the comparative gain of the primary sector is much too low, and is equal to 3.02337, while the capital investment in the tertiary sector is in extreme excess (0.87373).

Second, in terms of the employment situation, the relevant data of the 21 industries of the Ningxia Autonomous Region should contain 42 data points. However, because of lack of raw data, we only collected 38 of them. So, we can only base our discussion on these available statistics. Among all the industries, only one is basically in equilibrium, where the comparative gains of labor in "health, safety, and welfare" is equal to 1.022784. All the rest (37/38 ≥ 97%) are situated in extremely imbalanced states. In particular, the comparative gains of labor of the following 11 industries are all over 1. Especially, the comparative gain of labor of "finance" even reaches

Table 5.15 The GDPs, Total Investments in Fixed Properties, and Employment Situation of Ningxia's Three Economic Sectors and Industries (2011)

Sectors (i)	*Total Output of Sector (0.1 Billion Yuan)*	*Total Investment on Fixed Properties (0.1 Billion Yuan)*	*Employment (10,000 People)*
Total	2102.21	1648.48	339.5907
Primary sector	184.14	47.76	166.1608
Secondary sector	1056.15	827.16	55.3522
Tertiary sector	861.92	773.56	118.0777
Industries (*i*)			
Agriculture, forest, animal husbandry, fishery	184.14	**47.8**	166.1608
Mining	224.79	**124.8**	7.9383
Manufacturing	428.71	**423.2**	24.9668
Production/supply of electricity, gas, and water	163.29	**241.6**	4.5866
Construction	239.36	**28.1**	17.8605
Communication, transportation, storage, postal service	174.1	**105.5**	9.9762
Information transfer, computer, and software areas	66.08	**16.3**	1.8489
# Information Transfer			
Wholesale and retail	109.99	**41.6**	47.9987
Hotels and restaurants	37.15	**11.1**	8.706
Finance	134.18	**2.7**	3.4012
Real estate/housing	79.01	**416.6**	2.7176
Leasing and commercial services	19.74	**0.8**	5.2808

continued

Table 5.15 (continued) The GDPs, Total Investments in Fixed Properties, and Employment Situation of Ningxia's Three Economic Sectors and Industries (2011)

Sectors (i)	*Total Output of Sector (0.1 Billion Yuan)*	*Total Investment on Fixed Properties (0.1 Billion Yuan)*	*Employment (10,000 People)*
Scientific research, technology service, and geological survey	15.55	**1.3**	2.0237
Water conservation, environment, and public facility management	10.32	**96.8**	2.4094
# Public Facility Management			
Residents and other services	37.41	**1.5**	7.0066
Education	58.2	**27.1**	10.1288
Health, safety, and welfare	29	**11.6**	4.5803
Social, sports, and entertainment	8.27	**11.1**	1.567
Public administration and social organizations	82.92	**35.4**	10.0047

Source: The data on total investments of fixed properties are from *China Statistics Yearbook* (2012), and the rest are from *Ningxia Statistics Yearbook* (2012).

6.37284. That is (6.372874/0.179019 ≈) 36 times of the lowest value of the same indicator reached by "agriculture, forest, animal husbandry, fishery."

Finance	Information transfer, computer, and software areas	Production/supply of electricity, gas, and water	Real estate/housing
Mining	Communication, transportation, storage, postal service	Manufacturing	Construction
Public administration and social organizations	Scientific research, technology service, and geological survey	Health, safety, and welfare	

Table 5.16 The Structures and Comparative Gain Measures of Ningxia's Three Economic Sectors and Industries ($t = 2011$)

Sectors (i)	*Demand Structure* $d_i(t)$ *(%)*	*Employment Structure* $s_i^1(t)$ *(%)*	*Comparative Gain of Labor* $G_i^1(t)$	*Investment Structure* $s_i^2(t)$ *(%)*	*Comparative Gain of Capital* $G_i^2(t)$
Total	100	100	—	100	—
Primary sector	8.759353	48.92973	0.179019	2.897214	3.02337
Secondary sector	50.23999	16.29968	3.082268	50.17713	1.00125
Tertiary sector	41.00066	34.77059	1.179176	46.92565	0.87374
Industries (*i*)					
Agriculture, forest, animal husbandry, fishery	8.759353	48.92973	0.179019	2.899641	3.02084
Mining	10.69303	2.337608	4.574348	7.570611	1.41244
Manufacturing	20.39330	7.352027	2.773834	25.67213	0.79437
Production/supply of electricity, gas, and water	7.767540	1.350626	5.751067	14.65593	0.52999
Construction	11.38611	5.259420	2.164899	1.704601	6.67964
Communication, transportation, storage, postal service	8.281761	2.937713	2.819118	6.399835	1.29406
Information transfer, computer, and software areas	3.143359	0.544450	5.773452	0.98879	3.17900
# Information Transfer					
Wholesale and retail	5.232113	14.13428	0.370172	2.523537	2.07333
Hotels and restaurants	1.767188	2.563674	0.689318	0.673348	2.62448
Finance	6.382807	1.001559	6.372874	0.163787	38.9701
Real estate/ housing	3.758426	0.800257	4.696520	25.27177	0.14872
Leasing and commercial services	0.939012	1.555048	0.603847	0.04853	19.3493

continued

Table 5.16 (continued) The Structures and Comparative Gain Measures of Ningxia's Three Economic Sectors and Industries ($t = 2011$)

Sectors (i)	*Demand Structure $d_i(t)$ (%)*	*Employment Structure $s_i^1(t)$ (%)*	*Comparative Gain of Labor $G_i^1(t)$*	*Investment Structure $s_i^2(t)$ (%)*	*Comparative Gain of Capital $G_i^2(t)$*
Scientific research, technology service, and geological survey	0.739698	0.595923	1.241263	0.078861	9.37982
Water conservation, environment, and public facility management	0.490912	0.709501	0.691911	5.872076	0.08360
# Public Facility Management					
Residents and other services	1.779556	2.063248	0.862502	0.090993	19.5571
Education	2.768515	2.982649	0.928207	1.643939	1.68407
Health, safety, and welfare	1.379501	1.348771	1.022784	0.703679	1.96041
Social, sports, and entertainment	0.393396	0.461438	0.852543	0.673348	0.58424
Public administration and social organizations	3.944420	2.946105	1.338859	2.147433	1.83681

At the same time, the imbalance in terms of the comparative gains of capital seems to be way too severe. The comparative gains of capital in the following 14 industries are all over 1. In particular, the comparative gains of capital of "finance," "residents and other services," and "leasing and commercial services" are over 10. On the other hand, the value of this indicator of "water conservation, environment, and public facility management" is too low, equaling to merely 0.08360, which is only (0.08360/38.9701 ≈) 0.0022 when compared to that of "finance," (0.08360/19.5571 ≈) 0.0042 when compared to that of "residents and other services," (0.08360/19.3493 ≈) 0.0043 when compared to that of "leasing and commercial services," and (0.08360/9.37982 ≈) 0.009 when compared to that of "scientific research, technology service, and geological survey."

Finance	Residents and other services	Leasing and commercial services	Scientific research, technology service, and geological survey	Construction

Information transfer, computer, and software areas	Agriculture, forest, animal husbandry, fishery	Hotels and restaurants	Wholesale and retail	Health, safety, and welfare
Public administration and social organizations	Education	Mining	Communication, transportation, storage, postal service	

Therefore, in terms of the magnitudes of the comparative gains of labor, the industries of the Ningxia Autonomous Region with available data can be ranked as follows:

1	2	3	4
Finance	Information transfer, computer, and software areas	Production/supply of electricity, gas, and water	Real estate/ housing
5	6	7	8
Mining	Communication, transportation, storage, postal service	Manufacturing	Construction
9	10	11	12
Public administration and social organizations	Scientific research, technology service, and geological survey	Health, safety, and welfare	Education
13	14	15	16
Residents and other services	Social, sports, and entertainment	Water conservation, environment, and public facility management	Hotels and restaurants
17	18	19	
Leasing and commercial services	Wholesale and retail	Agriculture, forest, animal husbandry, fishery	

And, in terms of the magnitudes of the comparative gains of capital, the industries of the Ningxia Autonomous Region with available data can be ranked as follows:

1	2	3	4	5
Finance	Residents and other services	Leasing and commercial services	Scientific research, technology service, and geological survey	Construction

6	7	8	9	10
Information transfer, computer, and software areas	Agriculture, forest, animal husbandry, fishery	Hotels and restaurants	Wholesale and retail	Health, safety, and welfare
11	12	13	14	15
Public administration and social organizations	Education	Mining	Communication, transportation, storage, postal service	Manufacturing
16	17	18	19	
Social, sports, and entertainment	Production/supply of electricity, gas, and water	Real estate/housing	Water conservation, environment, and public facility management	

5.8.2 Investment and Employment in Ningxia's Economic Sectors and Industries

First, in terms of the comparative gains of labor, the values of this indicator for the following industries

Finance	Information transfer, computer, and software areas	Production/supply of electricity, gas, and water	Real estate/housing
Mining	Communication, transportation, storage, postal service	Manufacturing	Construction
Public administration and social organizations	Scientific research, technology service, and geological survey	Health, safety, and welfare	

are over 1, representing relatively optimal industries for employment. In all the other eight industries, there has been excessive supply of labor so that appropriate control on the increase of labor into these industries should be imposed. Specifically, the excessive labor in the following industries should be transferred into the previous 11 industries.

Agriculture, forest, animal husbandry, fishery	Wholesale and retail	Leasing and commercial services	Hotels and restaurants
Water conservation, environment, and public facility management	Social, sports, and entertainment	Residents and other services	Education

In other words, no matter whether the job seekers, who look for employment opportunities in the Ningxia Autonomous Region, are from western China, or other parts of the mainland China, or the rest of the world, they should first consider the secondary and the tertiary economic sectors with the relevant industries ranked as follows:

1	2	3	4
Finance	Information transfer, computer, and software areas	Production/supply of electricity, gas, and water	Real estate/housing
5	6	7	8
Mining	Communication, transportation, storage, postal service	Manufacturing	Construction
9	10	11	
Public administration and social organizations	Scientific research, technology service, and geological survey	Health, safety, and welfare	

Without special skills, don't look for employment opportunities in the primary economic sector and other industries, because those areas are experiencing excessive labor supplies.

That is the direction of employment.

Second, in terms of the comparative gains of capital, no matter whether the investors, who look for investment opportunities in the Ningxia Autonomous Region, are from western China, or other parts of the mainland China, or the rest of the world, they should consider the primary and the secondary economic sectors along with the following industries ranked from the most optimal to the least:

1	2	3	4	5
Finance	Residents and other services	Leasing and commercial services	Scientific research, technology service, and geological survey	Construction
6	7	8	9	10
Information transfer, computer, and software areas	Agriculture, forest, animal husbandry, fishery	Hotels and restaurants	Wholesale and retail	Health, safety, and welfare
11	12	13	14	
Public administration and social organizations	Education	Mining	Communication, transportation, storage, postal service	

At the same time, try not to invest additional capital in the tertiary economic sector and the other five industries:

Water conservation, environment, and public facility management	Real estate/ housing	Production/ supply of electricity, gas, and water	Social, sports, and entertainment	Manufacturing

because there has been a severe over supply of capital in these industries.

That is the direction of capital investment.

Third, by combining the two criteria discussed above, it follows that the industries in Ningxia Autonomous Region that are most optimal for both investment and employment are the following, because the comparative gains of labor and capital of each of these industries are over 1.

Finance	Information transfer, computer, and software areas	Construction	Mining
Scientific research, technology service, and geological survey	Public administration and social organizations	Health, safety, and welfare	Communication, transportation, storage, postal service

The worst choice is the industry of "water conservation, environment, and public facility management."

5.9 Chongqing City

Chongqing is a major city in central China and one of the four directly controlled municipalities (the other three are Beijing, Shanghai, and Tianjin) under the national government. The municipality was created in March 1997, succeeding the provincial city administration that was part of Sichuan province. As of April 29, 2011, the municipality had a population of 28,846,170 with roughly 6 or 7 million residing in the urbanized area. It represents the largest directly controlled municipality in China, and comprises 19 districts, 15 counties, and 4 autonomous counties. For more details, see Figure 5.9.

Geographically, Chongqing borders with Hubei in the east, Hunan in the southeast, Guizhou in the south, Sichuan in the west, and Shaanxi in the north. It is situated at the transitional area between the Qinghai-Tibet Plateau and the plain on the middle and lower reaches of the Yangtze River in the subtropical climate zone often swept by moist monsoons. It often rains in the night in late spring and early summer. The municipality reaches a maximum width of 470 km (290 miles) from east to west, and a maximum length of 450 km (280 miles) from north to south. A large area of the municipality is crisscrossed by rivers and mountains with the Daba Mountains standing in the north, the Wu Mountains in the east, the Wuling Mountains in the southeast, and the Dalou Mountains in the south. The whole region drops in elevation from north and south towards the Yangtze River valley with sharp rises and falls. The area is featured by typical karst landscape with stone forests, numerous collections of peaks, limestone caves and valleys found in

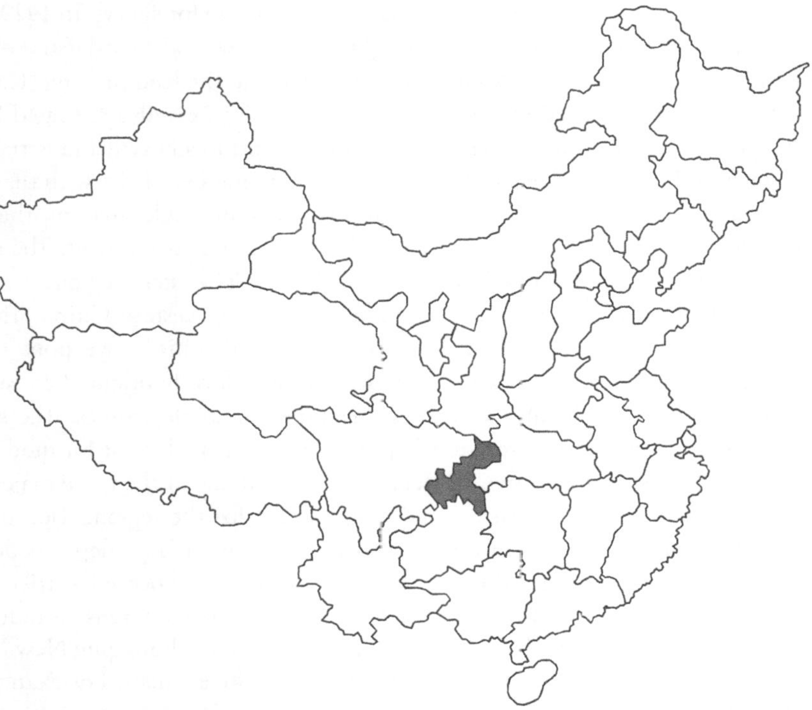

Figure 5.9 The geographic location of Chongqing.

many places. The Yangtze River runs through the whole municipality from west to east, covering a course of 665 km (413 miles), cutting through the Wu Mountains at three places, known as Three Gorges. Coming down from northwest and running through "the Jialing Lesser Three Gorges" of Libi, Wentang, and Guanyin, the Jialing River joins the Yangtze in Chongqing. The central urban area of Chongqing is a city with its unique features. It is built on mountains and partially surrounded by the Yangtze River and the Jialing River, making it known as a "mountain city" and a "city on rivers." With its particular topographical features, the municipality has the unique scenery of mountains, rivers, forests, springs, waterfalls, gorges, and caves. Chongqing has a monsoon-influenced humid subtropical climate. Its summers are long and very hot and humid with highs reaching 33–34°C (91–93°F) in the months of July and August in the urban area. And, winters are short and mild, but damp. Because of its location in the Sichuan Basin, it has only around 1055 hours of sunshine annually, which is lower than much of Northern Europe. On the other hand, Chongqing is covered by fog over 100 days a year so that it is known as a "fog city."

Historically, the tradition has associated Chongqing with the State of Ba, whose people established Chongqing during the Spring and Autumn period as their new capital. In 316 BC, the Ba was conquered by the State of Qin. It received its current name in 1189. In 1362, during the Yuan Dynasty, the Daxia Kingdom was established at Chongqing for a short time. In 1621, during the Ming Dynasty, another short-lived kingdom, known as Daliang, was established with Chongqing as its capital. In 1644, after the fall of the Ming Dynasty, Chongqing, along with the rest of Sichuan, was occupied by Zhang Xianzhong, who depopulated the region by massacring a large number of people. During the Qing Dynasty, immigration to Chongqing and Sichuan took place with the support of the Qing emperor. In 1890, the British Consulate General was opened in

Chongqing, making the city the first inland commerce port open to foreigners. In 1929, Chongqing became a municipality in the Republic of China. During the Second Sino-Japanese War (1937–1945), Chongqing was Chiang Kai-shek's provisional capital and the Kuomintang (KMT) government was moved to Chongqing. Many factories and universities were also relocated from eastern China to Chongqing, transforming this city from an inland port to a heavily industrialized city. In March 1997, the Eighth National People's Congress decided to merge the city with the neighboring Fuling, Wanzhou, and Qianjiang districts, making the resultant single division, the Chongqing Municipality, under the direct administrative control of the central government. The municipality immediately became the spearhead of China's effort to develop its western regions.

Economically, Chongqing is an important industrial area in western China. The industries of the municipality have been diversified from the traditional military weapons research and development. Because of its inland location, factories produce locally oriented consumer goods, such as processed food, cars, chemicals, textiles, machinery, and electronics. The municipality is China's third largest center for motor vehicle production and the largest for motorcycles. It is also one of the nine largest iron and steel centers in China and one of the three major aluminum producers. At the same time, agriculture remains significant for the region. Rice and fruits are the area's main produce. Natural resources are also abundant with large deposits of coal, natural gas, and minerals of more than 40 kinds. Recently, there has appeared a drive to move up the value chain by shifting toward high technology and knowledge-intensive industries. That resulted in the construction of new development zones, such as the Chongqing New North Zone. The city has also invested heavily in infrastructure to attract investment. For example, the network of roads and railways, connecting Chongqing to the rest of China, has been expanded and upgraded in order to reduce logistical costs. The nearby Three Gorges Dam, which is the world's largest, not only supplies Chongqing with power but also allows large ships to reach Chongqing's Yangtze River port. These infrastructure improvements have attracted numerous foreign direct investors ranging from car to finance and retailing. The nominal GDP of Chongqing in 2011 reached ¥1001.1 billion (US$158.9 billion) and the per capita GDP was ¥22,909 (US$3301).

For additional readings, refer to Ministry of Commerce (2013); Kapp (1974); MacKie (2005); Seetharaman (2013); and McIsaac (2000).

5.9.1 Current State of Investment and Employment Structures of Chongqing

All that is necessary for us to do is the computations of the gross domestic outputs of all the economic sectors and industries and the investment and employment situations (Table 5.17) of the Chongqing Municipality and the relevant comparative gains. By comparing the resultant comparative gains, we can then obtain the current states of investment and employment for the three economic sectors and industries. All the detailed computations are similar to what has been described earlier. The results are given in Table 5.18. Table 5.18 indicates the following facts:

First, in terms of the three major economic sectors, among the six relevant data points of the three sectors of the Chongqing Municipality, only one is basically in equilibrium, where the comparative gain of labor is equal to 0.972541, and the rest are situated in extreme instability. Specifically, the comparative gain of labor of the primary sector is way too low and equal to merely 0.221248. That is not even reaching one-tenth and one-fourth of that same indicator of the secondary and the tertiary sectors, respectively. In terms of the capital investment, the comparative gains of the primary and the secondary sectors are much too low, and are respectively equal to 2.32576 and 1.52813, while the capital investment in the tertiary sector is in extreme excess (0.60187).

Table 5.17 The GDPs, Total Investments in Fixed Properties, and Employment Situation of Chongqing's Three Economic Sectors and Industries (2011)

Sectors (i)	*Total Output of Sector (0.1 Billion Yuan)*	*Total Investment on Fixed Properties (0.1 Billion Yuan)*	*Employment (10,000 People)*
Total	10011.37	7685.8699	1585.16
Primary sector	844.52	278.7694	604.38
Secondary sector	5543.04	2784.7512	390.8
Tertiary sector	3623.81	4622.3493	589.98
Industries (*i*)			
Agriculture, forest, animal husbandry, fishery			
Mining		184.1855	28.11
Manufacturing		2069.4684	190.51
Production/supply of electricity, gas, and water		277.5541	8.54
Construction	852.58	253.5432	163.64
Communication, transportation, storage, postal service	456.25	833.5061	53.49
Information transfer, computer, and software areas	173.34	62.5201	10.57
# Information Transfer			
Wholesale and retail	747.3	79.9881	142.25
Hotels and restaurants	166.31	45.9208	88.22
Finance	704.66	2.5742	12.34
Real estate/housing	396.28	2408.0892	16.83
Leasing and commercial services	156.37	23.5491	26.31
Scientific research, technology service, and geological survey	90.22	14.9988	9.44
Water conservation, environment, and public facility management	37.71	778.58	7.45

continued

Table 5.17 (continued) The GDPs, Total Investments in Fixed Properties, and Employment Situation of Chongqing's Three Economic Sectors and Industries (2011)

Sectors (i)	Total Output of Sector (0.1 Billion Yuan)	Total Investment on Fixed Properties (0.1 Billion Yuan)	Employment (10,000 People)
# Public Facility Management			
Residents and other services	86.85	33.1431	133.83
Education	210.78	92.2153	39.88
Health, safety, and welfare	120.57	45.4422	17.42
Social, sports, and entertainment	48.68	58.9168	4.64
Public administration and social organizations	228.49	142.9055	27.31

Source: Data are from *Chongqing City's Statistics Yearbook 2012* (CLS, DOC).

Second, in terms of the employment situation, the relevant data of the 21 industries of the Chongqing Municipality should contain 42 data points. However, because of lack of raw data, we only collected 30 of them. So, we can only base our discussion on these available statistics. All the industries are situated in extremely imbalanced states. In particular, the comparative gains of labor of the following eight industries are all over 1. Especially, the comparative gain of labor in "finance" is even equal to 9.041574, which is more than (9.041574/0.102753 ≈) 88 times of the lowest value of this indicator that is reached by "residents and other services."

Finance	Real estate/housing	Information transfer, computer, and software areas	Social, sports, and entertainment
Scientific research, technology service, and geological survey	Communication, transportation, storage, postal service	Public administration and social organizations	Health, safety, and welfare

At the same time, the imbalance in terms of the comparative gains of capital seems to be way too severe. The comparative gains of capital in the following 11 industries are all over 1. In particular, the comparative gain of capital of "finance" is over 10, reaching the unbelievable level of 210.154, while the value of this indicator of "water conservation, environment, and public facility management" is too low, equaling to merely 0.03718, which is only (0.03718/210.154 ≈) 0.00018 when compared to that of "finance," (0.03718/17.17248 ≈) 0.0052 when compared to that of "wholesale and retail" (0.03718/5.09775 ≈) 0.0073 when compared to that of "leasing and commercial services," and (0.03718/4.61791 ≈) 0.0085 when compared to that of "scientific research, technology service, and geological survey.

Finance	Wholesale and retail	Leasing and commercial services	Scientific research, technology service, and geological survey
Hotels and restaurants	Construction	Information transfer, computer, and software areas	Health, safety, and welfare
Residents and other services	Education	Public administration and social organizations	

Table 5.18 The Structures and Comparative Gain Measures of Chongqing's Three Economic Sectors and Industries ($t = 2011$)

Sectors (i)	*Demand Structure $d_i(t)$ (%)*	*Employment Structure $s_i^1(t)$ (%)*	*Comparative Gain of Labor $G_i^1(t)$*	*Investment Structure $s_i^2(t)$ (%)*	*Comparative Gain of Capital $G_i^2(t)$*
Total	100	100	—	100	—
Primary sector	8.435609	38.12738	0.221248	3.627038	2.32576
Secondary sector	55.36745	24.65366	2.245810	36.23209	1.52813
Tertiary sector	36.19694	37.21896	0.972541	60.14087	0.60187
Industries (*i*)					
Mining		1.773323		2.396417	
Manufacturing		12.01835		26.92562	
Production/supply of electricity, gas, and water		0.538747		3.611226	
Construction	8.516117	10.32325	0.824946	3.298822	2.58156
Communication, transportation, storage, postal service	4.557318	3.374423	1.350548	10.84466	0.42024
Information transfer, computer, and software areas	1.731431	0.666810	2.596590	0.813442	2.12852
# Information Transfer					
Wholesale and retail	7.464513	8.973858	0.831806	1.040716	7.17248
Hotels and restaurants	1.661211	5.565369	0.298491	0.597470	2.78041
Finance	7.038597	0.778470	9.041574	0.033493	210.154

continued

Table 5.18 (continued) The Structures and Comparative Gain Measures of Chongqing's Three Economic Sectors and Industries ($t = 2011$)

Sectors (i)	*Demand Structure $d_i(t)$ (%)*	*Employment Structure $s_i^1(t)$ (%)*	*Comparative Gain of Labor $G_i^1(t)$*	*Investment Structure $s_i^2(t)$ (%)*	*Comparative Gain of Capital $G_i^2(t)$*
Real estate/housing	3.958299	1.061722	3.728187	31.33138	0.12634
Leasing and commercial services	1.561924	1.659769	0.941049	0.306395	5.09775
Scientific research, technology service, and geological survey	0.901175	0.595523	1.513249	0.195148	4.61791
Water conservation, environment, and public facility management	0.376672	0.469984	0.801456	10.13002	0.03718
# Public Facility Management					
Residents and other services	0.867514	8.442681	0.102753	0.431221	2.01176
Education	2.105406	2.515834	0.836862	1.199803	1.75479
Health, safety, and welfare	1.204331	1.098943	1.095899	0.591243	2.03695
Social, sports, and entertainment	0.486247	0.292715	1.661163	0.766560	0.63432
Public administration and social organizations	2.282305	1.722854	1.324723	1.859328	1.22749

Therefore, in terms of the magnitudes of the comparative gains of labor, the industries of Chongqing Municipality with available data can be ranked as follows:

1	2	3	4	5
Finance	Real estate/housing	Information transfer, computer, and software areas	Social, sports, and entertainment	Water conservation, environment, and public facility management
6	7	8	9	10
Communication, transportation, storage, postal service	Public administration and social organizations	Health, safety, and welfare	Leasing and commercial services	Education

11	12	13	14	15
Wholesale and retail	Construction	Water conservation, environment, and public facility management	Hotels and restaurants	Residents and other services

And, in terms of the magnitude of the comparative gains of capital, the industries of the Chongqing Municipality with available data can be ranked as follows:

1	2	3	4	5
Finance	Wholesale and retail	Leasing and commercial services	Scientific research, technology service, and geological survey	Hotels and restaurants
6	7	8	9	10
Construction	Information transfer, computer, and software areas	Health, safety, and welfare	Residents and other services	Education
11	12	13	14	15
Public administration and social organizations	Social, sports, and entertainment	Communication, transportation, storage, postal service	Real estate/ housing	Water conservation, environment, and public facility management

5.9.2 *Investment and Employment in Chongqing's Economic Sectors and Industries*

First, in terms of the comparative gains of labor, the values for this indicator for the following industries

Finance	Real estate/housing	Information transfer, computer, and software areas	Social, sports, and entertainment
Scientific research, technology service, and geological survey	Communication, transportation, storage, postal service	Public administration and social organizations	Health, safety, and welfare

are over 1, representing relatively optimal industries for employment. In all the other seven industries, there has been excessive supply of labor so that appropriate control on the increase of labor into these industries should be imposed. Specifically, the excessive labor in the following industries should be transferred into the previous eight industries.

In other words, no matter whether the job seekers, who look for employment opportunities in Chongqing Municipality, are from western China, or other parts of the mainland China, or the rest of the world, they should first consider the secondary economic sector with the relevant industries ranked as follows:

Residents and other services	Hotels and restaurants	Water conservation, environment, and public facility management	Construction
Wholesale and retail	Education	Leasing and commercial services	

1	2	3	4
Finance	Real estate/housing	Information transfer, computer, and software areas	Social, sports, and entertainment
5	6	7	8
Scientific research, technology service, and geological survey	Communication, transportation, storage, postal service	Public administration and social organizations	Health, safety, and welfare

Without special skills, don't look for employment opportunities in the primary and the tertiary economic sectors and other industries, because those areas are experiencing excessive labor supplies.

That is the direction of employment.

Second, in terms of capital, the following 11 industries have enjoyed the comparative gains of more than 1, representing relatively more optimal areas of additional investment.

Finance	Wholesale and retail	Leasing and commercial services	Scientific research, technology service, and geological survey
Hotels and restaurants	Construction	Information transfer, computer, and software areas	Health, safety, and welfare
Residents and other services	Education	Public administration and social organizations	

However, for the following four industries, excessive investment of capital has been given. Further capital investment in these industries should be curbed and redirected to the previous 11 industries.

Social, sports, and entertainment	Communication, transportation, storage, postal service	Real estate/housing	Water conservation, environment, and public facility management

In other words, no matter whether the investors, who look for investment opportunities in Chongqing Municipality, are from western China, or other parts of the mainland China, or the rest of the world, they should consider the following industries ranked from the most optimal to the least:

1	2	3	4	5
Finance	Residents and other services	Wholesale and retail	Information transfer, computer, and software areas	Leasing and commercial services
6	7	8	9	10
Public administration and social organizations	Education	Scientific research, technology service, and geological survey	Hotels and restaurants	Health, safety, and welfare

That is the direction of capital investment.

Third, by combining the two criteria discussed above, it follows that the industries in the Chongqing Municipality that are most optimal for both investment and employment are the following, because the comparative gains of labor and capital of each of these industries are over 1.

Finance	Scientific research, technology service, and geological survey	Information transfer, computer, and software areas	Health, safety, and welfare	Public administration and social organizations

The secondary choices are the industries of "leasing and commercial services" and "wholesale and retail." And the worst choices are the industries of "water conservation, environment, and public facility management" and "residents and other services."

5.10 Yunnan Province

Yunnan is located in the far southwest of the country at the far eastern edge of the Himalayan uplift with its capital at Kunming. It borders with Chinese provinces such as Tibet, Sichuan,

Guizhou, and Guangxi, and such countries as Myanmar in the southwest, Laos in the south, and Vietnam in the east. The eastern part of the province is a limestone plateau with karst topography and unnavigable rivers flowing through deep mountain gorges. The main surface formations of the plateau are the Lower Permian Maokou Formation, characterized by thick limestone deposits, the Lower Permian Qixia Formation, characterized by dolomitic limestones and dolostones, the Upper Permian basalts of the Ermeishan Formation, and the red sandstones, mudstones, siltstones, and conglomerates of the Mesozoic–Paleogene, including the Lufeng Formation and the Lunan Group (Lumeiyi, Xiaotun, and Caijiacong formations). The terrain of the province is largely mountainous, especially in the north and west, with a series of high mountain chains spreading across the land. The rugged, vertical terrain produces a wide range of flora and fauna, making the province known as a natural zoological and botanical garden. For more details, see Figure 5.10.

The province is the source of two rivers, the Xi River and the Yuan River and is drained by six major river systems: the Jinsha Jiang, the Pearl River, the Mekong, the Red River, the Nujiang, and the Irrawaddy. It has eight major lakes of areas of over 30 square kilometers (12 square miles), Dianchi Lake near Kunming, Fuxian Lake in Yuxi, Erhai Lake near Dali City, Lugu Lake in Ninglang, Qilu Lake in Tonghai County, Yangzong Lake in Yiliang County, Yilong Lake, and Xingyun Lake. Other than the diversity in its geographical features, Yunnan is also most diverse in China both biologically and culturally. It contains snow-capped mountains and truly tropical environments so that it supports an unusually full spectrum of species and vegetation types. This topographic range combined with tropical moisture sustains extremely high biodiversity and high degrees of endemism, making the region the richest botanically in the world's temperate regions. The fauna is also nearly diverse. Although Yunnan has less than 4% of the total land area of China,

Figure 5.10 The geographic location of Yunnan.

it contains about half of China's birds and mammals. It is home to, most notably, the Southeast Asian gaur, the tiger, and the Asian elephant.

Because of its location on south-facing mountain slopes, receiving the influence of both the Pacific and Indian oceans, Yunnan has a generally mild climate with pleasant and fair weather. Much of the area lies within the subtropical highland or humid subtropical zone with mild to warm winters, and tempered summers. In general, average temperatures in January range from 8 to 17°C (46–63°F); and average temperatures in July vary from 21 to 27°C (70–81°F) with an average annual rainfall ranging from 600 to 2300 mm (24 to 91 in.).

Historically, around the third century BC, the Chu general Zhuang Qiao entered the central area of Yunnan around present day Kunming from the upper Yangtze River and set himself up as "King of Dian." He brought an influx of Chinese influence into Yunnan, and started a long history of migration and cultural expansion. In 221 BC, the State of Qin established in Yunnan commanderies and counties. During the Han Dynasty, a series of military campaigns were dispatched against the Dian. During the time of Three Kingdoms, the dissolution of Chinese central authority led to increased autonomy for Yunnan. In the 320s, the Cuan clan migrated into Yunnan, while Cuan Chen named himself king and held authority from Lake Dian for the next four hundred plus years. In 738, the kingdom of Nanzhao was established in Yunnan by Pilucge, recognized by the imperial court of the Tang Dynasty. In 937, Duan Siping overthrew the Nanzhao and established the Kingdom of Dali, which was later conquered by the Mongol Empire in 1253 and made into a province. The Yunnan Province during the Yuan Dynasty also included significant portions of Upper Burma. During the Ming and Qing dynasties, the areas of Yunnan were administered under the native chieftain system. During the Qing Dynasty, an ethnic cleansing policy was adopted under the title of "washing off the Muslims." The province was transformed enormously by the war against Japan, and assumed great strategic significance, particularly as the Burma Road was constructed from Kunming to Lashio in Burma during this time.

Economically, the province's main source of wealth lies in its vast mineral resources. And mining is the leading industry in Yunnan. The potential value of the proven deposits in province is ¥3 trillion with 40% from fuel minerals, 7.3% from metallic minerals, and 52.7% from non-metallic minerals. The rich water resources of the province offer abundant hydro-energy. A series of dams on the Mekong has been constructed to develop the river into a waterway and source of power. Yunnan is one of the major production bases of copper, lead, zinc, tin, and aluminum in China. Gejiu is well known for its zinc reserves with the Yunxi brand refined tin being one of the main products, which is registered on the London Metal Exchange. Additionally, reserves of germanium, indium, zirconium, platinum, rock salt, sylvite, nickel, phosphate, mirabilite, arsenic, and blue asbestos are also high. And significant copper deposits are found at Dongchuan, iron ore at Wuding, and coal at Xuanwei and Kaiyuan. The economic policy to locate new industry in interior areas with substantial mineral wealth has led to major industrial development in Yunnan, especially in the Kunming area. And, the electricity industry is another important economic pillar of Yunnan, which plays a key role in the "West-East Electricity Transmission Project." Even so, Yunnan is one of China's relatively undeveloped provinces. For example, in 1994, about 7 million people lived below the poverty line. They were financially supported by the central government. With an input of ¥3.15 billion in 2002, the absolutely poor rural population in the province has been reduced from 4.05 million in 2000 to 2.86 million in 2002. The alleviation plan includes five projects aimed at improving infrastructure facilities, attempting at soil improvement, water conservation, electric power, roads, and "green belt" building. When the planned projects are fully functional, they will alleviate the shortages of grain, water, electric power, and roads. Although Yunnan lags behind the east coast of China in relation to socio-economic development, because of its geographic location the province has comparative advantages in regional and border trade with countries in Southeast Asia. The Lancang River is the waterway to Southeast Asia. The improved land transportation has strengthened the economic and trade co-operation among countries in the Greater Mekong subregion. The structure of Yunnan's natural resources determines the province's pillar industries: agriculture, tobacco, mining, hydro-electric power, and tourism. The secondary industry is currently the largest industrial tier in Yunnan, contributing over 45% of the GDP. The tertiary industry contributes 40% and agriculture 15%. The four pillar industries of the province include tobacco, agriculture/biology, mining, and tourism. The main manufacturing industries are iron and steel production and copper-smelting, commercial vehicles, chemicals, fertilizers, textiles, and optical instruments. Other than having trade contacts with over 70 countries and regions in the world, Yunnan has also established the Muse border trade zone along its border with Burma. In 2008, its total imports and exports reached US$9.6 billion. The nominal GDP of Yunnan in 2011 was ¥875.1 billion (US$138.92 billion), and its per capita GDP was ¥13,494 (US$1975).

For additional readings, refer to Ministry of Commerce (2013); National Bureau of Statistics of China (2013); Gernet (1996); Tian, Li, Zhang, and Guest (1998); Zhang et al. (2009); Forbes and Henley (2011); Booz (1998); McCarthy (2009); and Roberts (1837).

5.10.1 Current State of Investment and Employment Structures of Yunnan

All that is necessary for us to do is the computations of the gross domestic outputs of all the economic sectors and industries and the investment and employment situations (Table 5.19) of Yunnan Province and the relevant comparative gains. By comparing the resultant comparative gains, we can then obtain the current states of investment and employment for the three economic sectors and industries. All the detailed computations are similar to what has been described earlier. The results are given in Table 5.20.

5.11 Tibet Autonomous Region

Tibet is a plateau region in Asia, north-east of the Himalayas, and in the west of the Central China plain. It is the highest region on Earth with an average elevation of 4900 m (16,000 ft). The province has some of the world's tallest mountains. For example, Mount Everest, at 8848 m (29,029 ft), is the highest mountain on Earth, located on the border with Nepal. Several major rivers, such as the Yangtze River, Yellow River, Indus River, Mekong, Ganges, Salween, and the Yarlung Tsangpo River, originate in the Qinghai Province of the Tibetan Plateau. The Yarlung Tsangpo Grand Canyon, along the Yarlung Tsangpo River, is among the deepest and longest canyons in the world. The region has been called the "Water Tower" of Asia, and the central government is investing heavily in water projects in the area. For more details, see Figure 5.11.

The Indus and Brahmaputra rivers start at a lake in Western Tibet near Mount Kailash. The mountain is a holy pilgrimage site for both Hindus and Tibetans. The Hindus consider the mountain to be the abode of Lord Shiva, while the Tibetan names it as Khang Rinpoche. Tibet has numerous high-altitude lakes, such as Qinghai Lake, Lake Manasarovar, Namtso, Pangong Tso, Yamdrok Lake, Siling Co, Lhamo La-tso, Lumajangdong Co, Lake Puma Yumco, Lake Paiku, Lake Rakshastal, Dagze Co, and Dong Co. The Qinghai Lake is the largest lake in the People's Republic of China. The weather in Tibet is severely dry nine months out of the year, with average annual snowfall only 46 cm (18 in.) due to the rain shadow effect. Although the western passes

Table 5.19 The GDPs, Total Investments in Fixed Properties, and Employment Situation of Yunnan's Three Economic Sectors and Industries (2011)

Sectors (i)	*Total Output of Sector (0.1 Billion Yuan)*	*Total Investment on Fixed Properties (0.1 Billion Yuan)*	*Employment (10,000 People)*
Total	8893.12	7109.7	2857.24
Primary sector	1411.01	282.16	1697.24
Secondary sector	3780.32	2222.24	374.3
Tertiary sector	3701.79	4605.3	785.7
Industries (*i*)			
Agriculture, forest, animal husbandry, fishery		153.15	**11**
Mining		262.42	**18.4**
Manufacturing		895.5	**63.2**
Production/supply of electricity, gas, and water		792.61	**9.5**
Construction	786.02	6.13	**48.1**
Communication, transportation, storage, postal service	**217.22**	842.8	**13.6**
Information transfer, computer, and software areas		48.64	**4.1**
# Information Transfer			
Wholesale and retail	**932.21**	253.87	**19.4**
Hotels and restaurants	**278.2**	97.55	**7.4**
Finance	**456.23**	4.1	**9.7**
Real estate/housing	**222.31**	1272.72	**5.4**
Leasing and commercial services		21.49	**6.7**
Scientific research, technology service, and geological survey		33.28	**6.4**
Water conservation, environment, and public facility management		597.18	**5.5**

continued

Table 5.19 (continued) The GDPs, Total Investments in Fixed Properties, and Employment Situation of Yunnan's Three Economic Sectors and Industries (2011)

Sectors (i)	*Total Output of Sector (0.1 Billion Yuan)*	*Total Investment on Fixed Properties (0.1 Billion Yuan)*	*Employment (10,000 People)*
# Public facility management		338.4	
Residents and other services		14.46	**0.7**
Education		126.05	**52.9**
Health, safety, and welfare		81.54	**17.6**
Social, sports, and entertainment		98.63	**3.4**
Public administration and social organizations		115.49	**47**

Source: The data in bold are from *China Statistics Yearbook* (2012), the data on domestic GDPs and employments from *Yunnan Province's Statistics Yearbook* (2012), and the data on total investments of fixed properties are from *Statistics Report on Domestic Economies and Social Development of Yunnan Province* 2011, published by Yunnan Province's Statistics Bureau. What needs to be noted is that the data from the latter two sources do not agree with each other.

only receive small amounts of fresh snow each year, they remain traversable all year round. Low temperatures are prevalent throughout the western regions, where bleak desolation is unrelieved by any vegetation bigger than a low bush, and wind sweeps unchecked across the vast expanses of arid plain. Since the Indian monsoon exerts some influence on the eastern part of Tibet, the northern part is subject to high temperatures in the summer and intense cold in the winter.

Culturally, Tibet consists of several regions, such as Amdo in the northeast as parts of the Qinghai province, Gansu province, and Sichuan province, Kham in the southeast that encompasses parts of western Sichuan, northern Yunnan, southern Qinghai, and the eastern part of Tibet, Ü-Tsang that covers the central and western portion of Tibet. The Tibetan cultural influence extends to the neighboring states of Bhutan, Nepal, regions of India in addition to designated Tibetan autonomous areas in the adjacent Chinese provinces. Lhasa, Tibet's capital, contains two world heritage sites, the Potala Palace and Norbulingka, and a number of significant temples and monasteries, such as Jokhang and Ramoche Temple.

Historically, the Zhang Zhung culture made up of the people migrated from the Amdo region into what is now the region of Guge in western Tibet. It continued its dominance of the region until it was annexed by Songtsen Gampo in the seventh century. A unified Tibet began with the rule of Songtsän Gampo (604–650 BC), who founded the Tibetan Empire. In 640, he married Princess Wencheng, a niece of the powerful Chinese emperor of the Tang Dynasty. Under the next few Tibetan kings, Buddhism became established as the state religion and Tibetan power increased further over large areas of Central Asia with major inroads made into Chinese territory, reaching as far as the Tang's capital Chang'an (modern Xi'an) in late 763. However, the Tibetan occupation of Chang'an only lasted for 15 days before they were defeated by the Tang forces. In

Table 5.20 The Structures and Comparative Gain Measures of Yunnan's Three Economic Sectors and Industries ($t = 2011$)

Sectors (i)	*Demand Structure* $d_i(t)$ *(%)*	*Employment Structure* $s_i^1(t)$ *(%)*	*Comparative Gain of Labor* $G_i^1(t)$	*Investment Structure* $s_i^2(t)$ *(%)*	*Comparative Gain of Capital* $G_i^2(t)$
Total	100	100	—	100	—
Primary sector	15.86631	59.40138	0.267103	3.968663	3.99790
Secondary sector	42.50837	13.10005	3.244900	31.25645	1.35999
Tertiary sector	41.62532	27.49857	1.513727	64.77489	0.64262
Industries (*i*)					
Agriculture, forest, animal husbandry, fishery		0.384987		2.154099	
Mining		0.643978		3.691014	
Manufacturing		2.211925		12.59547	
Production/supply of electricity, gas, and water		0.332489		11.14829	
Construction	8.838518	1.683443	5.250263	0.086220	102.511
Communication, transportation, storage, postal service	2.442562	0.475984	5.131608	11.85423	0.20605
Information transfer, computer, and software areas		0.143495		0.684136	
# Information Transfer					
Wholesale and retail	10.48237	0.678977	15.43848	3.570755	2.93568
Hotels and restaurants	3.128261	0.258991	12.07854	1.372069	2.27996
Finance	5.130146	0.339488	15.11140	0.057668	88.9605
Real estate/ housing	2.499798	0.188994	13.22689	17.90118	0.13964
Leasing and commercial services		0.234492		0.302263	

continued

Table 5.20 (continued) The Structures and Comparative Gain Measures of Yunnan's Three Economic Sectors and Industries ($t = 2011$)

Sectors (i)	*Demand Structure $d_i(t)$ (%)*	*Employment Structure $s_i^1(t)$ (%)*	*Comparative Gain of Labor $C_i^1(t)$*	*Investment Structure $s_i^2(t)$ (%)*	*Comparative Gain of Capital $G_i^2(t)$*
Scientific research, technology service, and geological survey		0.223992		0.468093	
Water conservation, environment, and public facility management		0.192493		8.399511	
# Public Facility Management					
Residents and other services		0.024499		0.203384	
Education		1.851437		1.772930	
Health, safety, and welfare		0.615979		1.146884	
Social, sports, and entertainment		0.118996		1.387260	
Public administration and social organizations		1.644944		1.624400	

Note: The GDP data and the employment statistics come from *Yunnan Statistics Yearbook* (2012), while the data of investment on fixed properties from "Statistics Report on Domestic Economies and Social Development of Yunnan Province 2011", published by Yunnan Province's Statistics Bureau. Because these two sources disagree with each other and because there are too many missing data, the investment and employment situations of this province cannot be addressed.

747, the blockage of Tibet was loosened by General Gao Xianzhi in his attempt to re-establish direct communications between Central Asia and Kashmir. By 750, the Tibetans lost almost all of their central Asian possessions to the Chinese. However, with the start of the civil war in 755, Chinese influence decreased rapidly and Tibetan influence resumed. Tibet continued as a Central Asian empire until the mid-ninth century when a civil war over succession led to the collapse of imperial Tibet. The Yuan Dynasty ruled Tibet through a top-level administrative department, where Tibet retained nominal power over religious and regional political affairs, while the Yuan government managed a structural and administrative rule over the region. When the Yuan Dynasty was ended by the Ming, Tai Situ Changchub Gyaltsen founded between 1346 and 1354 the Phagmodrupa dynasty and sought to reduce Yuan influences. The Qing Dynasty put Amdo

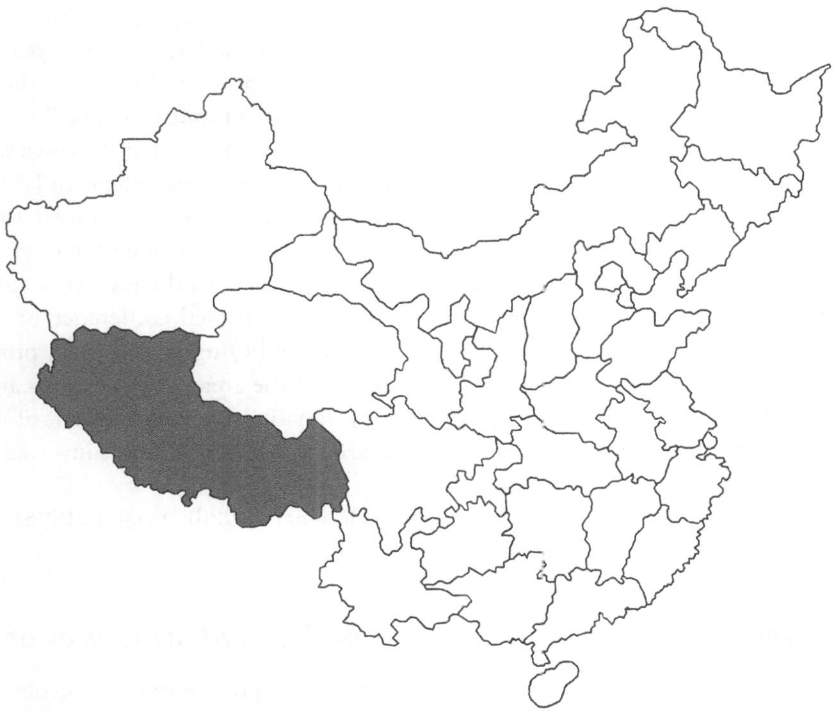

Figure 5.11 The geographic location of Tibet.

under its control in 1724. In 1834, the Sikh Empire invaded and annexed Ladakh. Seven years later when a Sikh army invaded western Tibet from Ladakh, the Sino-Sikh War broke out. The war ended with the signing of the Treaty of Chushul between the Chinese and Sikh empires. By the mid-nineteenth century, the Qing influence in Tibet was greatly weakened, making Qing's authority over Tibet more symbolic than real in the late nineteenth century. In 1910, the Qing government sent a military expedition to establish direct Manchu-Chinese rule. After the Xinhai Revolution (1911–12) toppled the Qing Dynasty, the new Republic of China restored the title of Dalai Lama. However, the Dalai Lama declared himself ruler of an independent Tibet. For the next 36 years, Tibet fought Chinese warlords for control of the areas along the upper reaches of the Yangtze River. In 1914, the Tibetan government signed the Simla Accord with Britain, ceding the South Tibet region to British India. The Chinese government denounced the agreement as illegal. In the 1930s and 1940s, the Republic of China expanded its reach into the territory. The People's Republic of China incorporated Tibet in 1950 and negotiated the Seventeen Point Agreement with the newly enthroned 14th Dalai Lama's government, affirming the People's Republic of China's sovereignty but granting the area autonomy. After the Dalai Lama government fled to Dharamsala, India, in 1959, it established a rival government-in-exile. Afterwards, the Central People's Government in Beijing renounced the agreement and began implementation of the halted social and political reforms. In 1962, China and India fought a brief war over the disputed South Tibet and Aksai Chin regions.

Economically, the economy of Tibet is dominated by subsistence agriculture. Because of the limited arable land, the primary occupation of the Tibetan Plateau is raising livestock, such as sheep, cattle, goats, camels, yaks, dzo, and horses. The main crops grown there are barley, wheat, buckwheat, rye, potatoes, and assorted fruits and vegetables. In recent years, tourism has become an increasingly important sector, and is actively promoted by the authorities. The Central National Government exempts Tibet from all taxation and provides 90% of Tibet's government expenditures. The Qingzang railway linking Tibet to Qinghai Province was opened in 2006, providing a direct link between the central plain of China and Tibet. In January 2007, the Chinese government published a report on the large mineral deposit under the Tibetan Plateau worth over $128 billion. On January 15, 2009, China announced the construction of Tibet's first expressway of 37.9 km long in southwestern Lhasa with the projected cost of ¥1.55 billion (US$227 million). During January 18–20, 2010, a national conference on Tibet and areas inhabited by Tibetans in Sichuan, Yunnan, Gansu, and Qinghai was held, producing an important substantial plan to improve the development of the area. The event was attended by all members of CPC Politburo Standing Committee, signaling the commitment of the leaders to the development of Tibet and ethnic Tibetan areas. Since 2001, China has invested over ¥310 billion (US$45.6 billion) in Tibet.

For additional readings, refer to Allen (2004); Dowman (1988); Feigon (1998); Goldstein (2007); Marks (1978); and Zhao et al. (2009).

5.11.1 Current State of Investment and Employment Structures of Tibet

All that is necessary for us to do is the computations of the gross domestic outputs of all the economic sectors and industries and the investment and employment situations (Table 5.21) of the Tibet Autonomous Region and the relevant comparative gains. By comparing the resultant comparative gains, we can then obtain the current states of investment and employment for the three economic sectors and industries. All the detailed computations are similar to what has been described earlier. The results are given in Table 5.22. Table 5.22 indicates the following facts:

First, in terms of the three major economic sectors, the six relevant data points of the three sectors of the Tibet Autonomous Region are all situated in extreme instability. Specifically, the comparative gain of labor of the primary sector is way too low and equal to merely 0.244173. That is not even reaching one-eleventh and one-fifth of that same indicator of the secondary and the tertiary sectors, respectively. In terms of the capital investment, the comparative gains of the primary and the secondary sectors are much too low, and are respectively equal to 2.20141 and 1.25620, while the capital investment in the tertiary sector is in extreme excess (0.79491).

Second, in terms of the employment situation, the relevant data of the 21 industries of the Tibet Autonomous Region should contain 42 data points. However, because of lack of raw data, we only collected 12 of them, all of which are situated in extremely imbalanced states. In particular, the comparative gains of labor of the following four industries are all over 1. Especially, the comparative gain of labor in "real estate/housing" is even equal to 19.36700, which is more than (19.36700/0.531809 ≈) 36 times of the lowest value of this indicator that is reached by "wholesale and retail."

Real estate/housing	Finance	Construction	Communication, transportation, storage, postal service

At the same time, the imbalance in terms of the comparative gains of capital seems to be way too severe. The comparative gains of capital in the following four industries are all over 1.

Table 5.21 The GDPs, Total Investments in Fixed Properties, and Employment Situation of Tibet's Three Economic Sectors and Industries (2011)

Sectors (i)	*Total Output of Sector (0.1 Billion Yuan)*	*Total Investment on Fixed Properties (0.1 Billion Yuan)*	*Employment (10,000 People)*
Total	605.83	549.27	185.55
Primary sector	74.47	**30.67**	93.41
Secondary sector	208.79	**150.69**	22.57
Tertiary sector	322.57	**367.91**	69.57
Industries (*i*)			
Agriculture, forest, animal husbandry, fishery		**24.2**	93.40
Mining		**15.3**	3.32
Manufacturing		**34.8**	2.78
Production/supply of electricity, gas, and water		**65.1**	0.99
Construction	**160.61**	**31.2**	15.48
Communication, transportation, storage, postal service	**23.95**	**146.1**	5.28
Information transfer, computer, and software areas		**6.0**	0.36
# Information Transfer			
Wholesale and retail	**34.25**	**3.4**	19.72
Hotels and restaurants	**17.75**	**13.4**	8.41
Finance	**31.7**	**0.5**	0.97
Real estate/housing	**17.44**	**52.1**	0.28
Leasing and commercial services		**1.7**	1.70
Scientific research, technology service, and geological survey		**0.8**	0.81
Water conservation, environment, and public facility management		**43.5**	0.24
# Public Facility Management			
Residents and other services		**1.8**	3.24

continued

Table 5.21 (continued) The GDPs, Total Investments in Fixed Properties, and Employment Situation of Tibet's Three Economic Sectors and Industries (2011)

Sectors (i)	Total Output of Sector (0.1 Billion Yuan)	Total Investment on Fixed Properties (0.1 Billion Yuan)	Employment (10,000 People)
Education		**15.9**	4.27
Health, safety, and welfare		**7.1**	1.65
Social, sports, and entertainment		**11.7**	1.91
Public administration and social organizations		**41.9**	10.02

Source: The data in bold are from *China Statistics Yearbook* (2012), and the rest are from *Tibet Statistics Yearbook* (2012).

In particular, the comparative gain of capital of "communication, transportation, storage, postal service" is very low, equaling to merely 0.14862, which is only (0.14862/57.4809 ≈) 0.0027 when compared to that of "finance" and (0.14862/9.13305 ≈) 0.016 when compared to that of "wholesale and retail."

Finance	Wholesale and retail	Construction	Hotels and restaurants

Therefore, in terms of the magnitude of the comparative gains of labor, the industries of the Tibet Autonomous Region with available data can be ranked as follows:

1	2	3
Real estate/housing	Finance	Construction
4	5	6
Communication, transportation, storage, postal service	Hotels and restaurants	Wholesale and retail

And, in terms of the magnitudes of the comparative gains of capital, the industries of the Tibet Autonomous Region with available data can be ranked as follows:

1	2	3
Finance	Wholesale and retail	Construction
4	5	6
Hotels and restaurants	Real estate/housing	Communication, transportation, storage, postal service

Table 5.22 The Structures and Comparative Gain Measures of Tibet's Three Economic Sectors and Industries ($t = 2011$)

Sectors (i)	*Demand Structure $d_i(t)$ (%)*	*Employment Structure $s_i^1(t)$ (%)*	*Comparative Gain of Labor $G_i^1(t)$*	*Investment Structure $s_i^2(t)$ (%)*	*Comparative Gain of Capital $G_i^2(t)$*
Total	100	100	—	100	—
Primary sector	12.29223	50.34223	0.244173	5.583785	2.20141
Secondary sector	34.46346	12.16384	2.833272	27.43464	1.25620
Tertiary sector	53.24431	37.49394	1.420078	66.98175	0.79491
Industries (*i*)					
Agriculture, forest, animal husbandry, fishery		50.33932		4.405856	
Mining		1.791431		2.78552	
Manufacturing		1.496524		6.335693	
Production/supply of electricity, gas, and water		0.535004		11.85212	
Construction	26.51074	8.343088	3.177569	5.680277	4.66716
Communication, transportation, storage, postal service	3.953254	2.843438	1.390308	26.59899	0.14862
Information transfer, computer, and software areas		0.462948		1.092361	
# Information Transfer					
Wholesale and retail	5.653401	10.63050	0.531809	0.619005	9.13305
Hotels and restaurants	2.929865	4.529830	0.646794	2.439606	1.20096
Finance	5.232491	0.520237	10.05790	0.09103	57.4809
Real estate/housing	2.878695	0.148639	19.36700	9.485334	0.30349
Leasing and commercial services		0.915387		0.309502	

continued

Table 5.22 (continued) The Structures and Comparative Gain Measures of Tibet's Three Economic Sectors and Industries ($t = 2011$)

Sectors (i)	*Demand Structure $d_i(t)$ (%)*	*Employment Structure $s_i^1(t)$ (%)*	*Comparative Gain of Labor $G_i^1(t)$*	*Investment Structure $s_i^2(t)$ (%)*	*Comparative Gain of Capital $G_i^2(t)$*
Scientific research, technology service, and geological survey		0.438049		0.145648	
Water conservation, environment, and public facility management		0.127297		7.919617	
# Public Facility Management					
Residents and other services		1.745998		0.327708	
Education		2.300404		2.894756	
Health, safety, and welfare		0.888332		1.292627	
Social, sports, and entertainment		1.028887		2.130104	
Public administration and social organizations		5.398060			

5.11.2 Investment and Employment in Tibet's Economic Sectors and Industries

First, in terms of the comparative gains of labor, the values of this indicator for the following industries

Real estate/housing	Finance	Construction	Communication, transportation, storage, postal service

are over 1, representing relatively optimal industries for employment. In the remaining two industries, there has been excessive supply of labor so that appropriate control on the increase of labor into these industries should be imposed. Specifically, the excessive labor in the following industries should be transferred into the previous four industries.

Hotels and restaurants	Wholesale and retail

In other words, no matter whether the job seekers, who look for employment opportunities in Tibet Autonomous Region, are from western China, or other parts of the mainland China, or the rest of the world, they should first consider the secondary and the tertiary economic sectors with the relevant industries ranked as follows:

1	2	3	4
Real estate/housing	Finance	Construction	Communication, transportation, storage, postal service

Without special skills, don't look for employment opportunities in the primary economic sector and other industries, because those areas are experiencing excessive labor supplies.

That is the direction of employment.

Second, in terms of capital, the following four industries have enjoyed the comparative gains of more than 1, representing relatively more optimal areas of additional investment.

Finance	Wholesale and retail	Construction	Hotels and restaurants

However, for the following two industries, excessive investment of capital has been given. Further capital investment in these industries should be curbed and redirected to the previous four industries.

Communication, transportation, storage, postal service	Real estate/housing

In other words, no matter whether the investors, who look for investment opportunities in Tibet Autonomous Region, are from western China, or other parts of the mainland China, or the rest of the world, they should consider the following industries ranked from the most optimal to the least:

Finance	Wholesale and retail	Construction	Hotels and restaurants

That is the direction of capital investment.

Third, by combining the two criteria discussed above, it follows that the economic sector and the industries in Tibet Autonomous Region that are most optimal for both investment and employment are the following, because the comparative gains of labor and capital of each of these industries are over 1.

Secondary sector	Finance	Construction

3

TRADITIONAL ADMINISTRATIVE AREAS AND MODERN SPECIAL ZONES

The geographic division of China has been changing with time. First, there is such a division as discussed in Section 2: the east, the central plain, and the west. Second, there appears a division of the following six major administrative areas: the northeast administrative area, the north administrative area, the east administrative area, the south administrative area, the southwest administrative area, and the northwest administrative area. Third, there are the current specific zones: Bohai Rim, Yangtze River Delta zone, and the Greater Pearl River Delta region.

We have studied the relevant situations of the second geographic division that consists of the east, central, and west. In this section, we will look at the third geographic division.

Chapter 6

Regions and Industries in the Six Administrative Areas

In this chapter, we look at the traditional division of China into the following six areas: Northeast, Northwest, North, East, South, and Southwest (Figure 6.1). Consequently, we are able to rank these areas in the order from the most optimal to the most difficult.

6.1 Investment and Employment in the Six Administrative Areas

6.1.1 Current Regional Economic Structures of the Six Areas

Similar to what we have done in the previous chapters, we compute one by one the vector of demand structure $d(t)$, the vector of employment structure $S^1(t)$, and the vector of investment structure $S^2(t)$ for each of the Chinese provinces, municipalities, and autonomous regions, by using the relevant statistics data of 2011 (Table 6.1). Then, we substitute the relevant numerical values into Equation 1.1 in Chapter 1 to calculate, respectively, the comparative gains of labor $G_i^1(t) = d_i(t)/S_i^1(t)$ and the comparative gains of capital $G_i^2(t) = d_i(t)/S_i^2(t)$, which lead to Table 6.2.

Table 6.2 indicates that the regional economic structures of the six major administrative areas of China are in severe imbalances, no matter whether we look at them by using the demand and supply of labor or the demand and supply of capital.

In fact, from the fifth column (the comparative gains of labor) and the seventh column (the comparative gains of capital) of the table, it follows that among all the 31 provinces, municipalities, and autonomous regions, only three of them have their comparative gains equal to 1, while all the rest are not equal to 1 and are situated in various states of severe imbalances.

If we classify regions with comparative gains $G_i^j(t)$ between 0.9500 and 1.0500 as regions in basic equilibrium, and the rest as in states of imbalance, then in terms of the six administrative areas, only five regions are in basic equilibrium while all the rest (7/12 = 58%) are situated in states of varying degrees of imbalance. In terms of the 31 provinces, municipalities, and autonomous

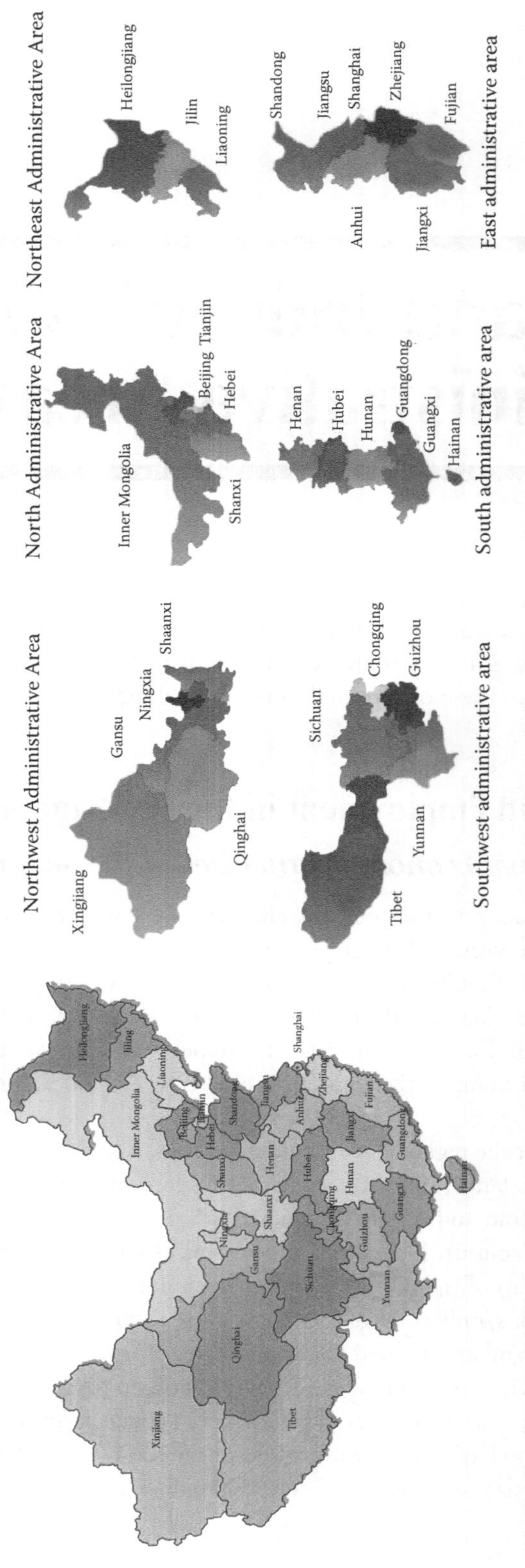

Figure 6.1 Traditional division of China into six areas.

Table 6.1 GDPs, Total Investments in Fixed Properties, and Employment Situations of All Provinces, Municipalities, and Autonomous Regions in China (2011)

Area	*Province/Municipality/ Autonomous Region (i)*	*Regional GDP (0.1 Billion Yuan)*	*Value of Fixed Properties (0.1 Billion Yuan)*	*Employment (10,000 People)*
North	1. Beijing	16251.93	5578.9	2143.9
	2. Tianjin	11307.28	7067.7	688
	3. Hebei	24515.76	16389.3	2157.4
	4. Shanxi	11237.55	7073.1	1225.7
	5. Inner Mongolia	14359.88	10366.2	1120.1
	Total	77672.40	46475.2	7334.6
Northeast	6. Liaoning	22226.70	17726.3	2675.8
	7. Jilin	10568.83	7441.7	1283.7
	8. Heilongjiang	12582.00	7475.4	1672.1
	Total	45377.53	32643.4	5631.6
East	9. Shanghai	19195.69	4962.1	2185
	10. Jiangsu	49110.27	26692.6	6451
	11. Zhejiang	32318.85	14185.3	4810.8
	12. Anhui	15300.65	12455.7	2028.8
	13. Fujian	17560.18	9910.9	2208.9
	14. Jiangxi	11702.82	9087.6	1942.4
	15. Shandong	45361.85	26749.7	4315.9
	Total	190550.31	104043.9	23943.8
South	16. Henan	26931.03	17769.0	2892.4
	17. Hubei	19632.26	12557.3	2701.7
	18. Hunan	19669.56	11880.9	2413.3
	19. Guangdong	53120.28	17069.2	5943.6
	20. Guangxi	11720.87	7990.7	1489.4
	21. Hainan	2522.66	1657.2	383.2
	Total	133686.66	68924.3	15823.6
Southwest	22. Chongqing	10011.37	7473.4	1580.5
	23. Sichuan	21026.68	14222.2	2647.7
	24. Guizhou	5701.84	4235.9	789.9
	25. Yunnan	8893.12	6191.0	1753.6
	26. Tibet	605.83	516.3	153.4
	Total	46238.84	32638.8	6925.1
Northwest	27. Shaanxi	12512.30	9431.1	1427.2
	28. Gansu	5020.37	3965.8	747

continued

Table 6.1 (continued) GDPs, Total Investments in Fixed Properties, and Employment Situations of All Provinces, Municipalities, and Autonomous Regions in China (2011)

Area	*Province/Municipality/ Autonomous Region (i)*	*Regional GDP (0.1 Billion Yuan)*	*Value of Fixed Properties (0.1 Billion Yuan)*	*Employments (10,000)*
	29. Qinghai	1670.44	1435.6	233.1
	30. Ningxia	2102.21	1644.7	321.3
	31. Xinjiang	6610.05	4632.1	771.5
	Total	27915.37	21109.3	3500.7
National total		521441.11	305834.9	63159.4

Source: Data are from *China Statistics Yearbook 2012* (XLS, DOC).

regions, other than 10 are in basic equilibrium, the rest (52/62 = 84%) are in states of varying degrees of imbalance. In particular, the degrees of imbalance in the demand-and-supply structures of the 10 provinces, municipalities, and autonomous regions of the west region (both southwest and northwest) and the following provinces, municipalities, and autonomous regions are extremely severe. The comparative gains of labor and capital of Tibet are the lowest, equaling, respectively, 0.4167 and 0.5, while the comparative gain of Tianjin is the highest, equaling 2.0183. This fact implies that Tianjin and Tibet represent the two extremes of production efficiency of China. At the same time, the comparative gain of the capital of Shanghai is the highest, equaling 2.313. This implies that Shanghai and Tibet are the two extremes of investment efficiency of China.

Beijing	Tianjin	Hebei	Inner Mongolia	Liaoning
Shanghai	Zhejiang	Anhui	Jiangxi	Henan
Guangdong	Guangxi	Hainan		

In terms of the six major administrative areas, the data of the table indicate that the supply and demand of both labor and capital are severely imbalanced. The mean of the comparative gains of labor of the five provinces and municipalities of Northern China is equal to 1.2920 ($\gg$1). So, this area is the best area of return on labor in China, representing the current-stage direction of development of China's social productivity. Therefore, additional supply of labor should be directed to this area. On the other hand, the comparative gain of capital of the Southern area is 1.138 ($\gg$1). So, this region is the area with the best return on capital. This implies that the supply of capital to Southern China is insufficient; and that additional supply of capital to this area should be strengthened. This indicates that it is very possible that this area will be the focus of next round of capital investment. As for the 10 provinces and municipalities of the Western area, other than the labor supply is basically in equilibrium (= 0.9747), the labor supply in the southwest region (0.8263$\ll$1) and the supplies of capital in the southwest (0.832$\ll$1) and the northwest (0.768$\ll$1) are severely in excess. They are regions in China with the worst labor productivity and returns on capital and are also the most economically deprived areas in China.

Because the comparative gains are defined as ratios of two percentage values, a comparative gain that is equal to 1 means equilibrium of supply and demand. For a chosen region, if its comparative gain of labor is greater than 1, it means that the relatively scarce labor of the region

Table 6.2 Regional Economic Structures and Comparative Gain Measures of the Provinces, Municipalities, and Autonomous Regions of China (2011)

Area	*Province/Municipality/ Autonomous Region (i)*	*Demand Structure $d_i(t)$ (%)*	*Employment Structure $s_i^1(t)$ (%)*	*Comparative Gain of Labor $G_i^1(t)$*	*Investment Structure $s_i^2(t)$ (%)*	*Comparative Gain of Capital $G_i^2(t)$*
North	1. Beijing	20.924	29.230	0.716	12.004	1.743
	2. Tianjin	14.558	9.380	1.552	15.207	0.957
	3. Hebei	31.563	29.414	1.073	35.265	0.895
	4. Shanxi	14.468	16.711	0.866	15.219	0.951
	5. Inner Mongolia	18.488	15.271	1.211	22.305	0.829
	Total	100	100	—	100	—
Northeast	6. Liaoning	48.982	47.514	1.031	54.303	0.902
	7. Jilin	23.291	22.795	1.022	22.797	1.022
	8. Heilongjiang	27.727	29.691	0.934	22.900	1.211
	Total	100	100	—	100	—
East	9. Shanghai	10.074	9.130	1.103	4.769	2.112
	10. Jiangsu	25.773	26.942	0.957	25.655	1.005
	11. Zhejiang	16.961	20.092	0.844	13.634	1.244
	12. Anhui	8.030	8.473	0.948	11.972	0.671
	13. Fujian	9.216	9.225	0.999	9.533	0.967
	14. Jiangxi	6.142	8.112	0.757	8.734	0.703
	15. Shandong	23.806	18.025	1.321	25.710	0.926
	Total	100	100	—	100	—

continued

Table 6.2 (continued) Regional Economic Structures and Comparative Gain Measures of the Provinces, Municipalities, and Autonomous Regions of China (2011)

Area	*Province/Municipality/ Autonomous Region (i)*	*Demand Structure $d_i(t)$ (%)*	*Employment Structure $s_i^1(t)$ (%)*	*Comparative Gain of Labor $G_i^1(t)$*	*Investment Structure $s_i^2(t)$ (%)*	*Comparative Gain of Capital $G_i^2(t)$*
South	16. Henan	20.145	18.279	1.102	25.780	0.781
	17. Hubei	14.685	17.074	0.860	18.219	0.806
	18. Hunan	14.713	15.251	0.965	17.238	0.854
	19. Guangdong	39.735	37.562	1.058	24.765	1.604
	20. Guangxi	8.767	9.413	0.931	11.593	0.756
	21. Hainan	1.887	2.422	0.779	2.404	0.785
	Total	100	100	—	100	—
Southwest	22. Chongqing	21.651	22.823	0.949	22.897	0.946
	23. Sichuan	45.474	38.233	1.189	43.575	1.044
	24. Guizhou	12.331	11.406	1.081	12.978	0.950
	25. Yunnan	19.233	25.322	0.760	18.968	1.014
	26. Tibet	1.310	2.215	0.591	1.582	0.828
	Total	100	100	—	100	—
Northwest	27. Shaanxi	44.822	40.769	1.099	44.677	1.003
	28. Gansu	17.984	21.339	0.843	18.787	0.957
	29. Qinghai	5.984	6.660	0.898	6.801	0.880
	30. Ningxia	7.531	9.178	0.821	7.791	0.967
	31. Xinjiang	23.679	22.038	1.074	21.943	1.079
	Total	100	100	—	100	—

has produced relatively abundant societal demand so that the productivity of labor of the region is relatively high and the comparative gain of labor is relatively large. For example, although the labor in Tianjin only amounts 1.09% of the national total, it has generated 2.2% of the societal demand when compared to the national total demand. Therefore, the comparative gain of labor of Tianjin is equal to 2.0183 (= 0.022/0.0109). This means that the labor productivity of Tianjin is way above those of the other 30 provinces, municipalities, and autonomous regions of China. On the other hand, the level of labor productivity in Tibet is the lowest with its comparative gain of labor equal to merely 0.500 (= 0.001/0.002). This means that the labor force in Tibet, which amounts to 0.2% of the national total, only generated 0.1% of the societal demand of China.

Therefore, in terms of the magnitudes of the comparative gains of labor, the 31 provinces, municipalities, and autonomous regions of China can be ranked as follows:

1	2	3	4	5	6	7	8
Tianjin	Inner Mongolia	Hebei	Shandong	Henan	Shanxi	Guangdong	Shanghai
9	10	11	12	13	14	15	16
Xinjiang	Shaanxi	Fujian	Liaoning	Hunan	Jilin	Sichuan	Guangxi
17	18	19	20	21	22	23	24
Jiangsu	Beijing	Heilongjiang	Hubei	Guizhou	Gansu	Anhui	Hainan
25	26	27	28	29	30	31	
Zhejiang	Qinghai	Ningxia	Chongqing	Jiangxi	Yunnan	Tibet	

Similarly, for any chosen region, if its comparative gain of capital is greater than 1, it means that the relatively small investment in the region has created a relatively large amount of societal demand. So, the return of capital investment in the region is relatively good. For example, the capital investment in Shanghai only amounts to 1.6% of the national total. However, this small percentage had led to 3.7% of the national demand. Therefore, the comparative gain of capital in Shanghai is equal to 2.313 (= 0.037/0.016). This implies that the return of capital investment in Shanghai has far outpaced those of all other 31 provinces, municipalities, and autonomous regions of China. On the other hand, the effect of capital investment in Tibet is the worst, where the capital investment amounts to 0.2% of the national total. However, the investment only generated 0.1% of the national demand.

Therefore, in terms of the magnitudes of the comparative gains of capital, the 31 provinces, municipalities, and autonomous regions of China can be ranked as follows:

1	2	3	4	5	6	7	8
Shanghai	Guangdong	Beijing	Liaoning	Zhejiang	Fujian	Jiangsu	Shandong
9	10	11	12	13	14	15	16
Heilongjiang	Hainan	Hunan	Tianjin	Shanxi	Hubei	Henan	Hebei
17	18	19	20	21	22	23	24
Xinjiang	Sichuan	Yunnan	Guangxi	Jilin	Inner Mongolia	Ningxia	Chongqing

25	26	27	28	29	30	31	
Guizhou	Shaanxi	Gansu	Jiangxi	Anhui	Qinghai	Tibet	

6.1.2 Investment and Employment Opportunities in the Six Administrative Areas

First, in terms of the comparative gains of labor, the six administrative areas are ordered as follows from the most optimal to the least:

1	2	3	4	5	6
North (1.2920)	South (1.0260)	Northeast (0.9753)	Northwest (0.9747)	East (0.9707)	Southwest (0.8263)

Therefore, based on the division of these six administrative areas, the labor forces should be moving toward the administrative areas of the north and the south. That represents the direction of employment.

Second, in terms of the comparative gains of capital, the six administrative areas are ordered as follows from the most optimal to the least:

1	2	3	4	5	6
South (1.138)	East (1.074)	North (0.980)	Northeast (0.832)	Southwest (0.832)	Northwest (0.768)

Therefore, based on the division of these six administrative areas, the future investment of capital should be directed toward the administrative areas of south and east. That represents the direction of investment.

Third, by combining the two criteria discussed above, we can order the six administrative areas of China from the most optimal to the least as follows:

1	2	3	4	5	6
South	North	East	Northeast	Northwest	Southwest

The following six sections will, respectively, discuss the investment and employment problems of these administrative areas.

6.2 The South Administrative Area

The south administrative area is depicted in Figure 6.2, which includes such provinces and autonomous regions as Henan, Hubei, Hunan, Guangdong, Guangxi, and Hainan.

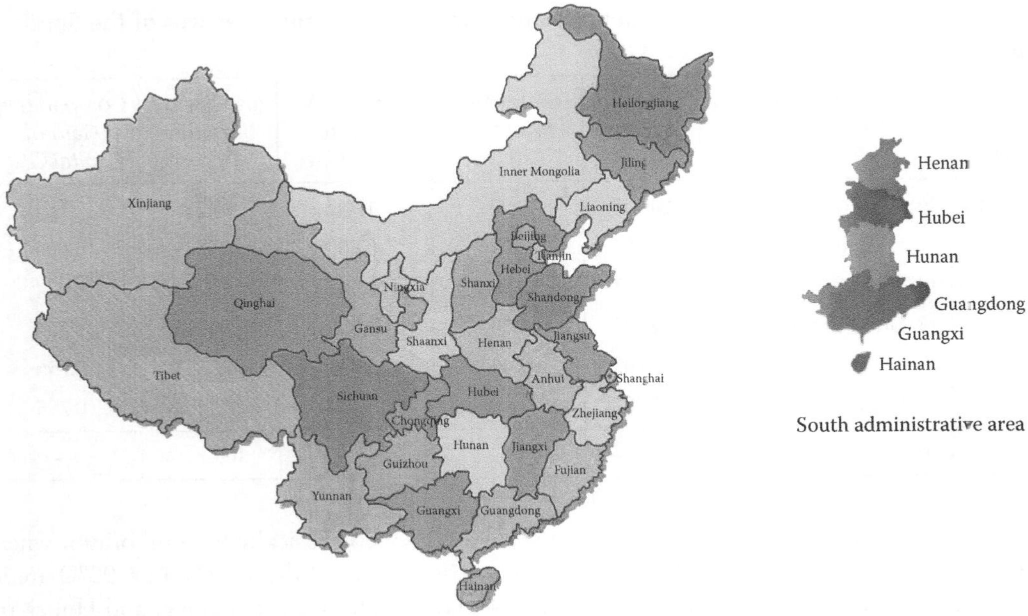

Figure 6.2 The composite of the south administrative area.

6.2.1 Current Regional Economic Structures of the South Area

In this section, we investigate the investment and employment problems of the Southern area. By employing the same methods as in the previous chapters, we obtain Tables 6.3 and 6.4.

Table 6.4 indicates that the regional economic structure of the Southern administrative areas is in severe imbalance, no matter whether we look at it by using the demand and supply of labor or the demand and supply of capital.

In fact, from the fifth column (the comparative gains of labor) and the seventh column (the comparative gains of capital) of the table, it follows that among all the 12 data points of the six

Table 6.3 GDPs, Total Investments in Fixed Properties, and Employment Situations in the South Administrative Area of China (2011)

Area	*Province/Municipality/ Aautonomous Region (i)*	*Regional GDP (0.1 Billion Yuan)*	*Value of Fixed Properties (0.1 Billion Yuan)*	*Employment (10,000 People)*
South	16. Henan	26931.03	17769.0	2892.4
	17. Hubei	19632.26	12557.3	2701.7
	18. Hunan	19669.56	11880.9	2413.3
	19. Guangdong	53120.28	17069.2	5943.6
	20. Guangxi	11720.87	7990.7	1489.4
	21. Hainan	2522.66	1657.2	383.2
	Total	133686.66	68924.3	15823.6

Source: Data are from *China Statistics Yearbook 2012* (XLS, DOC).

Table 6.4 Regional Economic Structures and Comparative Gain Measures of the South Administrative Area of China (2011)

Area	*Province/Municipality/ Autonomous Region (i)*	*Demand Structure $d_i(t)$ (%)*	*Employment Structure $s_i^1(t)$ (%)*	*Comparative Gain of Labor $G_i^1(t)$*	*Investment Structure $s_i^2(t)$ (%)*	*Comparative Gain of Capital $G_i^2(t)$*
South	16. Henan	20.145	18.279	1.102	25.780	0.781
	17. Hubei	14.685	17.074	0.860	18.219	0.806
	18. Hunan	14.713	15.251	0.965	17.238	0.854
	19. Guangdong	39.735	37.562	1.058	24.765	1.604
	20. Guangxi	8.767	9.413	0.931	11.593	0.756
	21. Hainan	1.887	2.422	0.779	2.404	0.785
	Total	100	100	—	100	—

provinces, municipalities, and autonomous regions, only one is basically in equilibrium, where the comparative gain of labor of Hunan is equal to 0.965, while all the rest (11/12 ≈ 92%) are in severe imbalances. In particular, the imbalances of Hainan, Henan, Guangdong, and Hubei are most severe with the comparative gain of labor of Hainan being the lowest, equaling 0.779, and the comparative gain of labor of Henan being the highest, equaling 1.102. On the other hand, the comparative gain of capital of Guangdong is the highest, equaling 1.604, and the value of this indicator of Guangxi is the lowest, equaling merely 0.779. This end implies that Guangdong and Guangxi represent the two extremes in the spectrum of the return of capital investment of the region. Specifically, other than the Guangdong Province, whose comparative gain of capital is far greater than 1, the values of this indicator of the other provinces are far less than 1. It implies that the capital investment in this region has been concentrated in Guangdong Province.

Therefore, in terms of the magnitudes of the comparative gains of labor, the six provinces of this region can be ranked as follows:

1	2	3	4	5	6
Henan	Guangdong	Hunan	Guangxi	Hubei	Hainan

and in terms of the magnitudes of the comparative gains of capital, the six provinces of this region can be ranked as follows:

1	2	3	4	5	6
Guangdong	Hunan	Hubei	Hainan	Henan	Guangxi

6.2.2 Investment and Employment Opportunities in the South Area

First, in terms of the comparative gain of labor, the future labor supplies of this area should be redirected toward Henan and Guangdong two provinces.

That is the direction of employment.

Second, in terms of the comparative gains of capital, the future investment capital should target the Guangdong Province. That is direction of investment.

In other words, in the coming years, no matter whether it is the supply of labor and capital in the Southern Administrative area, it should be directed toward the Guangdong Province.

Third, when these two criteria are combined, the only optimal province for both labor and investment is Guangdong.

6.3 The North Administrative Area

The north administrative area is depicted in Figure 6.3, which includes provinces, municipalities, and autonomous regions as Beijing, Tianjin, Hebei, Shanxi, and Inner Mongolia.

6.3.1 Current Regional Economic Structures of the North Area

In this section, we investigate the investment and employment problems of the Northern area. By employing the same methods as in the previous chapters, we obtain Tables 6.5 and 6.6.

Table 6.6 indicates that the regional economic structure of the Northern administrative areas is in severe imbalance, no matter whether we look at it by using the demand and supply of labor or the demand and supply of capital.

In fact, from the fifth column (the comparative gains of labor) and the seventh column (the comparative gains of capital) of the table, it follows that among all the 10 data points of the five provinces, municipalities, and autonomous regions of this area, only two are basically in equilibrium, where the comparative gains of capital of both Tianjin and Shanxi are respectively equal to 0.957 and 0.951, while all the rest (8/10 80%) are in imbalances. In particular, the imbalances of Beijing, Tianjin, and Inner Mongolia are most severe with the comparative gain of labor of Beijing being the lowest, equaling 0.716, and its comparative gain of capital being the highest, equaling 1.552, and the value of this indicator of Inner Mongolia is the lowest, equaling 0.829.

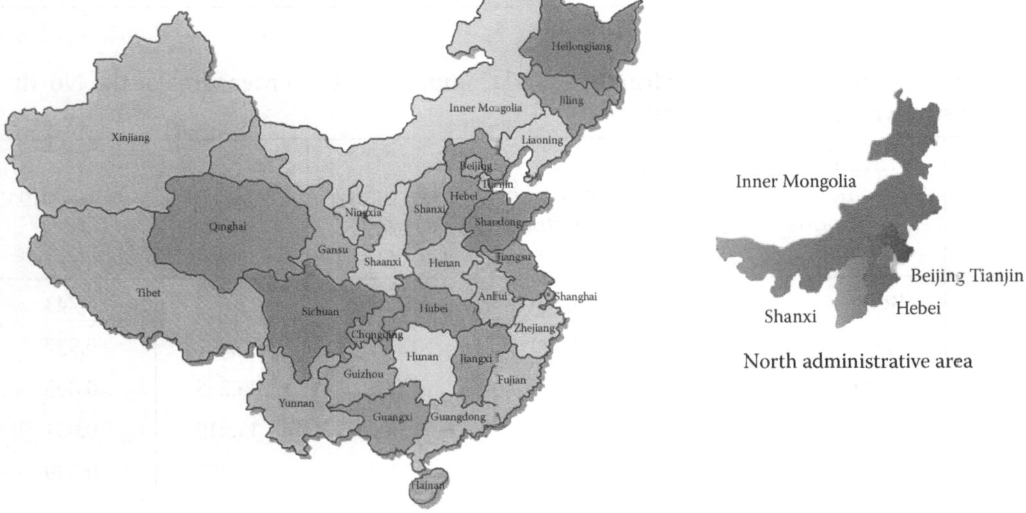

Figure 6.3 The composite of the north administrative area.

Table 6.5 GDPs, Total Investments in Fixed Properties, and Employment Situations of the North Administrative Area of China (2011)

Area	*Province/Municipality/ Autonomous Region (i)*	*Regional GDP (0.1 Billion Yuan)*	*Value of Fixed Properties (0.1 Billion Yuan)*	*Employment (10,000 People)*
North	1. Beijing	16251.93	5578.9	2143.9
	2. Tianjin	11307.28	7067.7	688
	3. Hebei	24515.76	16389.3	2157.4
	4. Shanxi	11237.55	7073.1	1225.7
	5. Inner Mongolia	14359.88	10366.2	1120.1
	Total	77672.40	46475.2	7334.6

Source: Data are from *China Statistics Yearbook 2012* (XLS, DOC).

Therefore, in terms of the magnitudes of the comparative gains of labor, the five provinces and municipalities of this region can be ranked as follows:

1	2	3	4	5
Tianjin	Inner Mongolia	Hebei	Shanxi	Beijing

and, in terms of the magnitudes of the comparative gains of capital, the five provinces and municipalities of this region can be ranked as follows:

1	2	3	4	5
Beijing	Tianjin	Shanxi	Hebei	Inner Mongolia

Table 6.6 Regional Economic Structures and Comparative Gain Measures of the North Administrative Area of China (2011)

Area	*Province/ Municipality/ Autonomous Region (i)*	*Demand Structure $d_i(t)$ (%)*	*Employment Structure $s_i^1(t)$ (%)*	*Comparative Gain of Labor $G_i^1(t)$*	*Investment structure $s_i^2(t)$ (%)*	*Comparative Gain of Capital $G_i^2(t)$*
North	1. Beijing	20.924	29.230	0.716	12.004	1.743
	2. Tianjin	14.558	9.380	1.552	15.207	0.957
	3. Hebei	31.563	29.414	1.073	35.265	0.895
	4. Shanxi	14.468	16.711	0.866	15.219	0.951
	5. Inner Mongolia	18.488	15.271	1.211	22.305	0.829
	Total	100	100	—	100	—

6.3.2 Investment and Employment Opportunities in the North Area

First, in terms of the comparative gain of labor, the future labor supplies of this area should be redirected as much as possible toward Tianjin, Inner Mongolia, and Hebei provinces.

That is the direction of employment.

Second, in terms of the comparative gains of capital, the future investment capital should target Beijing Municipality as much as possible. All other areas have experienced excessive investment of capital. That is direction of investment.

Third, when these two criteria are combined, the only optimal province for both labor and investment in this area is Tianjin. It is because its comparative gain of capital is basically in equilibrium, equaling 0.957, while its comparative gain of labor is the highest. That is different from the other four provinces and municipalities, each of which experiences severe excess in at least one of the comparative gains.

6.4 The East Administrative Area

The east administrative area is depicted in Figure 6.4, which includes such provinces, municipality, and autonomous regions as Shanghai, Jiangsu, Zhejiang, Anhui, Fujian, Jiangxi, and Shandong.

6.4.1 Current Regional Economic Structures of the East Area

In this section, we investigate the investment and employment problems of the Eastern area. By employing the same methods as in the previous chapters, we obtain Tables 6.7 and 6.8.

Table 6.8 indicates that the regional economic structure of the greater Pearl River Delta area is in severe imbalance, no matter whether we look at it by using the demand and supply of labor or the demand and supply of capital.

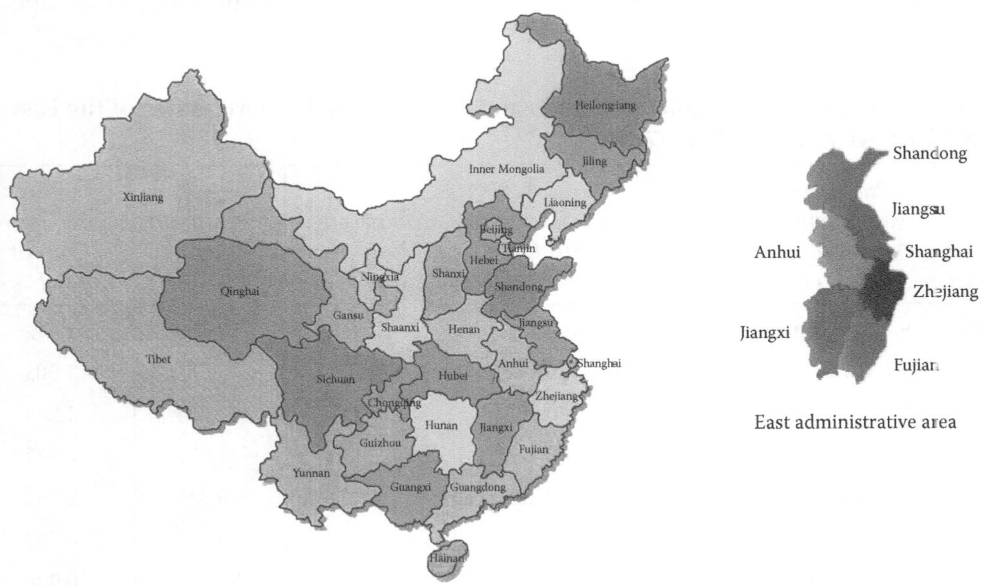

Figure 6.4 The composite of the east administrative area.

Table 6.7 GDPs, Total Investments in Fixed Properties, and Employment Situations of the East Administrative Area of China (2011)

Area	*Province/Municipality/ Autonomous Region (i)*	*Regional GDP (0.1 Billion Yuan)*	*Value of Fixed Properties (0.1 Billion Yuan)*	*Employment (10,000 People)*
East	9. Shanghai	19195.69	4962.1	2186
	10. Jiangsu	49110.27	26692.6	6451
	11. Zhejiang	32318.85	14185.3	4810.8
	12. Anhui	15300.65	12455.7	2028.8
	13. Fujian	17560.18	9910.9	2208.9
	14. Jiangxi	11702.82	9087.6	1942.4
	15. Shandong	45361.85	26749.7	4315.9
	Total	190550.31	104043.9	23943.8

Source: Data are from *China Statistics Yearbook 2012* (XLS, DOC).

In fact, from the fifth column (the comparative gains of labor) and the seventh column (the comparative gains of capital) of the table, it follows that among all the 14 data points of the seven provinces, municipalities, and autonomous regions of this area, only four are basically in equilibrium, where the comparative gains of labor and capital of both Jiangsu and Fujian are basically in equilibrium and are, respectively, equal to 0.957 and 1.005 for Jiangsu and 0.999 and 0.967 for Fujian, while all the rest (10/14 ≈ 71%) are in states of imbalances. In particular, the supply and demand imbalances of Shanghai, Shandong, Zhejiang, Jiangxi, and Anhui are most severe. The comparative gain of labor of Jiangxi is the lowest, equaling 0.757, and the comparative gain of labor of Shandong is the highest, equaling 1.321. Although the comparative gain of capital of

Table 6.8 Regional Economic Structures and Comparative Gain Measures of the East Administrative Area of China (2011)

Area	*Province/ Municipality/ Autonomous Region (i)*	*Demand Structure $d_i(t)$ (%)*	*Employment Structure $s_i^1(t)$ (%)*	*Comparative Gain of Labor $G_i^1(t)$*	*Investment Structure $s_i^2(t)$ (%)*	*Comparative Gain of Capital $G_i^2(t)$*
East	9. Shanghai	10.074	9.130	1.103	4.769	2.112
	10. Jiangsu	25.773	26.942	0.957	25.655	1.005
	11. Zhejiang	16.961	20.092	0.844	13.634	1.244
	12. Anhui	8.030	8.473	0.948	11.972	0.671
	13. Fujian	9.216	9.225	0.999	9.533	0.967
	14. Jiangxi	6.142	8.112	0.757	8.734	0.703
	15. Shandong	23.806	18.025	1.321	25.710	0.926
	Total	100	100	—	100	—

Shanghai is the highest, reaching 2.112, the value of this indicator for Anhui is the lowest, equaling merely 0.671.

Therefore, in terms of the magnitudes of the comparative gains of labor, the seven provinces and municipalities of this region can be ranked as follows:

1	2	3	4	5	6	7
Shandong	Shanghai	Fujian	Jiangsu	Anhui	Zhejiang	Jiangxi

and, in terms of the magnitudes of the comparative gains of capital, the seven provinces and municipalities of this region can be ranked as follows:

1	2	3	4	5	6	7
Shanghai	Zhejiang	Jiangsu	Fujian	Shandong	Jiangxi	Anhui

6.4.2 Investment and Employment Opportunities in the East Area

First, in terms of the comparative gain of labor, the future labor supplies of this area should be redirected as much as possible toward Shandong and Shanghai.

That is the direction of employment.

Second, in terms of the comparative gains of capital, the future investment capital should target Shanghai, Zhejiang, and Jiangsu. That is direction of investment.

Third, when these two criteria are combined, the only optimal place for both labor and investment in this area is Shanghai. It is because its comparative gains of labor and capital are way more than 1, while its comparative gain of capital is the highest. The next optional choices are Shandong and Jiangsu. The worst are Jiangxi and Anhui.

6.5 The Northeast Administrative Area

The northeast administrative area is depicted in Figure 6.5, which includes such provinces as Liaoning, Jilin, and Heilongjiang.

6.5.1 Current Regional Economic Structures of the Northeast Area

In this section, we investigate the investment and employment problems of the Northeast area. By employing the same methods as in the previous chapters, we obtain Tables 6.9 and 6.10.

Table 6.10 indicates that 50% of the regional economic structure of the Northeastern area of China is in severe imbalance, no matter whether we look at it from the angle of the demand and supply of labor or that of the demand and supply of capital.

In fact, from the fifth column (the comparative gains of labor) and the seventh column (the comparative gains of capital) of the table, it follows that among all the six data points of the three provinces of this area, three are basically in equilibrium, where the comparative gain of labor of Liaoning is equal to 1.031, and the comparative gains of labor and capital of Jilin Province are basically in in equilibrium, both of which are equal to 1.022. All the rest, which is exactly 50%, are in states of imbalances. In particular, the comparative gain of labor of Liaoning is the highest,

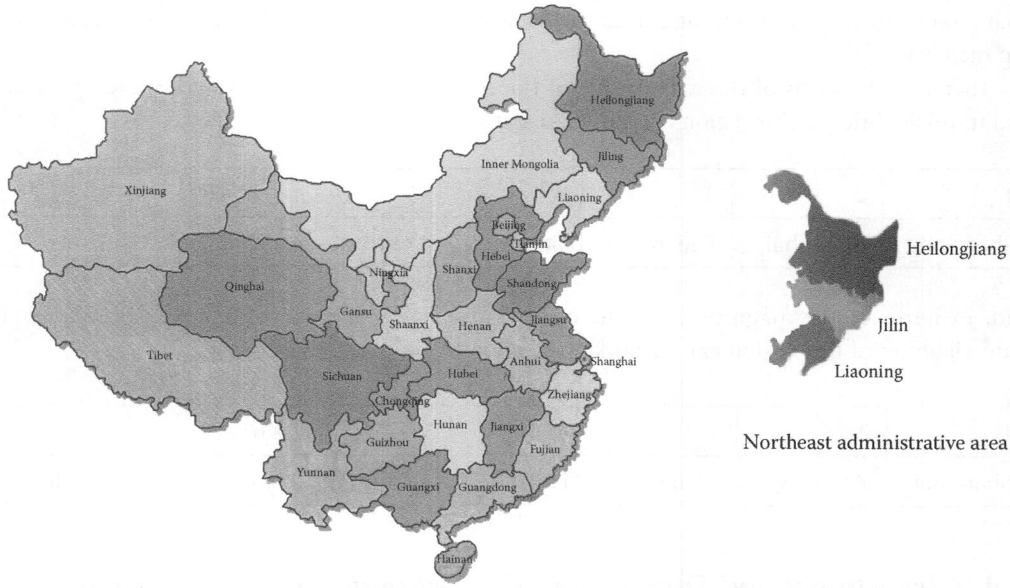

Figure 6.5 The composite of the northeast administrative area.

Table 6.9 GDPs, Total Investments in Fixed Properties, and Employment Situations of the Northeast Administrative Area of China (2011)

Area	*Province/Municipality/ Autonomous Region (i)*	*Regional GDP (0.1 Billion Yuan)*	*Value of Fixed Properties (0.1 Billion Yuan)*	*Employment (10,000 People)*
Northeast	6. Liaoning	22226.70	17726.3	2675.8
	7. Jilin	10568.83	7441,7	1283.7
	8. Heilongjiang	12582.00	7475.4	1672.1
	Total	45377.53	32643.4	5631.6

Source: Data are from *China Statistics Yearbook 2012* (XLS, DOC).

Table 6.10 Regional Economic Structures and Comparative Gain Measures of the Northeast Administrative Area of China (2011)

Area	*Province/ Municipality/ Autonomous Region (i)*	*Demand Structure $d_i(t)$ (%)*	*Employment Structure $s_i^1(t)$ (%)*	*Comparative Gain of Labor $G_i^1(t)$*	*Investment Structure $s_i^2(t)$ (%)*	*Comparative Gain of Capital $G_i^2(t)$*
Northeast	6. Liaoning	48.982	47.514	1.031	54.303	0.902
	7. Jilin	23.291	22.795	1.022	22.797	1.022
	8. Heilongjiang	27.727	29.691	0.934	22.900	1.211
	Total	100	100	—	100	—

equaling 1.031, while the value of this indicator for Heilongjiang is the lowest, equaling 0.934. In terms of the comparative gains of capital, the value for Heilongjiang is the highest, equaling 1.211, and the value of Liaoning is the lowest, equaling 0.902. Therefore, in terms of the magnitude of the comparative gains of labor, the three provinces of this region can be ranked as follows:

1	2	3
Liaoning	Jilin	Heilongjiang

and, in terms of the magnitudes of the comparative gains of capital, the three provinces of this region can be ranked as follows in the exact opposite order:

1	2	3
Heilongjiang	Jilin	Liaoning

6.5.2 Investment and Employment Opportunities in the Northeast Area

First, in terms of the comparative gain of labor, the future labor supplies of this area should be redirected as much as possible toward Liaoning and Jilin two provinces.

That is the direction of employment.

Second, in terms of the comparative gains of capital, the future investment capital should target Heilongjiang and Jilin. That is direction of investment.

Third, when these two criteria are combined, the only optimal place for both labor and investment in this area is Jilin. It is because its comparative gains of labor and capital are way more than 1.

6.6 The Northwest Administrative Area

The northwest administrative area is depicted in Figure 6.6, which includes such provinces and autonomous regions as Shaanxi, Gansu, Qinghai, Ningxia, and Xinjiang.

6.6.1 Current Regional Economic Structures of the Northwest Area

In this section, we investigate the investment and employment problems of the Northwestern area. By employing the same methods as in the previous chapters, we obtain Tables 6.11 and 6.12.

Table 6.12 indicates that 50% of the regional economic structure of the Northwestern area of China is in severe imbalance, no matter whether we look at it from the angle of the demand and supply of labor or that of the demand and supply of capital.

In fact, from the fifth column (the comparative gains of labor) and the seventh column (the comparative gains of capital) of the table, it follows that among all the 10 data points of the five provinces and municipalities of this area, only three are basically in equilibrium, where the comparative gains of labor of Shaanxi, Ningxia, and Gansu are in equilibrium and respectively equaling 1.003, 0.967, and 0.957. All the rest (7/10 = 70%) are situated in states of severe imbalances. In particular, the imbalances of supply and demand of Qinghai, Gansu, and Ningxia are very severe with the comparative gain of labor of Ningxia being the lowest, equaling 0.821 and the value of this indicator of

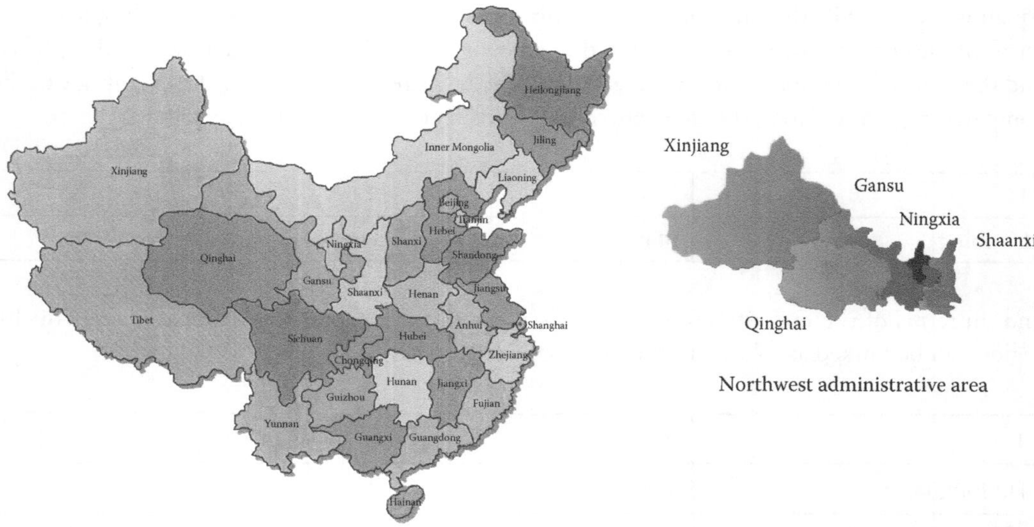

Figure 6.6 The composite of the northwest administrative area.

Shaanxi being the highest, equaling 1.099. In terms of the comparative gains of capital, the value of Xinjiang is the highest, equaling 1.079, and the value of Qinghai the lowest, equaling 0.880.

Therefore, in terms of the magnitude of the comparative gains of labor, the five provinces and autonomous regions of this area can be ranked as follows:

1	2	3	4	5
Shaanxi	Xinjiang	Qinghai	Gansu	Ningxia

and, in terms of the magnitude of the comparative gains of capital, the five provinces and autonomous regions of this area can be ranked as follows:

1	2	3	4	5
Xinjiang	Shaanxi	Ningxia	Gansu	Qinghai

6.6.2 Investment and Employment Opportunities in the Northwest Area

First, in terms of the comparative gain of labor, the future labor supplies of this area should be redirected as much as possible toward Shaanxi and Xinjiang provinces.

That is the direction of employment.

Second, in terms of the comparative gains of capital, the future investment capital should target Xinjiang and Shaanxi provinces. That is direction of investment.

Third, when these two criteria are combined, the most optimal place for both labor and investment in this area is Xinjiang with Shaanxi being the next best choice. It is because their comparative gains of labor and capital are way more than 1, while the values of these indicators of the other three provinces and regions are all smaller than 1.

Table 6.11 GDPs, Total Investments in Fixed Properties, and Employment Situations of the Northwest Administrative Area of China (2011)

Area	*Province/Municipality/ Autonomous Region (i)*	*Regional GDP (0.1 Billion Yuan)*	*Value of Fixed Properties (0.1 Billion Yuan)*	*Employment (10,000 People)*
Northwest	27. Shaanxi	12512.30	9431.1	1427.2
	28. Gansu	5020.37	3965.8	747
	29. Qinghai	1670.44	1435.6	233.1
	30. Ningxia	2102.21	1644.7	321.3
	31. Xinjiang	6610.05	4632.1	771.5
	Total	27915.37	21109.3	3500.7

Source: Data are from *China Statistics Yearbook 2012* (XLS, DOC).

Table 6.12 Regional Economic Structures and Comparative Gain Measures of the Northwest Administrative Area of China (2011)

Area	*Province/ Municipality/ Autonomous Region (i)*	*Demand Structure $d_i(t)$ (%)*	*Employment Structure $s_i^1(t)$ (%)*	*Comparative Gain of Labor $G_i^1(t)$*	*Investment Structure $s_i^2(t)$ (%)*	*Comparative Gain of Capital $G_i^2(t)$*
Northwest	27. Shaanxi	44.822	40.769	1.099	44.677	1.003
	28. Gansu	17.984	21.339	0.843	18.787	0.957
	29. Qinghai	5.984	6.660	0.898	6.801	0.880
	30. Ningxia	7.531	9.178	0.821	7.791	0.967
	31. Xinjiang	23.679	22.038	1.074	21.943	1.079
	Total	100	100	—	100	—

6.7 The Southwest Administrative Area

The southwest administrative area is depicted in Figure 6.7, which includes such provinces, municipality, and autonomous regions as Chongqing, Sichuan, Guizhou, Yunnan, and Tibet.

6.7.1 Current Regional Economic Structures of the Southwest Area

In this section, we investigate the investment and employment problems of the Southwestern area. By employing the same methods as in the previous chapters, we obtain Tables 6.13 and 6.14.

Table 6.14 indicates that the regional economic structure of the southwestern area of China is in severe imbalance, no matter whether we look at it from the angle of the demand and supply of labor or that of the demand and supply of capital.

In fact, from the fifth column (the comparative gains of labor) and the seventh column (the comparative gains of capital) of the table, it follows that among all the 10 data points of the five provinces, municipalities, and autonomous regions of this area, only three are basically

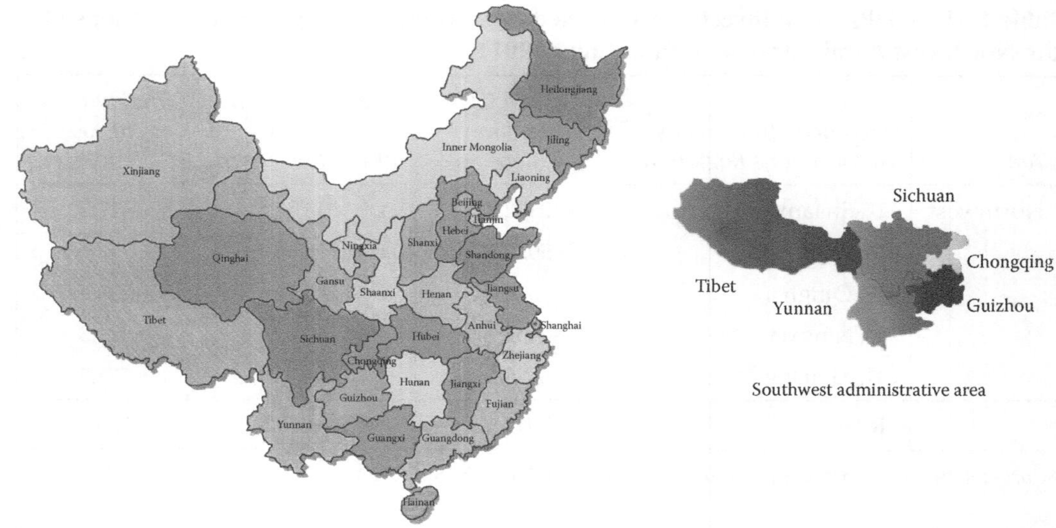

Figure 6.7 The composite of the southwest administrative area.

in equilibrium, where the comparative gains of labor of Yunnan, Sichuan, and Guizhou are located in between 0.95 and 1.05, equaling, respectively, 1.014, 1.044, and 0.950. And all the rest (7/10 = 70%) are situated in states of imbalances. In particular, the imbalances of supply and demand of Tibet, Yunnan, and Sichuan are very severe with the comparative gains of labor and capital of Tibet being the lowest, equaling, respectively, 0.591 and 0.828, and the values of these two indicators of Sichuan being the best, equaling, respectively, 1.189 and 1.044.

Therefore, in terms of the magnitude of the comparative gains of labor, the five provinces and autonomous regions of this area can be ranked as follows:

1	2	3	4	5
Sichuan	Guizhou	Chongqing	Yunnan	Tibet

Table 6.13 GDPs, Total Investments in Fixed Properties, and Employment Situations of the Southwest Administrative Area of China (2011)

Area	*Province/Municipality/ Autonomous Region (i)*	*Regional GDP (0.1 Billion Yuan)*	*Value of Fixed Properties (0.1 Billion Yuan)*	*Employment (10,000 People)*
Southwest	22. Chongqing	10011.37	7473.4	1580.5
	23. Sichuan	21026.68	14222.2	2647.7
	24. Guizhou	5701.84	4235.9	789.9
	25. Yunnan	8893.12	6191.0	1753.6
	26. Tibet	605.83	516.3	153.4
	Total	46238.84	32638.8	6925.1

Source: Data are from *China Statistics Yearbook 2012* (XLS, DOC).

Table 6.14 Regional Economic Structures and Comparative Gain Measures of the Southwest Administrative Area of China (2011)

Area	*Province/ Municipality/ Autonomous Region (i)*	*Demand Structure $d_i(t)$ (%)*	*Employment Structure $s_i^1(t)$ (%)*	*Comparative Gain of Labor $G_i^1(t)$*	*Investment Structure $s_i^2(t)$ (%)*	*Comparative Gain of Capital $G_i^2(t)$*
Southwest	22. Chongqing	21.651	22.323	0.949	22.897	0.946
	23. Sichuan	45.474	38.233	1.189	43.575	1.044
	24. Guizhou	12.331	11.406	1.081	12.978	0.950
	25. Yunnan	19.233	25.322	0.760	18.968	1.014
	26. Tibet	1.310	2.215	0.591	1.582	0.828
	Total	100	100	—	100	—

and, in terms of the magnitude of the comparative gains of capital, the five provinces and autonomous regions of this area can be ranked as follows:

1	2	3	4	5
Sichuan	Yunnan	Guizhou	Chongqing	Tibet

6.7.2 Investment and Employment Opportunities in the Southwest Area

First, in terms of the comparative gain of labor, the future labor supplies of this area should be redirected as much as possible toward Sichuan and Guizhou provinces.

That is the direction of employment.

Second, in terms of the comparative gains of capital, the future investment capital should target Sichuan and Yunnan provinces. That is direction of investment.

Third, when these two criteria are combined, the most optimal place for both labor and investment in this area is Sichuan. It is because its comparative gains of labor and capital are way more than 1 and the best. The next choice is Guizhou, whose comparative gain of labor is way above 1 and whose comparative gain of capital is also basically in equilibrium. The worst is Tibet.

Chapter 7

Regions in the Three Special Areas of Modern China

This chapter investigates the three particular economic areas: the Bohai Economic Rim, the Yangtze River Delta Economic Region, and the Greater Pearl River Delta Region, as so defined since the start of the recent economic reform in China. Our purpose is to determine the ranking of these areas according to their optimality in terms of investment and employment opportunities.

7.1 The Special Economic Areas

Figure 7.1 shows the relative locations of these three specific areas of economic development in China.

7.1.1 Current Regional Economic Structures of the Special Economic Areas

Similar to what we have done in the previous chapters, we compute one by one the vector of demand structure $d(t)$, the vector of employment structure $S^1(t)$, and the vector of investment structure $S^2(t)$ for each of the Chinese provinces, municipalities, and autonomous regions, by using the relevant statistics data of 2011 (Table 7.1). Then, we substitute the relevant numerical values into Equation 1.1 to calculate, respectively, the comparative gains of labor $G_i^1(t) = d_i(t)/S_i^1(t)$ and the comparative gains of capital $G_i^2(t) = d_i(t)/S_i^2(t)$, which lead to Table 7.2.

Table 7.2 indicates that the regional economic structures of the three particular economic areas of China are in severe imbalances, no matter whether we look at them by using the demand and supply of labor or the demand and supply of capital.

In fact, from the fifth column (the comparative gains of labor) and the seventh column (the comparative gains of capital) of the table, it follows that there are 54 data points for the 21 provinces, municipalities, and autonomous regions. However, only three of them are basically in equilibrium, where the comparative gain of labor of Liaoning Province is equal to 0.972, the

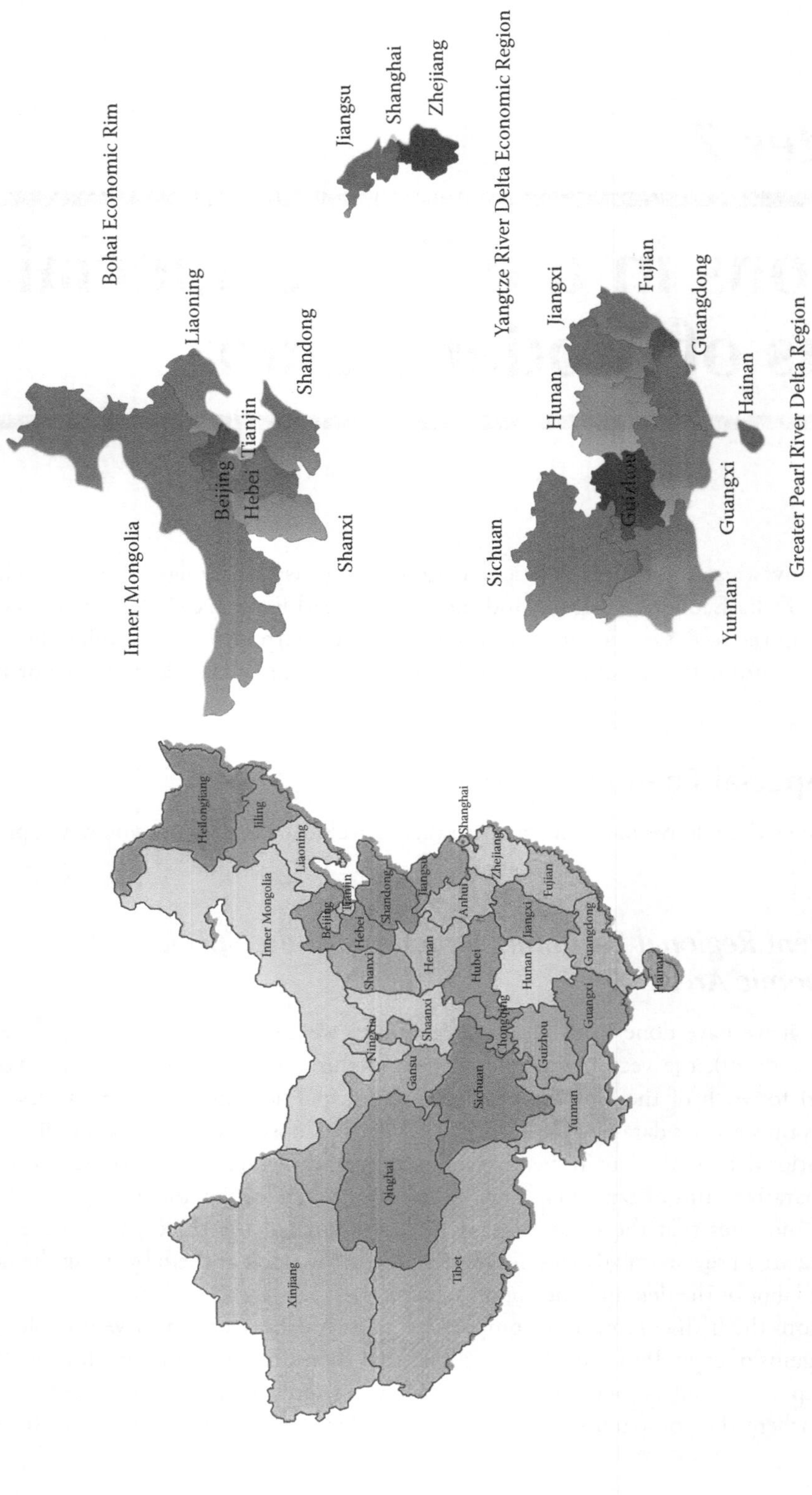

Figure 7.1 Relative locations of the three special areas of economic development.

Table 7.1 GDPs, Total Investments in Fixed Properties, and Employment Situations of the Three Special Economic Areas of China (2011)

Area	*Province/ Municipality/ Autonomous Region (i)*	*Regional GDP (0.1 Billion Yuan)*	*Value of Fixed Properties (0.1 Billion Yuan)*	*Employment (10,000 People)*
Bohai Economic Rim	1. Liaoning	22226.70	17726.3	2675.8
	2. Hebei	24515.76	16389.3	2157.4
	3. Beijing	16251.93	5578.9	2143.9
	4. Tianjin	11307.28	7067.7	688
	5. Shandong	45361.85	26749.7	4315.9
	6. Shanxi	11237.55	7073.1	1225.7
	7. Inner Mongolia	14359.88	10366.2	1120.1
	Total	145260.95	90951.2	14326.8
Yangtze River Delta Economic Zone	8. Jiangsu	49110.27	26692.6	6451
	9. Shanghai	19195.69	4962.1	2186
	10. Zhejiang	32318.85	14185.3	4810.8
	Total	100624.81	45840	13447.8
Greater Pearl River Delta Region	11. Fujian	17560.18	9910.9	2203.9
	12. Guangdong	53120.28	17069.2	5943.6
	13. Guangxi	11720.87	7990.7	1489.4
	14. Guizhou	5701.84	4235.9	789.9
	15. Hainan	2522.66	1657.2	383.2
	16. Hunan	19669.56	11880.9	2413.3
	17. Jiangxi	11702.82	9087.6	1942.4
	18. Sichuan	21026.68	14222.2	2647.7
	19. Yunnan	8893.12	6191.0	1753.6
	20. Macao	1642.89	#	33.5
	21. Hong Kong	15148.39	#	364.4
	Total	168709.29	82245.6	19969.9
National	Grand total	414594.34	219036.8	47744.5

Source: Data are from *China Statistics Yearbook 2012* (XLS, DOC).

comparative gain of capital of Jiangsu Province is equal to 0.973, and the comparative gain of labor of the Greater Pearl River Delta Region is equal to 0.973. And all the rest are not equal to 1 and are situated in various states of (severe) imbalances. In particular, the comparative gains of labor of Hong Kong and Macao, respectively, reached 4.789 and 5.657, while the comparative gains of capital of Beijing, Shanghai, and Guangdong were, respectively, equal to 2.140, 2.044,

Table 7.2 Regional Economic Structures and Comparative Gain Measures of the Three Special Economic Areas of China (2011)

Area	*Province/ Municipality/ Autonomous Region (i)*	*Demand Structure $d_i(t)$ (%)*	*Employment Structure $s_i^1(t)$ (%)*	*Comparative Gain of Labor $G_i^1(t)$*	*Investment Structure $s_i^2(t)$ (%)*	*Comparative Gain of Capital $G_i^2(t)$*
Bohai Economic Rim	1. Liaoning	5.361	5.604	0.975	8.093	0.662
	2. Hebei	5.913	4.519	1.308	7.482	0.790
	3. Beijing	3.920	4.490	0.873	2.547	2.149
	4. Tianjin	2.727	1.441	1.892	3.227	0.845
	5. Shandong	10.941	9.047	1.209	12.212	0.896
	6. Shanxi	2.710	2.567	1.056	3.229	0.839
	7. Inner Mongolia	3.464	2.346	1.477	4.733	0.732
	Total	35.037*	30.007*	1.168*	41.923*	0.836**
	8. Jiangsu	11.845	13.512	0.877	12.186	0.972
	9. Shanghai	4.630	4.581	1.174	2.265	2.044
	10. Zhejiang	7.795	10.076	0.774	6.476	1.204
	Total	24.271	28.166	0.862*	20.928	1.160**
Greater Pearl River Delta Region	11. Fujian	4.236	4.627	0.915	4.525	0.936
	12. Guangdong	12.813	12.449	1.029	7.973	1.607
	13. Guangxi	2.827	3.120	0.906	3.648	0.815
	14. Guizhou	1.375	1.654	0.831	1.933	0.856
	15. Hainan	0.608	0.803	0.757	0.757	0.803
	16. Hunan	4.744	5.055	0.938	5.424	0.932
	17. Jiangxi	2.623	4.068	0.649	4.149	0.680
	18. Sichuan	5.072	5.546	0.915	6.493	0.854
	19. Yunnan	2.145	3.673	0.584	2.826	1.230
	20. Macao	0.396	0.070	5.657	#	#
	21. Hong Kong	3.654	0.763	4.789	#	#
	Total	40.693	41.827	0.973*	37.549	1.084**
National	Grand total	1.000	1.000	–	1.000	–

Note 1: The data labeled with either * or ** are not the sums of the relevant data. Instead, they are the mean values of the relevant areas, where the data labeled with * stand for the mean values of the comparative gains of labor of relevant areas; and the data labeled with ** the mean values of the comparative gains of capital of relevant areas.

Note 2: The data cells occupied with # represent missing data. Therefore, for the Greater Pearl Delta River Region, we can only consider the employment situations; and the discussion on the investment opportunities we only look at the nine provinces without including the special administrative areas of Hong Kong and Macao.

and 1.607. This implies that in terms of these three particular economic areas, only one is basically in equilibrium and all the rest are situated in states of supply and demand imbalances. In terms of the 21 provinces, municipalities, and autonomous regions, other than 2 are basically in equilibrium, all the rest (40/42 ≥ 96%) are situated in states of supply and demand imbalances. More specifically, the 11 provinces, municipalities, and autonomous regions of the Greater Pearl River Delta Region and such other areas as Beijing, Tianjin, Inner Mongolia, Shanghai, and Zhejiang experience severe supply and demand imbalances. Here, the comparative gain of labor of Yunnan is the lowest, equaling 0.584, while the value of this indicator of Macao is the highest, equaling 5.657, which is nearly 10 times greater than the value of Yunnan. This implies that Macao and Yunnan represent the two extremes of the spectrum of the production efficiency of these three particular economic areas. In terms of the comparative gains of capital, Beijing is the highest, equaling 2.149, while the value of this indicator of Liaoning is the lowest, equaling merely 0.662. This means that Beijing and Liaoning are the two extremes in the spectrum of returns of capital investments in these three particular economic areas.

In terms of these three particular economic areas, the data of the table indicate that the supply and demand of both labor and capital are all severely imbalanced. The mean value of the comparative gains of labor of the seven provinces, municipalities, and autonomous regions of the Bohai Economic Rim is equal to 1.168 (≫1), representing the most optimal area of China in terms of labor efficiency and the development direction of societal labor forces in the current economic stage of these three particular economic areas. So, additional labor input should be given to this region. However, the mean value of the comparative gains of capital of the region is the lowest, equaling 0.836. This fact means that this region has experienced excessive supply of capital; the surplus should be redirected to other areas. On the other hand, the comparative gain of capital of the Yangtze River Delta Economic Region is equal to 1.160 (≫1), representing the best return of capital among the three particular economic areas. This implies that the supply of investment capital for the Yangtze River Delta Economic Region is relatively insufficient; and the capital input into this region should be strengthened in the next round of investment decisions. At the same time, the comparative gain of labor of the region is the lowest, equaling only 0.862, so that the excessive labor supply of the region should be redirected to the other areas.

Therefore, in terms of the magnitudes of the comparative gains of labor, the 21 provinces, municipalities, and autonomous regions of these three particular economic areas can be ranked as follows:

1	2	3	4	5	6	7	8
Macao	Hong Kong	Tianjin	Inner Mongolia	Hebei	Shandong	Shanghai	Shanxi
9	10	11	12	13	14	15	16
Guangdong	Liaoning	Hunan	Fujian	Sichuan	Guangxi	Jiangsu	Beijing
17	18	19	20	21			
Guizhou	Zhejiang	Hainan	Jiangxi	Yunnan			

And, in terms of the magnitudes of the comparative gains of capital, the 21 provinces, municipalities, and autonomous regions of these three particular economic areas can be ranked as follows:

1	2	3	4	5	6	7	8
Beijing	Shanghai	Guangdong	Yunnan	Zhejiang	Jiangsu	Fujian	Hunan
9	10	11	12	13	14	15	16
Shandong	Guizhou	Sichuan	Tianjin	Shanxi	Guangxi	Hainan	Hebei
17	18	19					
Inner Mongolia	Jiangxi	Liaoning					

7.1.2 Investment and Employment Opportunities in the Special Economic Areas

First, in terms of the comparative gains of labor, the resultant ordering of the three economic areas from the most optimal to the least is given below:

1	2	3
Bohai Economic Rim (1.168)	Greater Pearl River Delta Region (0.973)	Yangtze River Delta Economic Zone (0.862)

Therefore, the future flow of labor should be directed toward the Bohai Economic Rim. That is the direction of future employment.

Second, in terms of the comparative gains of capital, the resultant ordering of the three economic areas from the most optimal to the least is given below:

1	2	3
Yangtze River Delta Economic Zone (1.160)	Greater Pearl River Delta Region (1.084)	Bohai Economic Rim (0.863)

Therefore, the future flow of investment capital should be directed as much as possible toward the Yangtze River Delta Economic Zone and the Greater Pearl River Delta Region.

Third, by combining the two criteria discussed above, we can order the three economic areas of China from the most optimal to the least as follows:

1	2	3
Greater Pearl River Delta Region (0.973, 1.084)	Yangtze River Delta Economic Zone (0.862, 1.160)	Bohai Economic Rim (1.168, 0.863)

That is, the Greater Pearl River Delta Region is the most optimal area; the Yangtze River Delta Economic Zone is the second best area; and the Bohai Economic Rim the least optimal area.

The following three sections, respectively, discuss the investment and employment problems of these economic areas in terms of their economic sectors and industries.

7.2 The Greater Pearl River Delta Region: The Optimal Area

This section looks at the greater Pearl River Delta Region, Figure 7.2, which is composed of the following provinces, municipalities, and autonomous regions as Fujian, Guangdong, Guangxi, Guizhou, Hainan, Hunan, Jiangxi, Sichuan, Yunnan, Macao, and Hong Kong.

7.2.1 Current Regional Economic Structures of the Greater Pearl River Delta Region

The concept of the Pearl River Delta Region, which is the well-known 9 + 2 economic region, is formally proposed in July 2003 in China. This region covers the nine provinces and two municipalities of Southern China, Southeastern China, and Southwestern China. In particular, they are Fujian, Guangdong, Guangxi, Guizhou, Hainan, Hunan, Jiangxi, Sichuan, Yunnan, Hong Kong, and Macao. This Pearl River Delta Region covers one-fifth of the land mass of China and one-third of the national population.

This section investigates the investment and employment problems of this Pearl River Delta Region by using the same method as described in the previous chapters. The consequent results are shown in Tables 7.3 and 7.4.

Table 7.4 indicates that the regional economic structure of the greater Pearl River Delta Region is in severe imbalance, no matter whether we look at the structure by using the demand and supply of labor or the demand and supply of capital.

In fact, from the fifth column (the comparative gains of labor) and the seventh column (the comparative gains of capital) of the table, it follows that there are 20 data points for the 21 provinces, municipalities, and autonomous regions. However, only one of the data points is basically in equilibrium, where the comparative gain of labor of the Hunan Province is equal to 0.965. And all the rest are situated in states of various degrees of imbalances. In particular, the comparative gains of labor of Hong Kong and Macao, respectively, reached 4.789 and 5.657. This implies that in terms of the greater Pearl River Delta Region, only one is basically in equilibrium and all the

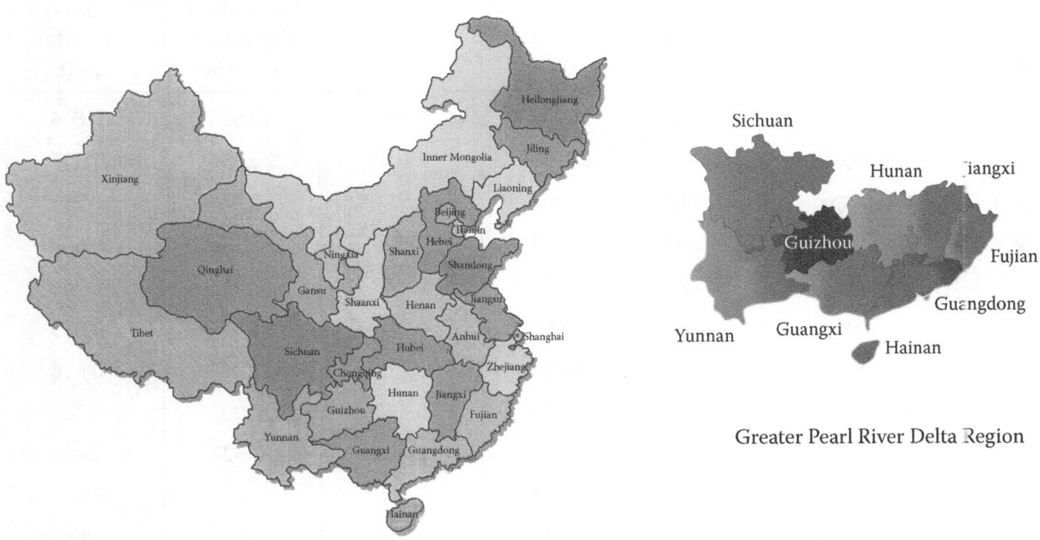

Figure 7.2 Composite of the greater Pearl River Delta Region.

Table 7.3 GDPs, Total Investments in Fixed Properties, and Employment Situations of the Greater Pearl River Delta Region of China (2011)

Area	*Province/Municipality/ Autonomous Region (i)*	*Regional GDP (0.1 Billion Yuan)*	*Value of Fixed Properties (0.1 Billion Yuan)*	*Employments (10,000)*
Greater Pearl River Delta Region	11. Fujian	17560.18	9910.9	2208.9
	12. Guangdong	53120.28	17069.2	5943.6
	13. Guangxi	11720.87	7990.7	1489.4
	14. Guizhou	5701.84	4235.9	789.9
	15. Hainan	2522.66	1657.2	383.2
	16. Hunan	19669.56	11880.9	2413.3
	17. Jiangxi	11702.82	9087.6	1942.4
	18. Sichuan	21026.68	14222.2	2647.7
	19. Yunnan	8893.12	6191.0	1753.6
	20. Macao	1642.89	#	33.5
	21. Hong Kong	15148.39	#	364.4
	Total	168709.29	82245.6#	19969.9

Source: Data are from *China Statistics Yearbook 2012* (XLS, DOC).

Table 7.4 Regional Economic Structure and Measures of Comparative Gains of the Greater Pearl River Delta Region of China (2011)

Area	*Province/ Municipality/ Autonomous Region (i)*	*Demand Structure $d_i(t)$ (%)*	*Employment Structure $s_i^1(t)$ (%)*	*Comparative Gain of Labor $G_i^1(t)$*	*Investment Structure $s_i^2(t)$ (%)*	*Comparative Gain of Capital $G_i^2(t)$*
Greater Pearl River Delta Region	Fujian	10.409	11.067	0.941	12.050	0.864
	Guangdong	31.486	29.763	1.058	20.754	1.517
	Guangxi	6.947	7.458	0.931	9.716	0.715
	Guizhou	3.380	3.955	0.855	5.150	0.656
	Hainan	1.495	1.919	0.779	2.015	0.742
	Hunan	11.659	12.085	0.965	14.446	0.875
	Jiangxi	6.937	9.727	0.713	11.049	0.628
	Sichuan	12.463	13.258	0.940	17.292	0.721
	Yunnan	5.271	8.781	0.600	7.527	0.700
	Macao	0.974	0.168	5.798	#	#
	Hong Kong	8.879	1.825	4.920	#	#
	Total	100	0.1651	—	100	—

rest (19/20 = 95%) are situated in states of supply and demand imbalances. Specifically, the supply and demand imbalances of Macao, Hong Kong, Yunnan, Jiangxi, Hainan, Guizhou, Guangxi, Sichuan, and Guangdong are the most severe, where the comparative gain of labor of Jiangxi is the lowest, equaling 0.600, while the value of the same indicator of Macao is the highest, equaling 5.798, which is nearly 10 times more than that of Jiangxi. This means that Macao and Yunnan are the two extremes on the spectrum of the productivity efficiency of the greater Pearl River Delta Region. On the other hand, the comparative gain of capital of Guangdong Province is the highest, equaling 1.517, while the value of the same indicator of Jiangxi is the lowest, equaling merely 0.628. This means that Guangdong and Jiangxi represent the two extremes on the spectrum of the capital investment efficiency of this region. Especially, other than Guangdong within this region that has its comparative gain of capital far greater than 1, the comparative gains of capital of all other eight provinces, municipalities, and autonomous regions are way smaller than 1. This implies that the capital investment in this region has been concentrated in Guangdong Province.

Therefore, in terms of the magnitudes of the comparative gains of labor, the 11 provinces, municipalities, and autonomous regions of this region can be ranked as follows:

1	2	3	4	5	6
Macao	Hong Kong	Guangdong	Hunan	Fujian	Sichuan
7	8	9	10	11	
Guangxi	Guizhou	Hainan	Jiangxi	Yunnan	

And, in terms of the magnitudes of the comparative gains of capital, the nine provinces, municipalities, and autonomous regions of this region, excluding Macao and Hong Kong, can be ranked as follows:

1	2	3	4	5	6	7	8	9
Guangdong	Hunan	Fujian	Hainan	Sichuan	Guangxi	Yunnan	Guizhou	Jiangxi

7.2.2 Investment and Employment Opportunities in the Greater Pearl River Delta Region

First, in terms of the comparative gains of labor, the future labor supply should be directed as much as possible toward Hong Kong, Macao, and Guangdong.

That is the direction of employment.

Second, in terms of the comparative gain of capital, the future capital investment in this region should be aimed at Guangdong Province.

That is the direction of investment.

In other words, in the coming years, no matter whether it is labor supply or capital investment, it should be directed toward Guangdong Province.

7.3 Yangtze River Delta Economic Zone: A Suboptimal Area

This section looks at the Yangtze River delta economic zone (Figure 7.3), which is composed of the following provinces and municipalities: Jiangsu, Shanghai, and Zhejiang.

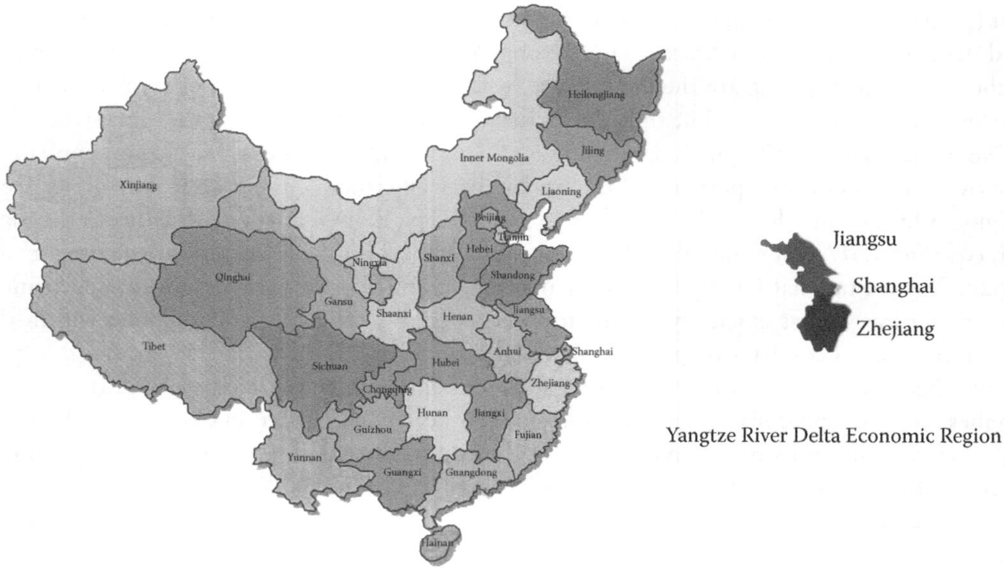

Figure 7.3 Composite of the Yangtze River delta economic zone.

7.3.1 Current Regional Economic Structures of the Yangtze River Delta Economic Zone

The planning of the Yangtze River Delta Region was initially and officially approved by the State Department of China on May 24, 2010, by implementing the legislative decision titled "The Guideline for the Reform and Economic Development of the Yangtze River Delta Region" of 2008. It represented a major implementation of the strategic decision to lift the overall strength of the Yangtze River Delta Region and to strengthen the national competitiveness of China in the international arena.

The Yangtze River Delta Region covers a land area of 21.07 square kilometers, including Shanghai Municipality, Jiangsu Province, and Zhejiang Province. This region enjoys its advantages of convenient location, rich natural resources, great economic foundation, relatively complete administrative structure, and well-developed education and cultural systems. It has been one of the most developed areas in China with the most overall competitiveness in China and occupies an important strategic position in the modernization of China.

This section investigates the investment and employment problems of this Yangtze River Delta Region by using the same method as described in the previous chapters. The consequent results are shown in Tables 7.5 and 7.6.

Table 7.6 indicates that the regional economic structure of the Yangtze River Delta Region is in severe imbalance, no matter whether we look at the structure by using the demand and supply of labor or the demand and supply of capital.

In fact, from the fifth column (the comparative gains of labor) and the seventh column (the comparative gains of capital) of the table, it follows that there are six data points for the three provinces and municipalities within this region. However, only two of the data points are basically in equilibrium, where the comparative gain of labor of Jiangsu Province is equal to 1.017, and the

Table 7.5 GDPs, Total Investments in Fixed Properties, and Employment Situations of Yangtze River Delta Economic Zone of China (2011)

Area	*Province/ Municipality/ Autonomous Region (i)*	*Regional GDP (0.1 Billion Yuan)*	*Value of Fixed Properties (0.1 Billion Yuan)*	*Employments (10,000)*
Yangtze River Delta Economic Zone	8. Jiangsu	49110.27	26692.6	6451
	9. Shanghai	19195.69	4962.1	2186
	10. Zhejiang	32318.85	14185.3	4810.8
	Total	100624.81	45840	13447.8

Source: Data are from *China Statistics Yearbook 2012* (XLS, DOC).

comparative gain of capital of Zhejiang is equal to 1.038. And all the rest are situated in states of various degrees of imbalances. This means that in terms of the Yangtze River Delta Region, other than two are basically in equilibrium, all the rest (2/6 = 67%) are in some states of supply and demand imbalances. In particular, the comparative gain of labor of Zhejiang is the lowest, equaling 0.898, and the comparative gain of capital of Jiangsu is the lowest, equaling 0.838. On the other hand, the values of these indicators of Shanghai are the highest, where the former is equal to 1.174 while the latter 1.762. This means that Shanghai and Zhejiang stand for the two extremes on the spectrum of productivity efficiency, and Shanghai and Jiangsu the two extremes on the spectrum of returns of capital investments in the Yangtze River Delta Region.

Therefore, in terms of the magnitudes of the comparative gains of labor, the three provinces and municipalities of the Yangtze River Delta Region can be ranked as follows:

1	2	3
Shanghai	Jiangsu	Zhejiang

And, in terms of the magnitudes of the comparative gains of capital, the three provinces and municipalities of the Yangtze River Delta Region can be ranked as follows:

1	2	3
Shanghai	Zhejiang	Jiangsu

Table 7.6 Regional Economic Structure and Measures of Comparative Gains of the Yangtze River Delta Region of China (2011)

Area	*Province/ Municipality/ Autonomous Region (i)*	*Demand Structure $d_i(t)$ (%)*	*Employment Structure $s_i^1(t)$ (%)*	*Comparative Gain of Labor $G_i^1(t)$*	*Investment Structure $s_i^2(t)$ (%)*	*Comparative Gain of Capital $G_i^2(t)$*
Yangtze River Delta Region	Jiangsu	48.805	47.971	1.017	58.230	0.838
	Shanghai	19.076	16.255	1.174	10.825	1.762
	Zhejiang	32.118	35.774	0.898	30.945	1.038
	Total	100	0.1651	—	100	—

7.3.2 Investment and Employment Opportunities in the Yangtze River Delta Economic Zone

First, in terms of the comparative gains of labor, the future labor supply within this region should be directed as much as possible toward Shanghai and Jiangsu. That is the direction of employment.

Second, in terms of the comparative gain of capital, the future capital investment in this region should be aimed at Shanghai and Zhejiang Province. That is the direction of investment.

Third, when these two criteria are jointly considered, the most optimal area in terms of the comparative gains of both labor and capital within this region is Shanghai, because the values of these indicators of Shanghai are both much greater than 1.

7.4 Bohai Economic Rim: A Relatively Promising Area

This section looks at the Bohai Economic Rim, Figure 7.4, which is composed of the following provinces, municipalities, and autonomous regions: Liaoning, Hebei, Beijing, Tianjin, Shandong, Shanxi, and Inner Mongolia.

7.4.1 Current Regional Economic Structures of the Bohai Economic Rim

The so-called Bohai Economic Rim stands for the greater economic region that is located around the Bohai Sea and parts of the region along the Yellow Sea. It is on the north of the Chinese coast of the Pacific Ocean. It is considered a golden coast of Northern China. Within the strategic development of the coastal areas of China, Bohai Economic Rim occupies an important position.

This economic rim includes two major municipalities, Beijing and Tianjin, which are under the direct control of the central government, and other five provinces: Liaoning, Hebei, Shanxi, Shandong, and Inner Mongolia. This rim covers a land mass of 112 square kilometers and a population of 0.26 billion people. This region contains 157 cities, which amounts to roughly one-fourth of the total number of cities in China. Among these cities, 13 of them are home to over 1 million residents.

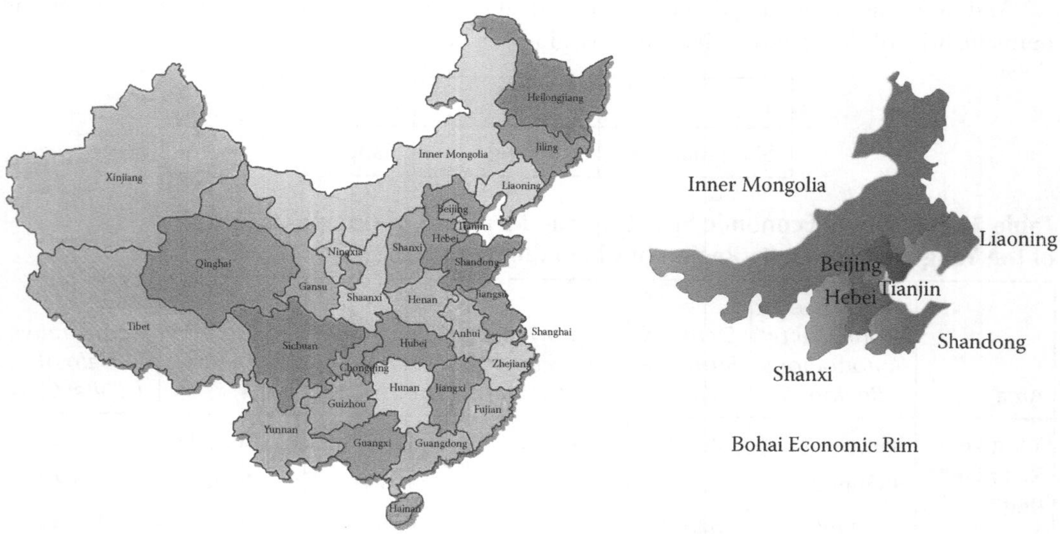

Figure 7.4 Composite of the Bohai Economic Rim.

Table 7.7 GDPs, Total Investments in Fixed Properties, and Employment Situations of Bohai Economic Rim of China (2011)

Area	*Province/Municipality/ Autonomous Region (i)*	*Regional GDP (0.1 Billion Yuan)*	*Value of Fixed Properties (0.1 Billion Yuan)*	*Employments (10,000)*
Bohai Economic Rim	1. Liaoning	22226.70	17726.3	2675.8
	2. Hebei	24515.76	16389.3	2157.4
	3. Beijing	16251.93	5578.9	2143.9
	4. Tianjin	11307.28	7067.7	688
	5. Shandong	45361.85	26749.7	4315.9
	6. Shanxi	11237.55	7073.1	1225.7
	7. Inner Mongolia	14359.88	10366.2	1120.1
	Total	145260.95	90951.2	14326.8

Source: Data are from *China Statistics Yearbook 2012* (XLS, DOC).

Additionally, the region can be seen as an economic rim consisting of Beijing, Tianjin, and Hebei as the core, and East Liao Peninsula and Shandong Peninsula as the wings. This region contains mainly Beijing, Tianjin, Hebei, Shandong, Liaoning, and is a 3 + 2 economic area that consists of three provinces and two municipalities. The regional GDP has reached 3800 billion Yuan, amounting to 28.2% of the national GDP.

This section investigates the investment and employment problems of this economic rim. By employing the same methods as described in the previous chapters, we obtain the results shown in Tables 7.7 and 7.8.

Table 7.8 indicates that the regional economic structure of the Bohai Economic Rim is in severe imbalance, no matter whether we look at the structure by using the demand and supply of labor or the demand and supply of capital.

Table 7.8 Regional Economic Structure and Measures of Comparative Gains of the Bohai Economic Rim of China (2011)

Area	*Province/ Municipality/ Autonomous Region (i)*	*Demand Structure $d_i(t)$ (%)*	*Employment Structure $s_i^1(t)$ (%)*	*Comparative Gain of Labor $G_i^1(t)$*	*Investment Structure $s_i^2(t)$ (%)*	*Comparative Gain of Capital $G_i^2(t)$*
Bohai Economic Rim	Liaoning	15.301	18.683	0.819	19.490	0.785
	Hebei	16.877	15.064	1.120	18.020	0.937
	Beijing	11.188	14.969	0.747	6.134	1.824
	Tianjin	7.784	4.804	1.620	7.771	1.002
	Shandong	31.228	30.135	1.036	29.411	1.062
	Shanxi	7.736	8.558	0.904	7.777	0.998
	Inner Mongolia	9.886	7.821	1.264	11.398	0.867
	Total	100	0.1651	—	100	—

In fact, from the fifth column (the comparative gains of labor) and the seventh column (the comparative gains of capital) of the table, it follows that there are 14 data points for the seven provinces and municipalities within this region. However, only three of the data points are basically in equilibrium, where the comparative gain of labor of Shandong Province is equal to 1.036, the comparative gains of capital of Tianjin and Shanxi are, respectively, equal to 1.002 and 0.998. And all the rest are situated in states of various degrees of imbalances. This means that in terms of the Bohai Economic Rim, other than three are basically in equilibrium, all the rest (11/14 = 78%) are in some states of supply and demand imbalances. In particular, the comparative gain of labor of Beijing is the lowest, equaling 0.747, and the comparative gain of capital of Liaoning is the lowest, equaling 0.785. On the other hand, the comparative gain of labor of Tianjin is the highest, equaling 1.620, and the comparative gain of capital of Beijing is the highest, equaling 1.762. This means that Tianjin and Beijing stand for the two extremes on the spectrum of productivity efficiency, and Beijing and Liaoning the two extremes on the spectrum of returns of capital investments in the Bohai Economic Rim.

Therefore, in terms of the magnitudes of the comparative gains of labor, the seven provinces and municipalities of the Bohai Economic Rim can be ranked as follows:

1	2	3	4	5	6	7
Tianjin	Inner Mongolia	Hebei	Shandong	Shanxi	Liaoning	Beijing

And, in terms of the magnitudes of the comparative gains of capital, the seven provinces and municipalities of the Bohai Economic Rim can be ranked as follows:

1	2	3	4	5	6	7
Beijing	Shandong	Tianjin	Shanxi	Hebei	Inner Mongolia	Liaoning

7.4.2 Investment and Employment Opportunities in the Bohai Economic Rim

First, in terms of the comparative gains of labor, the future labor supply within this region should be directed as much as possible toward Tianjin, Inner Mongolia, Hebei, Shandong provinces.

That is the direction of employment.

Second, in terms of the comparative gain of capital, the future capital investment in this region should be aimed at such three places as Beijing, Shandong, and Tianjin.

That is the direction of investment.

Third, when these two criteria are jointly considered, the most optimal areas in terms of the comparative gains of both labor and capital within this region are Tianjin and Shandong, because the values of the indicators of these two province and municipality are both much greater than 1. The secondary choices are Hebei and Inner Mongolia. Although Beijing enjoys the highest comparative gain of capital, its comparative gain of labor is the lowest, equaling merely 0.74. The last choice is Liaoning.

UPDATES 4

In Section 4, the reader will find all the relevant statistics that correspond to the previously analyzed tables in various chapters. All the detailed analyses are omitted in order to limit the length of this book.

Chapter 8

Updates with 2012 Statistics

In this chapter, we will analyze all the statistics from the year of 2012 that became available only recently before the deadline of submission of this manuscript. In order to limit the length of this book, we will only provide the statistics and the relevant numerical results without spelling out any of the related results. The interested reader can reference back to the previous chapters for how the corresponding results can be obtained.

8.1 Opportunities in East, Central, and West China

Table 8.1 The GDPs, Values of Fixed Properties, and Employments of Various Provinces, Municipalities, and Autonomous Regions (2012)

Area	*Province/Municipality/ Autonomous Region (i)*	*Regional GDP (0.1 Billion Yuan)*	*Value of Fixed Properties (0.1 Billion Yuan)*	*Employment (10,000 People)*
East China	1. Beijing	17879.4	7409.6	1795.8
	2. Tianjin	12893.9	9848.4	606.8
	3. Hebei	26575.0	15244.6	1792.3
	4. Liaoning	24846.4	15492.1	2218.1
	5. Shanghai	20181.7	7674.8	1781.6
	6. Jiangsu	54058.2	27258.1	5014.9
	7. Zhejiang	34665.3	15460.7	3838.8
	8. Fujian	19701.8	11304.8	1927.8
	9. Shandong	50013.2	27551.5	3564.0

continued

Table 8.1 (continued) The GDPs, Values of Fixed Properties, and Employments of Various Provinces, Municipalities, and Autonomous Regions (2012)

Area	*Province/Municipality/ Autonomous Region (i)*	*Regional GDP (0.1 Billion Yuan)*	*Value of Fixed Properties (0.1 Billion Yuan)*	*Employment (10,000 People)*
	10. Guangdong	57067.9	22871.9	5089.0
	11. Hainan	2855.5	2009.9	345.9
	Total	320738.3	162126.4	27974.5
Central China	12. Shanxi	12112.8	8223.9	1152.0
	13. Inner Mongolia	15880.6	13442.1	1034.5
	14. Jilin	11939.2	9136.2	1121.3
	15. Heilongjiang	13619.6	8143.7	1487.8
	16. Anhui	17212.1	8855.8	1745.7
	17. Jiangxi	12948.9	6513.7	1531.6
	18. Henan	29599.3	22060.0	2509.0
	19. Hubei	22250.5	12554.7	2380.0
	20. Hunan	22154.2	12488.8	2019.9
	21. Guangxi	13035.1	11068.5	1229.4
	Total	170752.3	112487.4	16211.2
West China	22. Chongqing	11409.6	6341.4	1437.1
	23. Sichuan	23872.8	12496.0	2323.5
	24. Guizhou	6852.2	4164.4	800.0
	25. Yunnan	10309.5	8576.4	1492.5
	26. Tibet	701.0	708.7	130.2
	27. Shaanxi	14453.7	9915.2	1211.9
	28. Gansu	5650.2	3298.1	657.0
	29. Qinghai	1893.5	1719.3	194.9
	30. Ningxia	2341.3	2086.9	262.9
	31. Xinjiang	7505.3	6792.2	753.6
	Total	84984.1	55098.6	9263.6
Entire Nation	Total	576479.7	329712.4	53449.8

Source: Data are from *China Statistical Abstract 2013* (XLS, DOC).

Table 8.2 The Regional Economic Structures and Measures of Comparative Gains of the 31 Provinces, Municipalities, and Autonomous Regions of East, Central, and West China (2012)

Area	Province/ Municipality/ Autonomous Region (i)	Demand Structure $d_i(t)$ (%)	Employment Structure $s_i^1(t)$ (%)	Comparative Gain of Labor $G_i^1(t)$	Investment Structure $s_i^2(t)$ (%)	Comparative Gain of Capital $G_i^2(t)$
East China	1. Beijing	3.10	3.36	0.92	2.25	1.38
	2. Tianjin	2.24	1.14	1.96	2.99	0.75
	3. Hebei	4.61	3.35	1.38	4.62	1.00
	4. Liaoning	4.31	4.15	1.04	4.70	0.92
	5. Shanghai	3.50	3.33	1.05	2.33	1.50
	6. Jiangsu	9.38	9.38	1.00	8.27	1.13
	7. Zhejiang	6.01	7.18	0.84	4.69	1.29
	8. Fujian	3.42	3.61	0.95	3.43	1.00
	9. Shandong	8.68	6.67	1.3	8.36	1.04
	10. Guangdong	9.90	9.52	1.04	6.94	1.43
	11. Hainan	0.50	0.63	0.79	0.61	0.82
	Total	55.64	52.32	1.06*	49.19	1.13**
Central China	12. Shanxi	2.10	2.16	0.97	2.49	0.84
	13. Inner Mongolia	2.75	1.94	1.42	4.08	0.67
	14. Jilin	2.07	2.10	0.99	2.77	0.75
	15. Heilongjiang	2.36	2.78	0.85	2.47	0.96
	16. Anhui	2.99	3.27	0.91	2.69	1.11
	17. Jiangxi	2.25	2.87	0.78	1.98	1.14
	18. Henan	5.13	4.69	1.09	6.69	0.77
	19. Hubei	3.86	4.45	0.87	3.81	1.01
	20. Hunan	3.84	3.78	1.02	3.79	1.01
	21. Guangxi	2.26	2.30	0.98	3.36	0.67
	Total	29.62	30.34	0.98*	34.13	0.87**
West China	22. Chongqing	1.98	2.69	0.74	1.92	1.03
	23. Sichuan	4.14	4.35	0.95	3.79	1.09
	24. Guizhou	1.19	1.50	0.79	1.26	0.94
	25. Yunnan	1.79	2.79	0.64	2.60	0.69
	26. Tibet	0.12	0.24	0.50	0.21	0.57
	27. Shaanxi	2.51	2.27	1.11	3.01	0.83
	28. Gansu	0.98	1.23	0.80	1.00	0.98

continued

Table 8.2 (continued) The Regional Economic Structures and Measures of Comparative Gains of the 31 Provinces, Municipalities, and Autonomous Regions of East, Central, and West China (2012)

Area	*Province/ Municipality/ Autonomous Region (i)*	*Demand Structure $d_i(t)$ (%)*	*Employment Structure $s_i^1(t)$ (%)*	*Comparative Gain of Labor $G_i^1(t)$*	*Investment Structure $s_i^2(t)$ (%)*	*Comparative Gain of Capital $G_i^2(t)$*
	29. Qinghai	0.33	0.36	0.92	0.52	0.63
	30. Ningxia	0.41	0.49	0.84	0.63	0.65
	31. Xinjiang	1.30	1.41	0.92	1.76	0.74
	Total	14.74	17.34	0.87*	16.70	0.88**

Source: Data are from *Shanghai Statistical Abstract 2013* (XLS, DOC). (* and **: These values are not the sums of the relevant data. Instead, they are the means of relevant regions, where * the means of the comparative gains of labor and ** the means of comparative gains of capital of relevant regions.)

Table 8.3 The Total Output Values, Investments on Fixed Assets, and State of Employments in China's Top Three Economic Sectors (2012)

Sectors (i)	*Total Output of Sector (0.1 Billion Yuan)*	*Total Investment on Fixed Properties (0.1 Billion Yuan)*	*Employment (10,000 People)*
Primary sector	52373.6	10996	375
Secondary sector	235162.0	157333	7249
Tertiary sector	231406.5	200292	7650
Total	518942.1	368621	15274

Source: Data are from *China Statistics Abstract 2013* (XLS, DOC).

Table 8.4 China's Economic Sectorial Structure and Relevant Measures of Comparative Gains (t = year of 2012)

Sectors (i)	*Demand Structure $d_i(t)$ (%)*	*Employment Structure $s_i^1(t)$ (%)*	*Comparative Gain of Labor $G_i^1(t)$*	*Investment Structure $s_i^2(t)$ (%)*	*Comparative Gain of Capital $G_i^2(t)$*
Primary sector	10.09	2.98	4.10	2.46	3.39
Secondary sector	45.32	42.68	0.95	47.46	1.06
Tertiary sector	44.59	54.34	0.89	50.08	0.82
Total	100	100	—	100	—

Source: Data from *China Statistics Abstract 2013* (XLS, DOC).

8.2 The East Region

Table 8.5 The GDPs, Values of Fixed Properties, and Employments of Various Provinces and Municipalities of East China (2012)

Area	Province/ Municipality/ Autonomous Region (i)	Regional GDP (0.1 Billion Yuan)	Value of Fixed Properties (0.1 Billion Yuan)	Employments (10,000 People)
East China	1. Beijing	17879.4	7409.6	1795.8
	2. Tianjin	12893.9	9848.4	606.8
	3. Hebei	26575.0	15244.6	1792.3
	4. Liaoning	24846.4	15492.1	2218.1
	5. Shanghai	20181.7	7674.8	1781.6
	6. Jiangsu	54058.2	27258.1	5014.9
	7. Zhejiang	34665.3	15460.7	3838.8
	8. Fujian	19701.8	11304.8	1927.8
	9. Shandong	50013.2	27551.5	3564.0
	10. Guangdong	57067.9	22871.9	5089.0
	11. Hainan	2855.5	2009.9	345.9
	Total	320738.3	162126.4	27974.5

Source: Data are from *China Statistical Abstract 2013* (XLS, DOC).

Note: *: These values are not the sums of the relevant data. Instead, they are the means of comparative gains of labor of relevant regions.

Table 8.6 The Regional Economic Structures and Measures of Comparative Gains of the Provinces and Municipalities of East China (2012)

Area	Province/ Municipality/ Autonomous Region (i)	Demand Structure $d_i(t)$ (%)	Employment Structure $s_i^1(t)$ (%)	Comparative Gain of Labor $G_i^1(t)$	Investment Structure $s_i^2(t)$ (%)	Comparative Gain of Capital $G_i^2(t)$
East China	1. Beijing	5.57	6.42	0.87	4.09	1.36
	2. Tianjin	4.02	2.17	1.85	6.07	0.66
	3. Hebei	8.29	6.41	1.29	9.46	0.88
	4. Liaoning	7.75	7.93	0.98	9.55	0.79
	5. Shanghai	6.29	6.37	0.99	4.26	1.48
	6. Jiangsu	16.85	17.93	0.94	16.81	1.02
	7. Zhejiang	10.82	13.72	0.79	9.53	1.16
	8. Fujian	6.14	6.89	0.89	6.97	0.88
	9. Shandong	15.59	12.74	1.22	16.99	0.89
	10. Guangdong	17.79	17.19	0.98	14.17	1.24
	11. Hainan	0.89	1.24	0.72	1.24	0.72
	Total	100	100	—	100	—

Table 8.7 The GDPs, Total Investments in Fixed Properties, and Employment Situation of Shanghai's Three Economic Sectors and Industries (2012)

Sectors (i)	*Total Output of Sector (0.1 Billion Yuan)*	*Total Investment on Fixed Properties (0.1 Billion Yuan)*	*Employment (10,000 People)*
Total	20181.72	5117.6	555.7
Primary sector	127.80	11	1.2
Secondary sector	7854.77	1245.5	270.5
Tertiary sector	12199.15	3861.1	284
Industries (*i*)			
Agriculture, forest, animal husbandry, fishery	127.80	11.0	1.2
Mining		0.4	0.1
Manufacturing		1080.6	218.7
Production/supply of electricity, gas, and water		163.0	5.8
Construction	757.01	1.5	48.9
Communication, transportation, storage, postal service	895.31	460.8	38.0
Information transfer, computer, and software areas		121.3	8.6
# Information Transfer			
Wholesale and retail	3291.93	62.1	59.8
Hotels and restaurants	298.40	33.3	20.1
Finance	2450.36	49.2	29.5
Real estate/housing	1147.04	2402.3	15.2
Leasing and commercial services		136.1	17.2
Scientific research, technology service, and geological survey		39.5	12.2
Water conservation, environment, and public facility management		338.3	6.2
# Public Facility Management			
Residents and other services		4.9	3.3
Education		50.9	28.0
Health, safety, and welfare		56.9	17.4

Table 8.7 (continued) The GDPs, Total Investments in Fixed Properties, and Employment Situation of Shanghai's Three Economic Sectors and Industries (2012)

Sectors (i)	*Total Output of Sector (0.1 Billion Yuan)*	*Total Investment on Fixed Properties (0.1 Billion Yuan)*	*Employment (10,000 People)*
Social, sports, and entertainment		90.6	5.1
Public administration and social organizations		14.9	20.5

Source: Data are from *Shanghai Statistical Yearbook 2013* (XLS, DOC).

Table 8.8 The Structures and Comparative Gain Measures of Shanghai's Three Economic Sectors and Industries ($t = 2012$)

Sectors (i)	*Demand Structure $d_i(t)$ (%)*	*Employment Structure $s_i^1(t)$ (%)*	*Comparative Gain of Labor $G_i^1(t)$*	*Investment Structure $s_i^2(t)$ (%)*	*Comparative Gain of Capital $G_i^2(t)$*
Total	100	100	—	100	—
Primary sector	0.63	0.22	2.86	0.21	3.00
Secondary sector	38.92	48.68	0.80	24.34	1.60
Tertiary sector	60.45	51.10	1.18	75.45	0.80
Industries (*i*)					
Agriculture, forest, animal husbandry, fishery	0.63	0.22	2.86	0.21	3.00
Mining		0.02		0.01	
Manufacturing		39.35		21.14	
Production/supply of electricity, gas, and water		1.04		3.19	
Construction	3.75	8.80	0.43	0.03	125
Communication, transportation, storage, postal service	4.44	6.84	0.65	9.00	0.49
Information transfer, computer, and software areas		1.55		2.37	
# Information Transfer					
Wholesale and retail	16.31	10.76	1.55	1.21	13.48

continued

Table 8.8 (continued) The Structures and Comparative Gain Measures of Shanghai's Three Economic Sectors and Industries ($t = 2012$)

Sectors (i)	*Demand Structure $d_i(t)$ (%)*	*Employment Structure $s_i^1(t)$ (%)*	*Comparative Gain of Labor $G_i^1(t)$*	*Investment Structure $s_i^2(t)$ (%)*	*Comparative Gain of Capital $G_i^2(t)$*
Hotels and restaurants	1.48	3.62	0.41	0.65	2.28
Finance	12.14	5.31	2.29	0.96	12.65
Real estate/housing	5.68	2.74	2.07	46.94	0.12
Leasing and commercial services		3.10		2.66	
Scientific research, technology service, and geological survey		2.20		0.77	
Water conservation, environment, and public facility management		1.12		6.61	
# Public Facility Management					
Residents and other services		0.59		0.10	
Education		5.04		0.99	
Health, safety, and welfare		3.13		1.11	
Social, sports, and entertainment		0.92		1.77	
Public administration and social organizations		3.69		0.29	

Table 8.9 The GDPs, Total Investments in Fixed Properties, and Employment Situation of Guangdong's Three Economic Sectors and Industries (2012)

Sectors (i)	*Total Output of Sector (0.1 Billion Yuan)*	*Total Investment on Fixed Properties (0.1 Billion Yuan)*	*Employment (10,000 People)*
Total	57067.92	18751.5	1304.0
Primary sector	2847.26	340.6	7.2
Secondary sector	27700.97	6062.9	672
Tertiary sector	26519.69	12348	624.7

Table 8.9 (continued) The GDPs, Total Investments in Fixed Properties, and Employment Situation of Guangdong's Three Economic Sectors and Industries (2012)

Sectors (i)	*Total Output of Sector (0.1 Billion Yuan)*	*Total Investment on Fixed Properties (0.1 Billion Yuan)*	*Employment (10,000 People)*
Industries (*i*)			
Agriculture, forest, animal husbandry, fishery	2847.26	340.6	7.2
Mining		82.6	3.4
Manufacturing		4923.9	540.9
Production/supply of electricity, gas, and water		1019.0	20.1
Construction	1890.90	37.4	107.6
Communication, transportation, storage, postal service	2367.46	1729.7	61.8
Information transfer, computer, and software areas	1618.09	341.5	18.6
# Information Transfer			
Wholesale and retail	6333.62	479.6	65.0
Hotels and restaurants	1308.40	390.5	32.3
Finance	3171.96	81.6	47.6
Real estate/housing	3643.87	6789.3	34.9
Leasing and commercial services	2132.42	175.0	37.0
Scientific research, technology service, and geological survey	597.72	118.2	21.3
Water conservation, environment, and public facility management	247.69	1526.4	14.3
# Public Facility Management			
Residents and other services	895.77	29.4	5.8
Education	1382.55	299.0	117.7
Health, safety, and welfare	853.78	140.8	56.1
Social, sports, and entertainment	329.95	153.9	9.5
Public administration and social organizations	1636.40	93.0	102.9

Source: Data are from *Statistical Yearbook of Guangdong Province 2013* (XLS, DOC).

Table 8.10 The Structures and Comparative Gain Measures of Guangdong's Three Economic Sectors and Industries ($t = 2012$)

Sectors (i)	*Demand Structure $d_i(t)$ (%)*	*Employment Structure $s_i^1(t)$ (%)*	*Comparative Gain of Labor $G_i^1(t)$*	*Investment Structure $s_i^2(t)$ (%)*	*Comparative Gain of Capital $G_i^2(t)$*
Total	100	100	—	100	—
Primary sector	4.99	0.55	9.00	1.82	2.75
Secondary sector	48.54	51.53	0.94	32.33	1.50
Tertiary sector	46.47	47.91	0.97	65.85	0.71
Industries (*i*)					
Agriculture, forest, animal husbandry, fishery	4.99	0.55	9.00	1.82	2.75
Mining		0.26		0.44	
Manufacturing		41.48		26.26	
Production/supply of electricity, gas, and water		1.54		5.43	
Construction	3.31	8.25	0.40	0.20	16.63
Communication, transportation, storage, postal service	4.15	4.74	0.88	9.22	0.45
Information transfer, computer, and software areas	2.84	1.43	1.99	1.82	1.57
# Information Transfer					
Wholesale and retail	11.10	4.98	2.23	2.56	4.34
Hotels and restaurants	2.29	4.76	0.93	2.08	1.10
Finance	5.56	3.65	1.52	0. 43	12.78
Real estate/housing	6.39	2.67	2.39	36.21	0.18
Leasing and commercial services	3.74	2.84	1.32	0. 93	4.00
Scientific research, technology service, and geological survey	1.05	1.63	0.64	0. 63	1.66
Water conservation, environment, and public facility management	0.43	1.10	0.40	8.14	0.05

Table 8.10 (continued) The Structures and Comparative Gain Measures of Guangdong's Three Economic Sectors and Industries ($t = 2012$)

Sectors (i)	*Demand Structure $d_i(t)$ (%)*	*Employment Structure $s_i^1(t)$ (%)*	*Comparative Gain of Labor $G_i^1(t)$*	*Investment Structure $s_i^2(t)$ (%)*	*Comparative Gain of Capital $G_i^2(t)$*
# Public Facility Management					
Residents and other services	1.57	0.45	3.50	0. 16	10.02
Education	2.42	9.03	0.27	1.59	1.52
Health, safety, and welfare	1.50	4.30	0.35	0. 75	1.99
Social, sports, and entertainment	0.58	0.73	0.80	0. 82	0.70
Public administration and social organizations	2.87	7.89	0.36	0. 50	5.78

Source: Data are from *China Statistics Abstract 2013* (XLS, DOC).

Table 8.11 The GDPs, Total Investments in Fixed Properties, and Employment Situation of Tianjin's Three Economic Sectors and Industries (2012)

Sectors (i)	*Total Output of Sector (0.1 Billion Yuan)*	*Total Investment on Fixed Properties (0.1 Billion Yuan)*	*Employment (10,000 People)*
Total	12893.88	7934.8	289.1
Primary sector	171.60	198.7	0.5
Secondary sector	6663.82	2851.5	162.6
Tertiary sector	6058.46	4884.6	125.9
Industries (*i*)			
Agriculture, forest, animal husbandry, fishery	171.60	198.7	0.5
Mining		199.0	6.9
Manufacturing		2319.9	120.2
Production/supply of electricity, gas, and water		283.9	4.4
Construction	540.76	48.6	31.1
Communication, transportation, storage, postal service	683.56	729.9	14.1
Information transfer, computer, and software areas		75.3	3.3

continued

Table 8.11 (continued) The GDPs, Total Investments in Fixed Properties, and Employment Situation of Tianjin's Three Economic Sectors and Industries (2012)

Sectors (i)	*Total Output of Sector (0.1 Billion Yuan)*	*Total Investment on Fixed Properties (0.1 Billion Yuan)*	*Employment (10,000 People)*
# Information Transfer			
Wholesale and retail	1680.33	243.9	17.5
Hotels and restaurants	222.18	70.5	6.9
Finance	1001.59	21.4	7.8
Real estate/housing	449.65	1824.6	5.5
Leasing and commercial services		588.4	5.4
Scientific research, technology service, and geological survey		48.9	8.3
Water conservation, environment, and public facility management		917.9	3.7
# Public Facility Management			
Residents and other services		52.9	10.9
Education		73.8	16.3
Health, safety, and welfare		63.0	8.8
Social, sports, and entertainment		90.7	1.9
Public administration and social organizations		83.5	15.5

Source: Data are from *China's Statistics Abstracts 2013* (XLS, DOC).

Table 8.12 The Structures and Comparative Gain Measures of Tianjin's Three Economic Sectors and Industries ($t = 2012$)

Sectors (i)	*Demand Structure $d_i(t)$ (%)*	*Employment Structure $s_i^1(t)$ (%)*	*Comparative Gain of Labor $G_i^1(t)$*	*Investment Structure $s_i^2(t)$ (%)*	*Comparative Gain of Capital $G_i^2(t)$*
Total	100	100	—	100	—
Primary sector	1.33	0.19	7.00	2.50	0.53
Secondary sector	51.68	56.25	0.92	35.94	1.44
Tertiary sector	46.99	43.56	1.08	61.56	0.76

Table 8.12 (continued) The Structures and Comparative Gain Measures of Tianjin's Three Economic Sectors and Industries ($t = 2012$)

Sectors (i)	*Demand Structure $d_i(t)$ (%)*	*Employment Structure $s_i^1(t)$ (%)*	*Comparative Gain of Labor $G_i^1(t)$*	*Investment Structure $s_i^2(t)$ (%)*	*Comparative Gain of Capital $G_i^2(t)$*
Industries (*i*)					
Agriculture, forest, animal husbandry, fishery	1.33	0.19	7.00	2.50	0.53
Mining		2.39		2.51	
Manufacturing		41.60		29.24	
Production/supply of electricity, gas, and water		1.51		3.58	
Construction	4.19	10.76	0.39	0.61	6.84
Communication, transportation, storage, postal service	5.30	4.87	1.09	9.20	0.58
Information transfer, computer, and software areas		1.13045		0.95	
# Information Transfer					
Wholesale and retail	13.03	6.04	2.16	3.07	4.24
Hotels and restaurants	1.72314	2.40	0.72	0.89	1.94
Finance	7.77	2.71	2.87	0.27	28.81
Real estate/housing	3.49	1.91	1.82	23.00	0.15
Leasing and commercial services		1.88		7.42	
Scientific research, technology service, and geological survey		2.86		0. 62	
Water conservation, environment, and public facility management		1.27		11.57	
# Public Facility Management					
Residents and other services		3.76		0.67	

continued

Table 8.12 (continued) The Structures and Comparative Gain Measures of Tianjin's Three Economic Sectors and Industries ($t = 2012$)

Sectors (i)	*Demand Structure $d_i(t)$ (%)*	*Employment Structure $s_i^1(t)$ (%)*	*Comparative Gain of Labor $G_i^1(t)$*	*Investment Structure $s_i^2(t)$ (%)*	*Comparative Gain of Capital $G_i^2(t)$*
Education		5.64		0.93	
Health, safety, and welfare		3.06		0.79	
Social, sports, and entertainment		0.66		1.14	
Public administration and social organizations		5.36		1.05	

Source: Data are from *China Statistics Abstract 2013* (XLS, DOC).

Table 8.13 The GDPs, Total Investments in Fixed Properties, and Employment Situation of Beijing's Three Economic Sectors and Industries (2012)

Sectors (i)	*Total Output of Sector (0.1 Billion Yuan)*	*Total Investment on Fixed Properties (0.1 Billion Yuan)*	*Employment (10,000 People)*
Total	17879.40	6112.4	717.4
Primary sector	150.20	127.3	2.5
Secondary sector	4059.27	643.40	166.50
Tertiary sector	13669.93	5341.70	548.30
Industries (*i*)			
Agriculture, forest, animal husbandry, fishery	150.20	127.3	2.5
Mining		4.3	6.9
Manufacturing		414.5	108.0
Production/supply of electricity, gas, and water		215.8	8.9
Construction	764.95	8.7	42.7
Communication, transportation, storage, postal service	816.31	696.4	57.8
Information transfer, computer, and software areas	1621.8	162.0	52.6
# Information Transfer			
Wholesale and retail	2229.77	27.8	68.6

Table 8.13 (continued) The GDPs, Total Investments in Fixed Properties, and Employment Situation of Beijing's Three Economic Sectors and Industries (2012)

Sectors (i)	*Total Output of Sector (0.1 Billion Yuan)*	*Total Investment on Fixed Properties (0.1 Billion Yuan)*	*Employment (10,000 People)*
Hotels and restaurants	373.06	55.3	32.0
Finance	2536.91	26.6	37.6
Real estate/housing	1244.17	3491.1	37.1
Leasing and commercial services	1340.60	36.6	61.4
Scientific research, technology service, and geological survey	1268.40	128.7	54.0
Water conservation, environment, and public facility management	101.30	402.9	9.4
# Public Facility Management			
Residents and other services	124.30	18.1	8.6
Education	681.80	95.0	44.4
Health, safety, and welfare	363.60	44.2	23.0
Social, sports, and entertainment	402.60	78.9	17.1
Public administration and social organizations	565.20	78.0	44.7

Source: Data are from *Beijing Statistical Yearbook 2013* (XLS, DOC).

Table 8.14 The Structures and Comparative Gain Measures of Beijing's Three Economic Sectors and Industries ($t = 2012$)

Sectors (i)	*Demand Structure $d_i(t)$ (%)*	*Employment Structure $s_i^1(t)$ (%)*	*Comparative Gain of Labor $G_i^1(t)$*	*Investment Structure $s_i^2(t)$ (%)*	*Comparative Gain of Capital $G_i^2(t)$*
Total	100	100	—	100	—
Primary sector	0.84	0.35	2.37	2.08	0.40
Secondary sector	22.70	23.21	0.98	10.53	2.16
Tertiary sector	76.46	76.44	1.00	87.39	0.87
Industries (*i*)					
Agriculture, forest, animal husbandry, fishery	0.84	0.35	2.37	2.08	0.40
Mining		0.96		0. 07	

continued

Table 8.14 (continued) The Structures and Comparative Gain Measures of Beijing's Three Economic Sectors and Industries ($t = 2012$)

Sectors (i)	*Demand Structure $d_i(t)$ (%)*	*Employment Structure $s_i^1(t)$ (%)*	*Comparative Gain of Labor $G_i^1(t)$*	*Investment Structure $s_i^2(t)$ (%)*	*Comparative Gain of Capital $G_i^2(t)$*
Manufacturing		15.05		6.78	
Production/supply of electricity, gas, and water		1.25		3.53	
Construction	4.28	5.96	0.72	0.14	30.09
Communication, transportation, storage, postal service	4.57	8.06	0.57	11.39	0.40
Information transfer, computer, and software areas	9.07	7.34	1.24	2.65	3.42
# Information Transfer					
Wholesale and retail	12.47	9.57	1.30	0.45	27.41
Hotels and restaurants	2.09	4.46	0.47	0.90	2.31
Finance	14.19	5.24	2.71	0.44	32.57
Real estate/housing	6.96	5.17	1.35	57.12	0.12
Leasing and commercial services	7.50	8.56	0.88	0.60	12.52
Scientific research, technology service, and geological survey	7.09	7.53	0.94	2.10	3.37
Water conservation, environment, and public facility management	0. 567	1.31	0.43	6.59	0.09
# Public Facility Management					
Residents and other services	0. 70	1.20	0.58	0.30	2.35
Education	3.81	6.19	0.62	1.55	2.45
Health, safety, and welfare	2.034	3.20	0.64	0.72	2.81
Social, sports, and entertainment	2.088371	1.45	1.44	0.93	2.24
Public administration and social organizations	3.258081	3.64	0.90	0.83	3.95

Source: Data are from *China Statistics Abstract 2013* (XLS, DOC).

Table 8.15 The GDPs, Total Investments in Fixed Properties, and Employment Situation of Shandong's Three Economic Sectors and Industries (2012)

Sectors (i)	*Total Output of Sector (0.1 Billion Yuan)*	*Total Investment on Fixed Properties (0.1 Billion Yuan)*	*Employment (10,000 People)*
Total	50013.24	31256.0	1110.2
Primary sector	4281.70	900.4	2.8
Secondary sector	25735.73	14479.9	631.7
Tertiary sector	19995.81	15875.7	475.7
Industries (*i*)			
Agriculture, forest, animal husbandry, fishery	4281.70	900.4	2.8
Mining		570.3	78.8
Manufacturing		12713.7	395.0
Production/supply of electricity, gas, and water		705.7	21.0
Construction	2937.40	490.1	136.8
Communication, transportation, storage, postal service	2516.19	1657.2	37.4
Information transfer, computer, and software areas		91.2	9.1
# Information Transfer			
Wholesale and retail	6507.40	1276.2	48.9
Hotels and restaurants	1059.47	444.8	15.2
Finance	1936.11	51.6	32.7
Real estate/housing	1984.49	7552.1	15.8
Leasing and commercial services		363.8	11.5
Scientific research, technology service, and geological survey		399.8	12.4
Water conservation, environment, and public facility management		1467.5	12.5
# Public Facility Management			
Residents and other services		391.7	3.9
Education		333.0	109.8

continued

Table 8.15 (continued) The GDPs, Total Investments in Fixed Properties, and Employment Situation of Shandong's Three Economic Sectors and Industries (2012)

Sectors (i)	*Total Output of Sector (0.1 Billion Yuan)*	*Total Investment on Fixed Properties (0.1 Billion Yuan)*	*Employment (10,000 People)*
Health, safety, and welfare		190.9	49.6
Social, sports, and entertainment		700.0	6.9
Public administration and social organizations		956.0	109.8

Source: Data are from *China Statistics Abstracts 2013* (XLS, DOC).

Table 8.16 The Structures and Comparative Gain Measures of Shandong's Three Economic Sectors and Industries (t = 2012)

Sectors (i)	*Demand Structure $d_i(t)$ (%)*	*Employment Structure $s_i^1(t)$ (%)*	*Comparative Gain of Labor $G_i^1(t)$*	*Investment Structure $s_i^2(t)$ (%)*	*Comparative Gain of Capital $G_i^2(t)$*
Total	100	100	—	100	—
Primary sector	8.56	0.25	34.56	2.88	2.97
Secondary sector	51.46	56.90	0.90	46.33	1.11
Tertiary sector	39.98	42.85	0.93	50.79	0.79
Industries (i)					
Agriculture, forest, animal husbandry, fishery	8.56	0.25	34.56	2.88	2.97
Mining		7.10		1.82	
Manufacturing		35.58		40.68	
Production/supply of electricity, gas, and water		1.89		2.26	
Construction	5.87	12.3257	0.48	1.57	3.75
Communication, transportation, storage, postal service	5.03	3.37	1.49	5.30	0.95
Information transfer, computer, and software areas		0.82		0.29	

Table 8.16 (continued) The Structures and Comparative Gain Measures of Shandong's Three Economic Sectors and Industries ($t = 2012$)

Sectors (i)	*Demand Structure* $d_i(t)$ *(%)*	*Employment Structure* $s_i^1(t)$ *(%)*	*Comparative Gain of Labor* $G_i^1(t)$	*Investment Structure* $s_i^2(t)$ *(%)*	*Comparative Gain of Capital* $G_i^2(t)$
# Information Transfer					
Wholesale and retail	13.01	4.41	2.95	4.08	3.19
Hotels and restaurants	2.12	1.37	1.55	1.42	1.49
Finance	3.87	2.95	1.31	0.17	23.44
Real estate/housing	3.97	1.42	2.79	24.16	0.16
Leasing and commercial services		0.01		0.01	
Scientific research, technology service, and geological survey		1.12		1.28	
Water conservation, environment, and public facility management		1.13		4.69	
# Public Facility Management					
Residents and other services		0. 55		1.25	
Education		9.89		1.07	
Health, safety, and welfare		4.47		0.61	
Social, sports, and entertainment		0.62		2.24	
Public administration and social organizations		9.89		3.06	

Source: Data are from *China Statistics Abstract 2013* (XLS, DOC).

Table 8.17 The GDPs, Total Investments in Fixed Properties, and Employment Situation of Hebei's Three Economic Sectors and Industries (2012)

Sectors (i)	*Total Output of Sector (0.1 Billion Yuan)*	*Total Investment on Fixed Properties (0.1 Billion Yuan)*	*Employment (10,000 People)*
Total	26575.01	19661.3	619.9
Primary sector	3186.66	804.4	5.5
Secondary sector	14003.57	9387	276.8
Tertiary sector	9384.78	9469.9	337.6
Industries (*i*)			
Agriculture, forest, animal husbandry, fishery	3186.66	804.4	5.5
Mining		620.5	28.8
Manufacturing		8008.9	145.4
Production/supply of electricity, gas, and water		713.4	21.2
Construction	1491.97	44.2	81.4
Communication, transportation, storage, postal service	2212.93	1543.3	24.3
Information transfer, computer, and software areas		88.8	6.5
# Information Transfer			
Wholesale and retail	2024.29	658.3	26.3
Hotels and restaurants	388.87	214.7	6.8
Finance	913.66	24.4	24.7
Real estate/housing	982.05	4656.5	6.8
Leasing and commercial services		210.8	5.2
Scientific research, technology service, and geological survey		110.4	12.5
Water conservation, environment, and public facility management		1200.2	11.3
# Public Facility Management			
Residents and other services		71.3	2.2
Education		208.9	89.9

Table 8.17 (continued) The GDPs, Total Investments in Fixed Properties, and Employment Situation of Hebei's Three Economic Sectors and Industries (2012)

Sectors (i)	*Total Output of Sector (0.1 Billion Yuan)*	*Total Investment on Fixed Properties (0.1 Billion Yuan)*	*Employment (10,000 People)*
Health, safety, and welfare		111.8	32.2
Social, sports, and entertainment		213.8	5.2
Public administration and social organizations		156.7	83.8

Source: All data are from *China Statistics Yearbook 2013*, and *Hebei Province's Statistics Yearbook 2013*.

Table 8.18 The Structures and Comparative Gain Measures of Hebei's Three Economic Sectors and Industries ($t = 2012$)

Sectors (i)	*Demand Structure $d_i(t)$ (%)*	*Employment Structure $s_i^1(t)$ (%)*	*Comparative Cain of Labor $G_i^1(t)$*	*Investment Structure $s_i^2(t)$ (%)*	*Comparative Gain of Capital $G_i^2(t)$*
Total	100	100	—	100	—
Primary sector	11.99	0.89	13.46	4.09	2.93
Secondary sector	52.69	44.65	1.18	47.74	1.10
Tertiary sector	35.31	54.46	0.65	48.17	0.73
Industries (*i*)					
Agriculture, forest, animal husbandry, fishery	11.99	0.89	13.46	4.09	2.93
Mining		4.65		3.16	
Manufacturing		23.46		40.73	
Production/supply of electricity, gas, and water		3.43		3.63	
Construction	5.61	13.13	0.43	0.22	24.95
Communication, transportation, storage, postal service	8.33	3.92	2.12	7.85	1.06
Information transfer, computer, and software areas		1.05		0.45	

continued

Table 8.18 (continued) The Structures and Comparative Gain Measures of Hebei's Three Economic Sectors and Industries ($t = 2012$)

Sectors (i)	*Demand Structure $d_i(t)$ (%)*	*Employment Structure $s_i^1(t)$ (%)*	*Comparative Gain of Labor $G_i^1(t)$*	*Investment Structure $s_i^2(t)$ (%)*	*Comparative Gain of Capital $G_i^2(t)$*
# Information Transfer					
Wholesale and retail	7.62	4.25	1.79	3.35	2.28
Hotels and restaurants	1.46	1.09403	1.34	1.09	1.34
Finance	3.44	3.98	0.86	0.12	27.72
Real estate/housing	3.70	1.10	3.35	23.68	0.16
Leasing and commercial services		0.83		1.07	
Scientific research, technology service, and geological survey		2.02		0.56	
Water conservation, environment, and public facility management		1.82		6.10	
# Public Facility Management					
Residents and other services		0.35		036	
Education		14.49		1.06	
Health, safety, and welfare		5.20		0.57	
Social, sports, and entertainment		0.84		1.09	
Public administration and social organizations		13.52		0.80	

Source: Data are from *China Statistics Abstract 2013* (XLS, DOC).

Table 8.19 The GDPs, Total Investments in Fixed Properties, and Employment Situation of Zhejiang's Three Economic Sectors and Industries (2012)

Sectors (i)	*Total Output of Sector (0.1 Billion Yuan)*	*Total Investment on Fixed Properties (0.1 Billion Yuan)*	*Employment (10,000 People)*
Total	34665.33	17649.4	1070.1
Primary sector	1667.88	200.1	0.8
Secondary sector	17316.32	6122.5	681.2
Tertiary sector	15681.13	11326.7	388.1
Industries (*i*)			
Agriculture, forest, animal husbandry, fishery	1667.88	200.1	0.8
Mining		33.0	1.4
Manufacturing		5333.7	372.5
Production/supply of electricity, gas, and water		727.9187	13.2999
Construction	1978.30	27.8674	294.0726
Communication, transportation, storage, postal service	1278.91	1349.7101	29.364
Information transfer, computer, and software areas	918.63	111.0178	14.0702
# Information Transfer			
Wholesale and retail	3684.34	322.9	39.3
Hotels and restaurants	655.74	212.8707	16.3226
Finance	2762.24	92.8641	36.3777
Real estate/housing	1927.93	6788.9	17.5
Leasing and commercial services	657.61	222.0	29.2
Scientific research, technology service, and geological survey	345.16	58.8944	17.4846
Water conservation, environment, and public facility management	161.61	1383.571	12.9023
# Public Facility Management			
Residents and other services	472.90	33.9	2.0
Education	877.64	199.8335	66.2361

continued

Table 8.19 (continued) The GDPs, Total Investments in Fixed Properties, and Employment Situation of Zhejiang's Three Economic Sectors and Industries (2012)

Sectors (i)	*Total Output of Sector (0.1 Billion Yuan)*	*Total Investment on Fixed Properties (0.1 Billion Yuan)*	*Employment (10,000 People)*
Health, safety, and welfare	589.94	125.2279	37.3721
Social, sports, and entertainment	254.15	207.7	6.7
Public administration and social organizations	1094.33	217.3	63.3

Source: Data are from *China Statistics Abstracts 2013* (XLS, DOC).

Table 8.20 The Structures and Comparative Gain Measures of Zhejiang's Three Economic Sectors and Industries ($t = 2012$)

Sectors (i)	*Demand Structure $d_i(t)$ (%)*	*Employment Structure $s_i^1(t)$ (%)*	*Comparative Gain of Labor $G_i^1(t)$*	*Investment Structure $s_i^2(t)$ (%)*	*Comparative Gain of Capital $G_i^2(t)$*
Total	100	100	—	100	—
Primary sector	4.81	0.08	62.92	1.13	4.24
Secondary sector	49.95	63.66	0.78	34.69	1.44
Tertiary sector	45.24	36.27	1.25	64.18	0.70
Industries (*i*)					
Agriculture, forest, animal husbandry, fishery	4.81	0.08	62.92	1.13	4.24
Mining		0.13		0.19	
Manufacturing		34.81		30.22	
Production/supply of electricity, gas, and water		1.24		4.12	
Construction	5.71	27.48	0.21	0.16	36.14
Communication, transportation, storage, postal service	3.69	2.74	1.34	7.65	0.48
Information transfer, computer, and software areas	2.65	1.31	2.02	0.63	4.21

Table 8.20 (continued) The Structures and Comparative Gain Measures of Zhejiang's Three Economic Sectors and Industries ($t = 2012$)

Sectors (i)	*Demand Structure $d_i(t)$ (%)*	*Employment Structure $s_i^1(t)$ (%)*	*Comparative Gain of Labor $G_i^1(t)$*	*Investment Structure $s_i^2(t)$ (%)*	*Comparative Gain of Capital $G_i^2(t)$*
# Information Transfer					
Wholesale and retail	10.63	3.67	2.90	1.83	5.81
Hotels and restaurants	1.89	1.53	1.24	1.21	1.57
Finance	7.97	3.40	2.34	0.53	15.14
Real estate/housing	5.56	1.63	3.41	38.47	0.14
Leasing and commercial services	1.90	2.73	0.69	1.26	1.51
Scientific research, technology service, and geological survey	1.00	1.63	0.62	0.33	2.98
Water conservation, environment, and public facility management	0.47	1.21	0.39	7.84	0.06
# Public Facility Management					
Residents and other services	1.36	0.19	7.20	0.19	7.20
Education	2.53	6.19	0.41	1.13	2.24
Health, safety, and welfare	1.70	3.49	0.49	0.71	2.40
Social, sports, and entertainment	0.73	0.63	1.16	1.18	0.62
Public administration and social organizations	3.16	5.91	0.53	1.23	2.56

Source: Data are from *China Statistics Abstract 2013* (XLS, DOC).

Table 8.21 The GDPs, Total Investments in Fixed Properties, and Employment Situation of Fujian's Three Economic Sectors and Industries (2012)

Sectors (i)	*Total Output of Sector (0.1 Billion Yuan)*	*Total Investment on Fixed Properties (0.1 Billion Yuan)*	*Employment (10,000 People)*
Total	19701.78	12439.94	637.86
Primary sector	1776.71	241.87	4.82
Secondary sector	10187.94	4600.80	428.90
Tertiary sector	7737.13	7597.20	204.20
Industries (*i*)			
Agriculture, forest, animal husbandry, fishery	1776.71	241.87	4.82
Mining	284.02	163.00	5.32
Manufacturing	7783.36	3765.51	292.38
Production/supply of electricity, gas, and water	474.56	621.90	9.48
Construction	1646.00	50.44	121.70
Communication, transportation, storage, postal service	1090.07	1441.95	18.64
Information transfer, computer, and software areas	404.35	143.28	5.33
# Information Transfer			
Wholesale and retail	1670.26	206.94	24.25
Hotels and restaurants	337.48	204.22	9.40
Finance	1015.37	30.14	14.75
Real estate/housing	1039.71	3643.37	10.18
Leasing and commercial services	378.18	149.63	6.00
Scientific research, technology service, and geological survey	124.72	22.49	6.54
Water conservation, environment, and public facility management	65.34	1072.58	5.00
# Public Facility Management			
Residents and other services	342.83	30.84	1.39
Education	364.38	194.88	46.69
Health, safety, and welfare	244.06	72.96	18.11

Table 8.21 (continued) The GDPs, Total Investments in Fixed Properties, and Employment Situation of Fujian's Three Economic Sectors and Industries (2012)

Sectors (i)	*Total Output of Sector (0.1 Billion Yuan)*	*Total Investment on Fixed Properties (0.1 Billion Yuan)*	*Employment (10,000 People)*
Social, sports, and entertainment	191.43	186.48	3.55
Public administration and social organizations	468.95	197.46	34.33

Source: The data are from *China Statistics Yearbook 2013* and *Fujian Province's Statistics Yearbook 2013.*

Table 8.22 The Structures and Comparative Gain Measures of Fujian's Three Economic Sectors and Industries ($t = 2012$)

Sectors (i)	*Demand Structure $d_i(t)$ (%)*	*Employment Structure $s_i^1(t)$ (%)*	*Comparative Gain of Labor $G_i^1(t)$*	*Investment Structure $s_i^2(t)$ (%)*	*Comparative Gain of Capital $G_i^2(t)$*
Total	100	100	—	100	—
Primary sector	9.02	0.76	11.94	1.94	4.64
Secondary sector	51.71	67.24	0.77	36.98	1.40
Tertiary sector	39.27	32.00	1.23	61.07	0.64
Industries (*i*)					
Agriculture, forest, animal husbandry, fishery	9.02	0.76	11.94	1.94	4.64
Mining	1.44	0.83	1.73	1.31	1.10
Manufacturing	39.51	45.84	0.86	30.27	1.31
Production/supply of electricity, gas, and water	2.41	1.49	1.62	5.00	0.48
Construction	8.35	19.08	0.44	0.41	20.61
Communication, transportation, storage, postal service	5.53	2.92	1.89	11.60	0.48
Information transfer, computer, and software areas	2.05	0.84	2.46	1.15	1.78

continued

Table 8.22 (continued) The Structures and Comparative Gain Measures of Fujian's Three Economic Sectors and Industries ($t = 2012$)

Sectors (i)	*Demand Structure $d_i(t)$ (%)*	*Employment Structure $s_i^1(t)$ (%)*	*Comparative Gain of Labor $G_i^1(t)$*	*Investment Structure $s_i^2(t)$ (%)*	*Comparative Gain of Capital $G_i^2(t)$*
# Information Transfer					
Wholesale and retail	8.48	3.80	2.23	1.66	5.10
Hotels and restaurants	1.71	1.47	1.16	1.64	1.04
Finance	5.15	2.31	2.23	0.24	21.27
Real estate/housing	5.28	1.60	3.31	29.29	0.18
Leasing and commercial services	1.92	0.94	2.04	1.20	1.60
Scientific research, technology service, and geological survey	0.63	1.02	0.62	0.18	3.50
Water conservation, environment, and public facility management	0.33	0.78	0.42	8.62	0.04
# Public Facility Management					
Residents and other services	1.74	0.22	8.00	0.25	7.02
Education	1.85	7.32	0.25	1.57	1.18
Health, safety, and welfare	1.24	2.84	0.44	0.59	2.11
Social, sports, and entertainment	0.97	0.56	1.74	1.50	0.65
Public administration and social organizations	2.38	5.38	0.44	1.59	1.50

Table 8.23 The GDPs, Total Investments in Fixed Properties, and Employment Situation of Hainan's Three Economic Sectors and Industries (2012)

Sectors (i)	*Total Output of Sector (0.1 Billion Yuan)*	*Total Investment on Fixed Properties (0.1 Billion Yuan)*	*Employment (10,000 People)*
Total	2855.54	2145.4	90.1
Primary sector	711.54	40.8	11.9
Secondary sector	804.47	394.4	18.5
Tertiary sector	1339.53	1710.2	59.5
Industries (*i*)			
Agriculture, forest, animal husbandry, fishery	711.54	40.8	11.9
Mining		30.2	1.0
Manufacturing		191.6	9.5
Production/supply of electricity, gas, and water		114.7	2.1
Construction	283.32	57.8	6.0
Communication, transportation, storage, postal service	133.40	143.2	4.5
Information transfer, computer, and software areas		34.9	1.0
# Information Transfer			
Wholesale and retail	300.52	47.9	5.4
Hotels and restaurants	98.96	187.0	6.2
Finance	130.69	0.6	2.8
Real estate/housing	238.11	1030.8	4.2
Leasing and commercial services		15.7	2.0
Scientific research, technology service, and geological survey		5.0	1.7
Water conservation, environment, and public facility management		154.6	2.6
# Public Facility Management			
Residents and other services		4.8	0.2
Education		31.1	12.3

continued

Table 8.23 (continued) The GDPs, Total Investments in Fixed Properties, and Employment Situation of Hainan's Three Economic Sectors and Industries (2012)

Sectors (i)	*Total Output of Sector (0.1 Billion Yuan)*	*Total Investment on Fixed Properties (0.1 Billion Yuan)*	*Employment (10,000 People)*
Health, safety, and welfare		21.7	4.8
Social, sports, and entertainment		20.2	1.1
Public administration and social organizations		12.9	10.8

Source: The data are from Hainan Province's statistical report of gross domestic products 2013, *China Statistics Yearbook 2013*, and *Hainan Province's Statistical Yearbook 2013*.

Table 8.24 The Structures and Comparative Gain Measures of Hainan's Three Economic Sectors and Industries ($t = 2012$)

Sectors (i)	*Demand Structure $d_i(t)$ (%)*	*Employment Structure $s_i^1(t)$ (%)*	*Comparative Gain of Labor $G_i^1(t)$*	*Investment Structure $s_i^2(t)$ (%)*	*Comparative Gain of Capital $G_i^2(t)$*
Total	100	100	—	100	—
Primary sector	24.92	13.21	1.89	1.90	13.10
Secondary sector	28.17	20.53	1.37	18.38	1.53
Tertiary sector	46.91	66.16	0.71	79.72	0.59
Industries (*i*)					
Agriculture, forest, animal husbandry, fishery	24.92	13.21	1.89	1.90	13.10
Mining		1.09		1.41	
Manufacturing		10.54		8.93	
Production/supply of electricity, gas, and water		2.35		5.35	
Construction	9.92	6.61	1.50	2.69	3.68
Communication, transportation, storage, postal service	4.67	5.01	0.93	6.67	0.70
Information transfer, computer, and software areas		1.11		1.63	

Table 8.24 (continued) The Structures and Comparative Gain Measures of Hainan's Three Economic Sectors and Industries ($t = 2012$)

Sectors (i)	*Demand Structure $d_i(t)$ (%)*	*Employment Structure $s_i^1(t)$ (%)*	*Comparative Gain of Labor $G_i^1(t)$*	*Investment Structure $s_i^2(t)$ (%)*	*Comparative Gain of Capital $G_i^2(t)$*
# Information Transfer					
Wholesale and retail	10.52	5.96	1.77	2.23	4.71
Hotels and restaurants	3.47	6.90	0.50	8.72	0.40
Finance	4.58	3.07	1.49	0.03	177.88
Real estate/housing	8.34	4.71	1.77	48.05	0.17
Leasing and commercial services		2.19		0.73	
Scientific research, technology service, and geological survey		1.92		0.23	
Water conservation, environment, and public facility management		2.85		7.21	
# Public Facility Management					
Residents and other services		0.24		0.22	
Education		13.63		1.45	
Health, safety, and welfare		5.38		1.01	
Social, sports, and entertainment		1.25		0.94	
Public administration and social organizations		11.98		0.60	

Table 8.25 The GDPs, Total Investments in Fixed Properties, and Employment Situation of Jiangsu's Three Economic Sectors and Industries (2012)

Sectors (i)	*Total Output of Sector (0.1 Billion Yuan)*	*Total Investment on Fixed Properties (0.1 Billion Yuan)*	*Employment (10,000 People)*
Total	54058.22	30854.2	830.9
Primary sector	3418.29	251.9	8.6
Secondary sector	27121.95	15797.5	454.9
Tertiary sector	23517.98	14804.9	367.5
Industries (*i*)			
Agriculture, forest, animal husbandry, fishery	3418.29	251.9	8.6
Mining	327.45	85.6	12.8
Manufacturing	22393.82	14792.5	359.7
Production/supply of electricity, gas, and water	1187.20	839.9	12.9
Construction	3213.48	79.4	69.5
Communication, transportation, storage, postal service	2352.40	1397.1	30.6
Information transfer, computer, and software areas	1103.84	266.0	10.8
# Information Transfer			
Wholesale and retail	5704.66	713.2	33.5
Hotels and restaurants	1045.21	475.7	11.9
Finance	3136.51	93.3	29.4
Real estate/housing	2992.82	7746.8	8.2
Leasing and commercial services	1415.19	678.2	12.3
Scientific research, technology service, and geological survey	612.53	324.8	12.8
Water conservation, environment, and public facility management	321.98	1897.5	13.3
# Public Facility Management			
Residents and other services	685.95	154.6	1.3
Education	1420.47	301.7	88.8

Table 8.25 (continued) The GDPs, Total Investments in Fixed Properties, and Employment Situation of Jiangsu's Three Economic Sectors and Industries (2012)

Sectors (i)	*Total Output of Sector (0.1 Billion Yuan)*	*Total Investment on Fixed Properties (0.1 Billion Yuan)*	*Employment (10,000 People)*
Health, safety, and welfare	731.58	156.6	41.1
Social, sports, and entertainment	302.99	329.1	6.1
Public administration and social organizations	1691.85	270.3	67.4

Source: The data are from Statistics Report of Jiangsu Province's Domestic Economy and Social Development 2013, *China Statistics Yearbook 2013*, and *Jiangsu Province's Statistics Yearbook 2013*.

Table 8.26 The Structures and Comparative Gain Measures of Jiangsu's Three Economic Sectors and Industries ($t = 2012$)

Sectors (i)	*Demand Structure $d_i(t)$ (%)*	*Employment Structure $s_i^1(t)$ (%)*	*Comparative Gain of Labor $G_i^1(t)$*	*Investment Structure $s_i^2(t)$ (%)*	*Comparative Gain of Capital $G_i^2(t)$*
Total	100	100	—	100	—
Primary sector	6.32	1.04	6.11	0.82	7.75
Secondary sector	50.17	54.75	0.92	51.20	0.98
Tertiary sector	43.51	44.23	0.98	47.98	0.91
Industries (*i*)					
Agriculture, forest, animal husbandry, fishery	6.32	1.04	6.11	0.82	7.75
Mining	0.61	1.54	0.39	0.28	2.18
Manufacturing	41.43	43.29	0.96	47.94	0.86
Production/supply of electricity, gas, and water	2.20	1.55	1.42	2.72	0.81
Construction	5.94	8.36	0.71	0.26	23.09
Communication, transportation, storage, postal service	4.35	3.68	1.18	4.53	0.96

continued

Table 8.26 (continued) The Structures and Comparative Gain Measures of Jiangsu's Three Economic Sectors and Industries ($t = 2012$)

Sectors (i)	*Demand Structure $d_i(t)$ (%)*	*Employment Structure $s_i^1(t)$ (%)*	*Comparative Gain of Labor $G_i^1(t)$*	*Investment Structure $s_i^2(t)$ (%)*	*Comparative Gain of Capital $G_i^2(t)$*
Information transfer, computer, and software areas	2.04	1.30	1.57	0.86	2.37
# Information Transfer					
Wholesale and retail	10.55	4.03	2.62	2.31	4.57
Hotels and restaurants	1.93	1.44	1.35	1.54	1.25
Finance	5.80	3.53	1.64	0.30	19.20
Real estate/housing	5.54	1.00	5.54	25.11	0.22
Leasing and commercial services	2.62	1.48	1.77	2.20	1.19
Scientific research, technology service, and geological survey	1.13	1.54	0.74	1.05	1.08
Water conservation, environment, and public facility management	0.60	1.60	0.37	6.15	0.10
# Public Facility Management					
Residents and other services	1.27	0.15	8.36	0.50	2.53
Education	2.63	10.69	0.25	0.98	2.69
Health, safety, and welfare	1.35	4.95	0.27	0.51	2.67
Social, sports, and entertainment	0.56	0.73	0.77	1.07	0.53
Public administration and social organizations	3.13	8.11	0.39	0.88	3.57

Table 8.27 The GDPs, Total Investments in Fixed Properties, and Employment Situation of Liaoning's Three Economic Sectors and Industries (2012)

Sectors (i)	*Total Output of Sector (0.1 Billion Yuan)*	*Total Investment on Fixed Properties (0.1 Billion Yuan)*	*Employment (10,000 People)*
Total	24846.43	21836.3	598.7
Primary sector	2155.82	601.4	25.2
Secondary sector	13230.49	9422.3	278.2
Tertiary sector	9460.12	11812.6	295.3
Industries (*i*)			
Agriculture, forest, animal husbandry, fishery	2155.82	601.4	25.2
Mining		683.7	32.6
Manufacturing		7493.1	168.3
Production/supply of electricity, gas, and water		776.2	16.4
Construction	1625.42	469.4	60.9
Communication, transportation, storage, postal service	1297.18	1070.1	32.8
Information transfer, computer, and software areas		133.1	9.0
# Information Transfer			
Wholesale and retail	2191.19	763.1	21.1
Hotels and restaurants	487.49	428.5	7.2
Finance	969.37	63.9	22.5
Real estate/housing	1050.03	6006.7	12.4
Leasing and commercial services		369.8	10.2
Scientific research, technology service, and geological survey		161.2	15.8
Water conservation, environment, and public facility management		1677.4	15.6
# Public Facility Management			
Residents and other services		210.1	2.7

continued

Table 8.27 (continued) The GDPs, Total Investments in Fixed Properties, and Employment Situation of Liaoning's Three Economic Sectors and Industries (2012)

Sectors (i)	*Total Output of Sector (0.1 Billion Yuan)*	*Total Investment on Fixed Properties (0.1 Billion Yuan)*	*Employment (10,000 People)*
Education		196.2	57.8
Health, safety, and welfare		161.9	30.4
Social, sports, and entertainment		282.6	5.1
Public administration and social organizations		287.9	52.8

Source: The data are from *China Statistics Yearbook 2013* and *Liaoning Province's Statistics Yearbook 2013*.

Table 8.28 The Structures and Comparative Gain Measures of Liaoning's Three Economic Sectors and Industries ($t = 2012$)

Sectors (i)	*Demand Structure $d_i(t)$ (%)*	*Employment Structure $s_i^1(t)$ (%)*	*Comparative Gain of Labor $G_i^1(t)$*	*Investment Structure $s_i^2(t)$ (%)*	*Comparative Gain of Capital $G_i^2(t)$*
Total	100	100	—	100	—
Primary sector	8.68	4.20	2.06	2.75	3.15
Secondary sector	53.25	46.47	1.15	43.15	1.23
Tertiary sector	38.07	49.32	0.77	54.10	0.70
Industries (*i*)					
Agriculture, forest, animal husbandry, fishery	8.68	4.20	2.06	2.75	3.15
Mining		5.44		3.13	
Manufacturing		28.11		34.31	
Production/supply of electricity, gas, and water		2.75		3.55	
Construction	6.54	1.02	6.43	2.15	3.04
Communication, transportation, storage, postal service	5.22	5.48	0.95	4.90	1.07
Information transfer, computer, and software areas		1.50		0.61	

Table 8.28 (continued) The Structures and Comparative Gain Measures of Liaoning's Three Economic Sectors and Industries (t = 2012)

Sectors (i)	*Demand Structure* $d_i(t)$ *(%)*	*Employment Structure* $s_i^1(t)$ *(%)*	*Comparative Gain of Labor* $G_i^1(t)$	*Investment Structure* $s_i^2(t)$ *(%)*	*Comparative Gain of Capital* $G_i^2(t)$
# Information Transfer					
Wholesale and retail	8.82	3.52	2.50	3.49	2.52
Hotels and restaurants	1.96	1.21	1.62	1.96	1.00
Finance	3.90	3.76	1.04	0.29	13.34
Real estate/housing	4.23	2.07	2.05	27.51	0.15
Leasing and commercial services		1.70		1.69	
Scientific research, technology service, and geological survey		2.65		0.74	
Water conservation, environment, and public facility management		2.60		7.68	
# Public Facility Management					
Residents and other services		0.45		0.96	
Education		9.65		0.90	
Health, safety, and welfare		5.08		0.74	
Social, sports, and entertainment		0.84		1.29	
Public administration and social organizations	8.68	4.20	2.06	2.75	3.15

8.3 The Central Plane

Table 8.29 The GDPs, Values of Fixed Properties, and Employments of Various Provinces, Municipalities, and Autonomous Regions of Central China (2012)

Area	*Province/ Municipality/ Autonomous Region (i)*	*Regional GDP (0.1 Billion Yuan)*	*Value of Fixed Properties (0.1 Billion Yuan)*	*Employment (10,000 People)*
Central China	12. Shanxi	12112.8	8223.9	1152.0
	13. Inner Mongolia	15880.6	13442.1	1034.5
	14. Jilin	11939.2	9136.2	1121.3
	15. Heilongjiang	13619.6	8143.7	1487.8
	16. Anhui	17212.1	8855.8	1745.7
	17. Jiangxi	12948.9	6513.7	1531.6
	18. Henan	29599.3	22060.0	2509.0
	19. Hubei	22250.5	12554.7	2380.0
	20. Hunan	22154.2	12488.8	2019.9
	21. Guangxi	13035.1	11068.5	1229.4
	Total	170752.3	112487.4	16211.2

Source: The data are from *China Statistics Abstracts 2013* (XLS, DOC).

Table 8.30 The Regional Economic Structures and Measures of Comparative Gains of the Provinces, Municipalities, and Autonomous Regions of Central China (2012)

Area	*Province/ Municipality/ Autonomous Region (i)*	*Demand Structure $d_i(t)$ (%)*	*Employment Structure $s_i^1(t)$ (%)*	*Comparative Gain of Labor $G_i^1(t)$*	*Investment Structure $s_i^2(t)$ (%)*	*Comparative Gain of Capital $G_i^2(t)$*
Central China	12. Shanxi	7.09	7.11	1.00	7.31	0.97
	13. Inner Mongolia	9.30	6.38	1.46	11.95	0.78
	14. Jilin	6.99	6.92	1.01	8.12	0.86
	15. Heilongjiang	7.98	9.18	0.87	7.24	1.10
	16. Anhui	10.08	10.77	0.94	7.87	1.28
	17. Jiangxi	7.58	9.45	0.80	5.79	1.31
	18. Henan	17.33	15.48	1.12	19.61	0.88
	19. Hubei	13.03	14.68	0.89	11.16	1.17
	20. Hunan	12.97	12.46	1.04	11.10	1.17
	21. Guangxi	7.63	7.58	1.01	9.84	0.80
	Total	100	100	—	100	—

Table 8.31 The GDPs, Total Investments in Fixed Properties, and Employment Situation of Henan's Three Economic Sectors and Industries (2012)

Sectors (i)	*Total Output of Sector (0.1 Billion Yuan)*	*Total Investment on Fixed Properties (0.1 Billion Yuan)*	*Employment (10,000 People)*
Total	29599.31	21450.0	881.2
Primary sector	3769.54	825.9	5.8
Secondary sector	16672.20	11041.8	429.3
Tertiary sector	9157.57	9582.3	446
Industries (*i*)			
Agriculture, forest, animal husbandry, fishery	3769.54	825.9	5.8
Mining		690.8	63.0
Manufacturing		9782.3	218.3
Production/supply of electricity, gas, and water		555.9	22.5
Construction	1654.64	12.8	125.6
Communication, transportation, storage, postal service	1151.91	927.9	30.9
Information transfer, computer, and software areas	347.89	46.8	6.3
# Information Transfer			
Wholesale and retail	1877.82	526.7	42.6
Hotels and restaurants	898.36	217.2	10.1
Finance	1013.60	14.8	23.3
Real estate/housing	1040.70	5363.6	13.1
Leasing and commercial services	312.05	84.7	11.3
Scientific research, technology service, and geological survey	197.59	75.5	13.1
Water conservation, environment, and public facility management	83.95	1497.2	13.1
# Public Facility Management			
Residents and other services	268.86	123.1	1.6

continued

Table 8.31 (continued) The GDPs, Total Investments in Fixed Properties, and Employment Situation of Henan's Three Economic Sectors and Industries (2012)

Sectors (i)	*Total Output of Sector (0.1 Billion Yuan)*	*Total Investment on Fixed Properties (0.1 Billion Yuan)*	*Employment (10,000 People)*
Education	814.45	278.1	118.9
Health, safety, and welfare	320.01	160.0	46.1
Social, sports, and entertainment	100.91	182.3	7.1
Public administration and social organizations	729.47	84.2	108.4

Source: Data are from *Henan Province's Statistical Yearbook 2013.*

Table 8.32 The Structures and Comparative Gain Measures of Henan's Three Economic Sectors and Industries ($t = 2012$)

Sectors (i)	*Demand Structure $d_i(t)$ (%)*	*Employment Structure $s_i^1(t)$ (%)*	*Comparative Gain of Labor $G_i^1(t)$*	*Investment Structure $s_i^2(t)$ (%)*	*Comparative Gain of Capital $G_i^2(t)$*
Total	100	100	–	100	–
Primary sector	12.74	0.66	19.26	3.85	3.31
Secondary sector	56.32	48.72	1.16	51.48	1.09
Tertiary sector	30.94	50.61	0.61	44.67	0.69
Industries (*i*)					
Agriculture, forest, animal husbandry, fishery	12.74	0.66	19.26	3.85	3.31
Mining		7.15		3.22	
Manufacturing		24.77		45.60	
Production/supply of electricity, gas, and water		2.55		2.59	
Construction	5.59	14.25	0.39	0.06	93.77
Communication, transportation, storage, postal service	3.89	3.51	1.11	4.33	0.90
Information transfer, computer, and software areas	1.17533	0.72	1.63	0.22	5.39

Table 8.32 (continued) The Structures and Comparative Gain Measures of Henan's Three Economic Sectors and Industries ($t = 2012$)

Sectors (i)	*Demand Structure* $d_i(t)$ *(%)*	*Employment Structure* $s_i^1(t)$ *(%)*	*Comparative Gain of Labor* $G_i^1(t)$	*Investment Structure* $s_i^2(t)$ *(%)*	*Comparative Gain of Capital* $G_i^2(t)$
# Information Transfer					
Wholesale and retail	6.34	4.83	1.31	2.46	2.58
Hotels and restaurants	3.04	1.15	2.65	1.01	3.00
Finance	3.42	2.65	1.29	0.07	49.50
Real estate/housing	3.52	1.49	2.36	25.01	0.14
Leasing and commercial services	1.05	1.28	0.82	0.39	2.67
Scientific research, technology service, and geological survey	0.67	1.49	0.45	0.35	1.90
Water conservation, environment, and public facility management	0.28	1.49	0.19	6.98	0.04
# Public Facility Management					
Residents and other services	0.91	0.18	5.01	0.57	1.58
Education	2.75	13.49	0.20	1.30	2.12
Health, safety, and welfare	1.08	5.23	0.21	0.75	1.45
Social, sports, and entertainment	0. 34	0.81	0.42	0.85	0.40
Public administration and social organizations	2.46	12.30	0.20	0.39	6.28

Source: Data is from *China Statistics Abstract (2012)* (XLS, DOC).

Table 8.33 The GDPs, Total Investments in Fixed Properties, and Employment Situation of Shanxi's Three Economic Sectors and Industries (2012)

Sectors (i)	*Total Output of Sector (0.1 Billion Yuan)*	*Total Investment on Fixed Properties (0.1 Billion Yuan)*	*Employment (10,000 People)*
Total	12112.83	8863.3	436.0
Primary sector	698.32	381.4	2.8
Secondary sector	6731.56	4146.7	210.5
Tertiary sector	4682.95	4335.2	222.7
Industries (*i*)			
Agriculture, forest, animal husbandry, fishery	698.32	381.4	2.8
Mining		1581.7	90.3
Manufacturing		1941.6	70.3
Production/supply of electricity, gas, and water		606.5	11.1
Construction	708.01	16.9	38.8
Communication, transportation, storage, postal service	847.44	1013.4	22.3
Information transfer, computer, and software areas		35.6	5.3
# Information Transfer			
Wholesale and retail	991.08	195.9	19.2
Hotels and restaurants	299.67	60.7	6.9
Finance	639.61	2.0	15.8
Real estate/housing	301.88	1856.7	2.6
Leasing and commercial services		25.9	5.3
Scientific research, technology service, and geological survey		34.9	5.9
Water conservation, environment, and public facility management		742.0	7.6
# Public Facility Management			
Residents and other services		15.3	0.6
Education		168.5	49.9

Table 8.33 (continued) The GDPs, Total Investments in Fixed Properties, and Employment Situation of Shanxi's Three Economic Sectors and Industries (2012)

Sectors (i)	*Total Output of Sector (0.1 Billion Yuan)*	*Total Investment on Fixed Properties (0.1 Billion Yuan)*	*Employment (10,000 People)*
Health, safety, and welfare		54.3	17.1
Social, sports, and entertainment		83.1	4.5
Public administration and social organizations		46.8	59.6

Source: Data are from *China Statistics Yearbook 2013.*

Table 8.34 The Structures and Comparative Gain Measures of Shanxi's Three Economic Sectors and Industries ($t = 2012$)

Sectors (i)	*Demand Structure $d_i(t)$ (%)*	*Employment Structure $s_i^1(t)$ (%)*	*Comparative Gain of Labor $G_i^1(t)$*	*Investment Structure $s_i^2(t)$ (%)*	*Comparative Gain of Capital $G_i^2(t)$*
Total	100	100	—	100	—
Primary sector	5.77	0.65	8.90	4.30	1.34
Secondary sector	55.57	48.28	1.15	46.79	1.19
Tertiary sector	38.66	51.08	0.76	48.91	0.79
Industries (i)					
Agriculture, forest, animal husbandry, fishery	5.77	0.65	8.90	4.30	1.34
Mining		20.70		17.85	
Manufacturing		16.12		21.91	
Production/supply of electricity, gas, and water		2.55		6.84	
Construction	5.85	8.89	0.66	0.19	30.57
Communication, transportation, storage, postal service	7.00	5.12	1.37	11.43	0.61
Information transfer, computer, and software areas		1.21		0.40	

continued

Table 8.34 (continued) The Structures and Comparative Gain Measures of Shanxi's Three Economic Sectors and Industries ($t = 2012$)

Sectors (i)	*Demand Structure $d_i(t)$ (%)*	*Employment Structure $s_i^1(t)$ (%)*	*Comparative Gain of Labor $G_i^1(t)$*	*Investment Structure $s_i^2(t)$ (%)*	*Comparative Gain of Capital $G_i^2(t)$*
# Information Transfer					
Wholesale and retail	8.18	4.40	1.86	2.21	3.70
Hotels and restaurants	2.47	1.59	1.56	0.68	3.61
Finance	5.28	3.62	1.46	0.02	231.83
Real estate/housing	2.49	0.60	4.16	0.21	0.12
Leasing and commercial services		1.23		0.29	
Scientific research, technology service, and geological survey		1.36		0.39	
Water conservation, environment, and public facility management		1.75		8.37	
# Public Facility Management					
Residents and other services		0.15		0.17	
Education		11.45		1.90	
Health, safety, and welfare		3.93		0.61	
Social, sports, and entertainment		1.02		0.94	
Public administration and social organizations		13.68		0.53	

Table 8.35 The GDPs, Total Investments in Fixed Properties, and Employment Situation of Hunan's Three Economic Sectors and Industries (2012)

Sectors (i)	*Total Output of Sector (0.1 Billion Yuan)*	*Total Investment on Fixed Properties (0.1 Billion Yuan)*	*Employment (10,000 People)*
Total	22154.23	14523.2	567.5
Primary sector	3004.21	499.2	2.4
Secondary sector	10506.42	6092.6	253.6
Tertiary sector	8643.60	7931.4	311.4
Industries (*i*)			
Agriculture, forest, animal husbandry, fishery	3004.21	499.2	2.4
Mining		524.6	15.2
Manufacturing		4948.1	128.2
Production/supply of electricity, gas, and water		465.5	15.7
Construction	1367.92	154.4	94.5
Communication, transportation, storage, postal service	1077.65	1122.3	23.5
Information transfer, computer, and software areas		59.6	7.1
# Information Transfer			
Wholesale and retail	1849.04	417.8	18.7
Hotels and restaurants	460.70	183.9	9.5
Finance	579.76	44.1	20.7
Real estate/housing	568.52	3319.2	9.9
Leasing and commercial services		296.6	8.8
Scientific research, technology service, and geological survey		116.8	9.5
Water conservation, environment, and public facility management		1413.2	9.7
# Public Facility Management			
Residents and other services		84.5	1.5

continued

Table 8.35 (continued) The GDPs, Total Investments in Fixed Properties, and Employment Situation of Hunan's Three Economic Sectors and Industries (2012)

Sectors (i)	*Total Output of Sector (0.1 Billion Yuan)*	*Total Investment on Fixed Properties (0.1 Billion Yuan)*	*Employment (10,000 People)*
Education		212.7	72.0
Health, safety, and welfare		111.9	34.8
Social, sports, and entertainment		113.5	5.0
Public administration and social organizations		435.2	80.7

Source: The data is from *Abstract of China Statistics 2013* (XLS, DOC).

Table 8.36 The Structures and Comparative Gain Measures of Hunan's Three Economic Sectors and Industries ($t = 2012$)

Sectors (i)	*Demand Structure $d_i(t)$ (%)*	*Employment Structure $s_i^1(t)$ (%)*	*Comparative Gain of Labor $G_i^1(t)$*	*Investment Structure $s_i^2(t)$ (%)*	*Comparative Gain of Capital $G_i^2(t)$*
Total	100	100	—	100	—
Primary sector	13.56	0.43	31.84	3.44	3.94
Secondary sector	47.42	44.69	1.06	41.95	1.13
Tertiary sector	39.02	54.87	0.71	54.61	0.71
Industries (*i*)					
Agriculture, forest, animal husbandry, fishery	13.56	0.43	31.84	3.44	3.94
Mining		2.68		3.61	
Manufacturing		22.59		34.07	
Production/supply of electricity, gas, and water		2.76		3.21	
Construction	6.17	16.67	0.37	1.06	5.81
Communication, transportation, storage, postal service	4.86	4.15928	1.17	7.73	0.63
Information transfer, computer, and software areas		1.24		0. 41	

Table 8.36 (continued) The Structures and Comparative Gain Measures of Hunan's Three Economic Sectors and Industries ($t = 2012$)

Sectors (i)	*Demand Structure $d_i(t)$ (%)*	*Employment Structure $s_i^1(t)$ (%)*	*Comparative Gain of Labor $G_i^1(t)$*	*Investment Structure $s_i^2(t)$ (%)*	*Comparative Gain of Capital $G_i^2(t)$*
# Information Transfer					
Wholesale and retail	8.35	3.30	2.53	2.88	2.90
Hotels and restaurants	2.08	1.68	1.24	1.27	1.64
Finance	2.62	3.65	0.72	0.30	8.63
Real estate/housing	2.57	1.75	1.46	22.85	0.11
Leasing and commercial services		1.54		2.04	
Scientific research, technology service, and geological survey		1.67		0.80	
Water conservation, environment, and public facility management		1.70		9.73	
# Public Facility Management					
Residents and other services		0.27		0.58	
Education		12.68		1.46	
Health, safety, and welfare		6.13		0.77	
Social, sports, and entertainment		0. 87		0.78	
Public administration and social organizations		14.22		3.00	

Table 8.37 The GDPs, Total Investments in Fixed Properties, and Employment Situation of Inner Mongolia's Three Economic Sectors and Industries (2012)

Sectors (i)	*Total Output of Sector (0.1 Billion Yuan)*	*Total Investment on Fixed Properties (0.1 Billion Yuan)*	*Employment (10,000 People)*
Total	15880.58	11875.7	270.8
Primary sector	1448.58	591.7	25.0
Secondary sector	8801.50	6068.8	94.4
Tertiary sector	5630.50	5215.2	151.4
Industries (*i*)			
Agriculture, forest, animal husbandry, fishery	1448.58	591.7	25.0
Mining		1116.4	21.5
Manufacturing		3829.4	42.7
Production/supply of electricity, gas, and water		1036.5	12.0
Construction	1065.71	86.4	18.2
Communication, transportation, storage, postal service	1185.30	1148.0	16.9
Information transfer, computer, and software areas		83.1	4.4
# Information Transfer			
Wholesale and retail	1415.47	284.4	8.0
Hotels and restaurants	433.94	104.7	2.9
Finance	502.01	25.2	10.8
Real estate/housing	385.22	1866.4	1.8
Leasing and commercial services		79.6	3.2
Scientific research, technology service, and geological survey		53.1	4.9
Water conservation, environment, and public facility management		996.0	7.8
# Public Facility Management			
Residents and other services		36.0	0.7
Education		111.0	34.8

Table 8.37 (continued) The GDPs, Total Investments in Fixed Properties, and Employment Situation of Inner Mongolia's Three Economic Sectors and Industries (2012)

Sectors (i)	*Total Output of Sector (0.1 Billion Yuan)*	*Total Investment on Fixed Properties (0.1 Billion Yuan)*	*Employment (10,000 People)*
Health, safety, and welfare		67.5	12.9
Social, sports, and entertainment		125.7	3.3
Public administration and social organizations		234.5	39.0

Source: The data from *Inner Mongolia's Statistics Yearbook 2013.*

Table 8.38 The Structures and Comparative Gain Measures of Inner Mongolia's Three Economic Sectors and Industries ($t = 2012$)

Sectors (i)	*Demand Structure $d_i(t)$ (%)*	*Employment Structure $s_i^1(t)$ (%)*	*Comparative Gain of Labor $G_i^1(t)$*	*Investment Structure $s_i^2(t)$ (%)*	*Comparative Gain of Capital $G_i^2(t)$*
Total	100	100	—	100	—
Primary sector	9.12	9.23	0.99	4.98	1.83
Secondary sector	55.42	34.86	1.59	51.10	1.08
Tertiary sector	35.46	55.91	0.63	43.91	0.81
Industries (*i*)					
Agriculture, forest, animal husbandry, fishery	9.12	9.23	0.99	4.98	1.83
Mining		7.93		9.40	
Manufacturing		15.76		32.25	
Production/supply of electricity, gas, and water		4.42		8.73	
Construction	6.71	6.74	1.00	0.73	9.22
Communication, transportation, storage, postal service	7.46	6.26	1.19	9.67	0.77
Information transfer, computer, and software areas		1.63		0.70	

continued

Table 8.38 (continued) The Structures and Comparative Gain Measures of Inner Mongolia's Three Economic Sectors and Industries ($t = 2012$)

Sectors (i)	*Demand Structure $d_i(t)$ (%)*	*Employment Structure $s_i^1(t)$ (%)*	*Comparative Gain of Labor $G_i^1(t)$*	*Investment Structure $s_i^2(t)$ (%)*	*Comparative Gain of Capital $G_i^2(t)$*
# Information Transfer					
Wholesale and retail	8.91	2.95	3.03	2.39	3.72
Hotels and restaurants	2.73	1.07	2.56	0.88	3.10
Finance	3.16	3.99	0.79	0.21	14.93
Real estate/housing	2.43	0.68	3.56	15.72	0.15
Leasing and commercial services		1.16		0.67	
Scientific research, technology service, and geological survey		1.80		0.45	
Water conservation, environment, and public facility management		2.89		8.39	
# Public Facility Management					
Residents and other services		0.27		0.30	
Education		12.85		0.93	
Health, safety, and welfare		7.55		0.57	
Social, sports, and entertainment		1.21		1.059	
Public administration and social organizations		14.42		1.97	

Table 8.39 The GDPs, Total Investments in Fixed Properties, and Employment Situation of Heilongjiang's Three Economic Sectors and Industries (2012)

Sectors (i)	*Total Output of Sector (0.1 Billion Yuan)*	*Total Investment on Fixed Properties (0.1 Billion Yuan)*	*Employment (10,000 People)*
Total	13691.58	9694.7	471.0
Primary sector	2113.66	757.3	93.3
Secondary sector	6037.61	4318.5	156
Tertiary sector	5540.31	4619	221.7
Industries (*i*)			
Agriculture, forest, animal husbandry, fishery	2113.66	757.3	93.3
Mining		596.0	41.3
Manufacturing		3026.9	63.3
Production/supply of electricity, gas, and water		519.6	15.7
Construction	796.96	176.0	35.6
Communication, transportation, storage, postal service	598.78	519.3	25.6
Information transfer, computer, and software areas		124.6	6.1
# Information Transfer			
Wholesale and retail	1339.05	329.6	15.9
Hotels and restaurants	367.40	91.5	4.6
Finance	485.11	31.6	16.0
Real estate/housing	522.27	2004.9	5.9
Leasing and commercial services		84.2	4.9
Scientific research, technology service, and geological survey		83.5	11.9
Water conservation, environment, and public facility management		787.8	10.2
# Public Facility Management			
Residents and other services		54.8	4.4

continued

Table 8.39 (continued) The GDPs, Total Investments in Fixed Properties, and Employment Situation of Heilongjiang's Three Economic Sectors and Industries (2012)

Sectors (i)	*Total Output of Sector (0.1 Billion Yuan)*	*Total Investment on Fixed Properties (0.1 Billion Yuan)*	*Employment (10,000 People)*
Education		103.4	47.0
Health, safety, and welfare		78.6	21.0
Social, sports, and entertainment		135.7	4.2
Public administration and social organizations		189.7	44.1

Source: The data from *China Statistics Abstract 2013* (XLS, DOC).

Table 8.40 The Structures and Comparative Gain Measures of Heilongjiang's Three Economic Sectors and Industries ($t = 2012$)

Sectors (i)	*Demand Structure $d_i(t)$ (%)*	*Employment Structure $s_i^1(t)$ (%)*	*Comparative Gain of Labor $G_i^1(t)$*	*Investment Structure $s_i^2(t)$ (%)*	*Comparative Gain of Capital $G_i^2(t)$*
Total	100	100	—	100	—
Primary sector	15.44	19.80	0.78	7.81	1.98
Secondary sector	44.10	33.12	1.33	44.54	0.99
Tertiary sector	40.46	47.07	0.86	47.64	0.85
Industries (*i*)					
Agriculture, forest, animal husbandry, fishery	15.44	19.80	0.78	7.81	1.98
Mining		8.77		6.15	
Manufacturing		13.00		0.31	
Production/supply of electricity, gas, and water		3.34		0.05	
Construction	5.82	7.56	0.77	8.15	0.71
Communication, transportation, storage, postal service	4.37	5.44	0.80	5.36	0.81
Information transfer, computer, and software areas		1.30		1.29	

Table 8.40 (continued) The Structures and Comparative Gain Measures of Heilongjiang's Three Economic Sectors and Industries ($t = 2012$)

Sectors (i)	*Demand Structure* $d_i(t)$ (%)	*Employment Structure* $s_i^1(t)$ (%)	*Comparative Gain of Labor* $G_i^1(t)$	*Investment Structure* $s_i^2(t)$ (%)	*Comparative Gain of Capital* $G_i^2(t)$
# Information Transfer					
Wholesale and retail	9.78	3.33	2.90	3.40	2.88
Hotels and restaurants	2.68	0.97	2.77	0.94	2.85
Finance	3.54	3.40	1.04	0.33	10.73
Real estate/housing	3.81	1.24	3.07	20.68	0.18
Leasing and commercial services		1.04		0.87	
Scientific research, technology service, and geological survey		2.53		0.86	
Water conservation, environment, and public facility management		2.17		8.13	
# Public Facility Management					
Residents and other services		0.93		0.57	
Education		9.97		1.07	
Health, safety, and welfare		4.46		0.81	
Social, sports, and entertainment		0.90		1.40	
Public administration and social organizations		9.36		1.96	

Table 8.41 The GDPs, Total Investments in Fixed Properties, and Employment Situation of Jilin's Three Economic Sectors and Industries (2012)

Sectors (i)	*Total Output of Sector (0.1 Billion Yuan)*	*Total Investment on Fixed Properties (0.1 Billion Yuan)*	*Employment (10,000 People)*
Total	11939.24	9511.5	285.5
Primary sector	1412.11	379.8	15.7
Secondary sector	6376.77	5161.2	108
Tertiary sector	4150.36	3970.5	161.8
Industries (*i*)			
Agriculture, forest, animal husbandry, fishery	1412.11	379.8	15.7
Mining		547.8	16.6
Manufacturing		4150.3	64.8
Production/supply of electricity, gas, and water		413.9	8.9
Construction	794.29	49.1	17.6
Communication, transportation, storage, postal service	462.13	547.3	15.9
Information transfer, computer, and software areas		69.7	4.7
# Information Transfer			
Wholesale and retail	986.46	410.9	9.5
Hotels and restaurants	240.70	104.7	3.0
Finance	244.63	14.5	10.9
Real estate/housing	240.86	1545.3	5.1
Leasing and commercial services		48.1	4.9
Scientific research, technology service, and geological survey		59.7	7.6
Water conservation, environment, and public facility management		760.5	8.2
# Public Facility Management			
Residents and other services		59.6	0.9
Education		66.1	37.2

Table 8.41 (continued) The GDPs, Total Investments in Fixed Properties, and Employment Situation of Jilin's Three Economic Sectors and Industries (2012)

Sectors (i)	*Total Output of Sector (0.1 Billion Yuan)*	*Total Investment on Fixed Properties (0.1 Billion Yuan)*	*Employment (10,000 People)*
Health, safety, and welfare		69.6	16.3
Social, sports, and entertainment		83.4	3.6
Public administration and social organizations		131.2	33.9

Source: The data is from *China Statistics Abstract 2013* (XLS, DOC).

Table 8.42 The Structures and Comparative Gain Measures of Jilin's Three Economic Sectors and Industries ($t = 2012$)

Sectors (i)	*Demand Structure $d_i(t)$ (%)*	*Employment Structure $s_i^1(t)$ (%)*	*Comparative Gain of Labor $G_i^1(t)$*	*Investment Structure $s_i^2(t)$ (%)*	*Comparative Gain of Capital $G_i^2(t)$*
Total	100	100	–1	100	—
Primary sector	11.83	5.50	2.15	3.99	2.96
Secondary sector	53.41	37.83	1.41	54.26	0.98
Tertiary sector	34.76	56.68	0.61	41.74	0.83
Industries (*i*)					
Agriculture, forest, animal husbandry, fishery	11.83	5.50	2.15	3.99	2.96
Mining		5.83		5.76	
Manufacturing		22.71		43.63	
Production/supply of electricity, gas, and water		3.12		4.35	
Construction	6.65	6.18	1.08	0.52	12.38
Communication, transportation, storage, postal service	3.87	5.58	0.69	5.75	0.67
Information transfer, computer, and software areas		1.66		0.73	

continued

Table 8.42 (continued) The Structures and Comparative Gain Measures of Jilin's Three Economic Sectors and Industries ($t = 2012$)

Sectors (i)	*Demand Structure* $d_i(t)$ *(%)*	*Employment Structure* $s_i^1(t)$ *(%)*	*Comparative Gain of Labor* $G_i^1(t)$	*Investment Structure* $s_i^2(t)$ *(%)*	*Comparative Gain of Capital* $G_i^2(t)$
# Information Transfer					
Wholesale and retail	8.26	3.34	2.48	4.32	1.91
Hotels and restaurants	2.02	1.06	1.91	1.10	1.83
Finance	2.05	3.83	0.54	0.15	13.41
Real estate/housing	2.02	1.80	1.12	16.25	0.12
Leasing and commercial services		1.70		0.51	
Scientific research, technology service, and geological survey		2.66		0.63	
Water conservation, environment, and public facility management		2.86		8.00	
# Public Facility Management					
Residents and other services		0.31		0.63	
Education		13.03		0.70	
Health, safety, and welfare		5.70		0.73	
Social, sports, and entertainment		1.28		0.88	
Public administration and social organizations		11.88		1.38	

Source: Data are from *China Statistics Abstract (2012)* (XLS, DOC).

Table 8.43 The GDPs, Total Investments in Fixed Properties, and Employment Situation of Hubei's Three Economic Sectors and Industries (2012)

Sectors (i)	*Total Output of Sector (0.1 Billion Yuan)*	*Total Investment on Fixed Properties (0.1 Billion Yuan)*	*Employment (10,000 People)*
Total	22250.45	15578.3	598.0
Primary sector	2848.77	531.4	9.7
Secondary sector	11193.10	6977.1	287.8
Tertiary sector	8208.58	8069.9	300.5
Industries (*i*)			
Agriculture, forest, animal husbandry, fishery	2848.77	531.4	9.7
Mining		283.7	12.0
Manufacturing		6250.1	160.6
Production/supply of electricity, gas, and water		402.9	18.2
Construction	1457.95	40.4	97.1
Communication, transportation, storage, postal service	934.96	1266.6	24.4
Information transfer, computer, and software areas		86.4	5.5
# Information Transfer			
Wholesale and retail	1687.52	409.8	30.1
Hotels and restaurants	512.02	193.8	9.2
Finance	870.36	45.0	16.3
Real estate/housing	692.82	3510.6	10.0
Leasing and commercial services		227.7	6.2
Scientific research, technology service, and geological survey		81.4	13.5
Water conservation, environment, and public facility management		1231.4	9.8

continued

Table 8.43 (continued) The GDPs, Total Investments in Fixed Properties, and Employment Situation of Hubei's Three Economic Sectors and Industries (2012)

Sectors (i)	*Total Output of Sector (0.1 Billion Yuan)*	*Total Investment on Fixed Properties (0.1 Billion Yuan)*	*Employment (10,000 People)*
# Public Facility Management			
Residents and other services		52.5	1.5
Education		148.8	71.8
Health, safety, and welfare		131.2	34.2
Social, sports, and entertainment		163.4	6.3
Public administration and social organizations		521.2	61.5

Source: The data are from *China Statistics Abstract 2013*.

Table 8.44 The Structures and Comparative Gain Measures of Hubei's Three Economic Sectors and Industries ($t = 2012$)

Sectors (i)	*Demand Structure $d_i(t)$ (%)*	*Employment Structure $s_i^1(t)$ (%)*	*Comparative Gain of Labor $G_i^1(t)$*	*Investment Structure $s_i^2(t)$ (%)*	*Comparative Gain of Capital $G_i^2(t)$*
Total	100	100	—	100	—
Primary sector	12.80	1.62	7.89	3.41	3.75
Secondary sector	50.31	48.13	1.05	44.79	1.12
Tertiary sector	36.89	50.25	0.73	51.80	0.71
Industries (*i*)					
Agriculture, forest, animal husbandry, fishery	12.80	1.62	7.89	3.41	3.75
Mining		2.00		1.82	
Manufacturing		26.85		40.12	
Production/supply of electricity, gas, and water		3.04		2.59	
Construction	6.55	16.23	0.40	0.26	25.27
Communication, transportation, storage, postal service	4.20	4.08	1.03	8.13	0.52

Table 8.44 (continued) The Structures and Comparative Gain Measures of Hubei's Three Economic Sectors and Industries ($t = 2012$)

Sectors (i)	*Demand Structure $d_i(t)$ (%)*	*Employment Structure $s_i^1(t)$ (%)*	*Comparative Gain of Labor $G_i^1(t)$*	*Investment Structure $s_i^2(t)$ (%)*	*Comparative Gain of Capital $G_i^2(t)$*
Information transfer, computer, and software areas		0.94		0.55	
# Information Transfer					
Wholesale and retail	7.58	5.0388	1.51	2.63	2.88
Hotels and restaurants	2.30	1.55	1.49	1.24	1.85
Finance	3.91	2.73	1.43	0.29	13.54
Real estate/housing	3.11	1.67	1.87	22.54	0.14
Leasing and commercial services		1.03		1.46	
Scientific research, technology service, and geological survey		2.25		0.52	
Water conservation, environment, and public facility management		1.64		7.90	
# Public Facility Management					
Residents and other services		0.25		0.34	
Education		12.01		0.96	
Health, safety, and welfare		5.72		0.84	
Social, sports, and entertainment		1.06		1.05	
Public administration and social organizations		10.28		3.35	

Table 8.45 The GDPs, Total Investments in Fixed Properties, and Employment Situation of Guangxi's Three Economic Sectors and Industries (2012)

Sectors (i)	*Total Output of Sector (0.1 Billion Yuan)*	*Total Investment on Fixed Properties (0.1 Billion Yuan)*	*Employment (10,000 People)*
Total	13035.10	9808.6	358.0
Primary sector	2172.37	370.5	9.9
Secondary sector	6247.43	4051.4	129.5
Tertiary sector	4615.30	5386.7	218.6
Industries (*i*)			
Agriculture, forest, animal husbandry, fishery	2172.37	370.5	9.9
Mining	251.21	322.0	5.3
Manufacturing	4544.12	3238.8	71.9
Production/supply of electricity, gas, and water	483.93	462.8	9.8
Construction	968.17	27.7	42.5
Communication, transportation, storage, postal service	625.57	925.8	18.5
Information transfer, computer, and software areas	198.78	92.8	4.4
# Information Transfer			
Wholesale and retail	982.18	269.4	12.8
Hotels and restaurants	364.46	161.3	4.8
Finance	573.05	23.0	11.7
Real estate/housing	489.43	2209.6	5.4
Leasing and commercial services	221.96	135.3	8.9
Scientific research, technology service, and geological survey	74.95	34.0	9.3
Water conservation, environment, and public facility management	33.61	995.7	8.8
# Public Facility Management			
Residents and other services	128.66	58.5	0.7
Education	310.01	175.6	60.8

Table 8.45 (continued) The GDPs, Total Investments in Fixed Properties, and Employment Situation of Guangxi's Three Economic Sectors and Industries (2012)

Sectors (i)	*Total Output of Sector (0.1 Billion Yuan)*	*Total Investment on Fixed Properties (0.1 Billion Yuan)*	*Employment (10,000 People)*
Health, safety, and welfare	202.93	80.9	26.0
Social, sports, and entertainment	61.88	113.0	3.3
Public administration and social organizations	347.84	111.7	43.0

Source: The data are from *China Statistics Abstracts 2013*.

Table 8.46 The Structures and Comparative Gain Measures of Guangxi's Three Economic Sectors and Industries ($t = 2012$)

Sectors (i)	*Demand Structure $d_i(t)$ (%)*	*Employment Structure $s_i^1(t)$ (%)*	*Comparative Gain of Labor $G_i^1(t)$*	*Investment Structure $s_i^2(t)$ (%)*	*Comparative Gain of Capital $G_i^2(t)$*
Total	100	100	—	100	—
Primary sector	16.67	2.77	6.01	3.78	4.41
Secondary sector	47.93	36.17	1.32	41.30	1.16
Tertiary sector	35.41	61.06	0.58	54.92	0.64
Industries (*i*)					
Agriculture, forest, animal husbandry, fishery	16.67	2.77	6.01	3.78	4.41
Mining	1.93	1.48	1.31	3.28	0.59
Manufacturing	34.86	20.07	1.74	33.02	1.06
Production/supply of electricity, gas, and water	3.71	2.75	1.35	4.72	0.79
Construction	7.43	11.88	0.63	0.28	26.27
Communication, transportation, storage, postal service	4.80	5.16	0.93	9.44	0.51
Information transfer, computer, and software areas	1.52	1.24	1.23	0.95	1.61

continued

Table 8.46 (continued) The Structures and Comparative Gain Measures of Guangxi's Three Economic Sectors and Industries ($t = 2012$)

Sectors (i)	*Demand Structure* $d_i(t)$ *(%)*	*Employment Structure* $s_i^1(t)$ *(%)*	*Comparative Gain of Labor* $G_i^1(t)$	*Investment Structure* $s_i^2(t)$ *(%)*	*Comparative Gain of Capital* $G_i^2(t)$
# Information Transfer					
Wholesale and retail	7.53	3.59	2.10	2.75	2.74
Hotels and restaurants	2.80	1.33	2.10	1.64	1.70
Finance	4.40	3.26	1.35	0.23	18.73
Real estate/housing	3.75	1.51	2.49	22.53	0.17
Leasing and commercial services	1.70	2.49	0.68	1.38	1.23
Scientific research, technology service, and geological survey	0.57	2.61	0.22	0.35	1.66
Water conservation, environment, and public facility management	0.26	2.47	0.10	10.15	0.03
# Public Facility Management					
Residents and other services	0.99	0.21	4.77	0.60	1.66
Education	2.38	17.00	0.14	1.79	1.33
Health, safety, and welfare	1.56	7.27	0.21	0.83	1.89
Social, sports, and entertainment	0.47	0.92	0.52	1.15	0.41
Public administration and social organizations	2.67	12.01	0.22	1.14	2.34

Table 8.47 The GDPs, Total Investments in Fixed Properties, and Employment Situation of Anhui's Three Economic Sectors and Industries (2012)

Sectors (i)	*Total Output of Sector (0.1 Billion Yuan)*	*Total Investment on Fixed Properties (0.1 Billion Yuan)*	*Employment (10,000 People)*
Total	17212.05	15425.8	436.3
Primary sector	2178.73	379.4	5.4
Secondary sector	9404.84	6955.5	205.4
Tertiary sector	5628.48	8090.9	226.1
Industries (*i*)			
Agriculture, forest, animal husbandry, fishery	2178.73	379.4	5.4
Mining		389.8	34.6
Manufacturing		6072.5	90.9
Production/supply of electricity, gas, and water		436.6	9.9
Construction	1379.00	56.6	69.9
Communication, transportation, storage, postal service	650.21	585.8	16.3
Information transfer, computer, and software areas	190.70	72.8	4.5
# Information Transfer			
Wholesale and retail	1223.97	358.1	15.9
Hotels and restaurants	268.95	215.4	4.3
Finance	617.62	62.6	16.8
Real estate/housing	665.76	4459.1	6.9
Leasing and commercial services	351.90	151.8	3.7
Scientific research, technology service, and geological survey	132.70	141.3	8.2
Water conservation, environment, and public facility management	106.90	1273.6	8.1
# Public Facility Management			
Residents and other services	165.80	46.4	0.6

continued

Table 8.47 (continued) The GDPs, Total Investments in Fixed Properties, and Employment Situation of Anhui's Three Economic Sectors and Industries (2012)

Sectors (i)	*Total Output of Sector (0.1 Billion Yuan)*	*Total Investment on Fixed Properties (0.1 Billion Yuan)*	*Employment (10,000 People)*
Education	412.90	204.0	63.3
Health, safety, and welfare	247.50	107.7	25.3
Social, sports, and entertainment	92.40	158.0	3.6
Public administration and social organizations	501.20	254.4	48.5

Source: The data are from *China Statistics Abstracts 2013*.

Table 8.48 The Structures and Comparative Gain Measures of Anhui's Three Economic Sectors and Industries (t = 2012)

Sectors (i)	*Demand Structure $d_i(t)$ (%)*	*Employment Structure $s_i^1(t)$ (%)*	*Comparative Gain of Labor $G_i^1(t)$*	*Investment Structure $s_i^2(t)$ (%)*	*Comparative Gain of Capital $G_i^2(t)$*
Total	100	100	—	100	—
Primary sector	12.66	1.23	10.28	2.46	5.15
Secondary sector	54.64	47.02	1.16	45.09	1.21
Tertiary sector	32.70	51.76	0.63	52.45	0.62
Industries (*i*)					
Agriculture, forest, animal husbandry, fishery	12.66	1.23	10.28	2.46	5.15
Mining		7.92		2.53	
Manufacturing		20.82		39.37	
Production/supply of electricity, gas, and water		2.26		2.83	
Construction	8.01	16.01	0.50	0.37	21.84
Communication, transportation, storage, postal service	3.78	3.74	1.01	3.80	0.99
Information transfer, computer, and software areas	1.11	1.04	1.06	0.47	2.35

Table 8.48 (continued) The Structures and Comparative Gain Measures of Anhui's Three Economic Sectors and Industries ($t = 2012$)

Sectors (i)	*Demand Structure* $d_i(t)$ *(%)*	*Employment Structure* $s_i^1(t)$ *(%)*	*Comparative Gain of Labor* $G_i^1(t)$	*Investment Structure* $s_i^2(t)$ *(%)*	*Comparative Gain of Capital* $G_i^2(t)$
# Information Transfer					
Wholesale and retail	7.11	3.65	1.95	2.32	3.06
Hotels and restaurants	1.56	0.99	1.57	1.40	1.12
Finance	3.59	3.85	0.93	0.41	8.84
Real estate/housing	3.87	1.58	2.45	28.91	0.13
Leasing and commercial services	2.04	0.84	2.43	0.98	2.08
Scientific research, technology service, and geological survey	0.77	1.88	0.41	0.93	0.84
Water conservation, environment, and public facility management	0.62	1.85	0.34	8.26	0.08
# Public Facility Management					
Residents and other services	0.96	0.13	7.47	0.30	3.20
Education	2.40	0.14	0.17	1.32	1.81
Health, safety, and welfare	1.44	5.80	0.25	0.70	2.06
Social, sports, and entertainment	0.54	0.82	0.65	1.02	0.52
Public administration and social organizations	2.91	11.11	0.26	1.65	1.77

Table 8.49 The GDPs, Total Investments in Fixed Properties, and Employment Situation of Jiangxi's Three Economic Sectors and Industries (2012)

Sectors (i)	*Total Output of Sector (0.1 Billion Yuan)*	*Total Investment on Fixed Properties (0.1 Billion Yuan)*	*Employment (10,000 People)*
Total	12948.88	10774.2	385.8
Primary sector	1520.23	346.5	11.6
Secondary sector	6942.59	6026.2	189.3
Tertiary sector	4486.06	4401.4	184.9
Industries (*i*)			
Agriculture, forest, animal husbandry, fishery	1520.23	346.5	11.6
Mining		265.5	10.1
Manufacturing		5363.1	102.0
Production/supply of electricity, gas, and water		294.3	10.5
Construction	1114.39	103.3	66.6
Communication, transportation, storage, postal service	630.56	474.1	12.9
Information transfer, computer, and software areas		50.6	4.3
# Information Transfer			
Wholesale and retail	935.94	411.4	16.7
Hotels and restaurants	310.84	254.1	4.3
Finance	413.07	24.7	10.6
Real estate/housing	421.83	1679.6	4.5
Leasing and commercial services		121.9	2.9
Scientific research, technology service, and geological survey		38.5	5.4
Water conservation, environment, and public facility management		827.2	5.9
# Public Facility Management			
Residents and other services		69.8	0.6
Education		137.8	44.2

Table 8.49 (continued) The GDPs, Total Investments in Fixed Properties, and Employment Situation of Jiangxi's Three Economic Sectors and Industries (2012)

Sectors (i)	*Total Output of Sector (0.1 Billion Yuan)*	*Total Investment on Fixed Properties (0.1 Billion Yuan)*	*Employment (10,000 People)*
Health, safety, and welfare		74.6	21.1
Social, sports, and entertainment		97.8	3.9
Public administration and social organizations		139.3	47.6

Source: The data is from *China Statistics Abstracts 2013.*

Table 8.50 The Structures and Comparative Gain Measures of Jiangxi's Three Economic Sectors and Industries (t = 2012)

Sectors (i)	*Demand Structure $d_i(t)$ (%)*	*Employment Structure $s_i^1(t)$ (%)*	*Comparative Gain of Labor $G_i^1(t)$*	*Investment Structure $s_i^2(t)$ (%)*	*Comparative Gain of Capital $G_i^2(t)$*
Total	100	100	—	100	—
Primary sector	11.74	3.02	3.89	3.22	3.65
Secondary sector	53.62	49.07	1.09	55.93	0.96
Tertiary sector	34.64	47.93	0.72	40.85	0.85
Industries (*i*)					
Agriculture, forest, animal husbandry, fishery	11.74	3.02	3.89	3.22	3 65
Mining		6.23		2.46	
Manufacturing		26.45		49.78	
Production/supply of electricity, gas, and water		2.73		2.73	
Construction	8.61	17.27	0.50	0.96	8.98
Communication, transportation, storage, postal service	4.87	3.34	1.46	4.40016	1.11
Information transfer, computer, and software areas		1.13		0.47	

continued

Table 8.50 (continued) The Structures and Comparative Gain Measures of Jiangxi's Three Economic Sectors and Industries ($t = 2012$)

Sectors (i)	*Demand Structure $d_i(t)$ (%)*	*Employment Structure $s_i^1(t)$ (%)*	*Comparative Gain of Labor $G_i^1(t)$*	*Investment Structure $s_i^2(t)$ (%)*	*Comparative Gain of Capital $G_i^2(t)$*
# Information Transfer					
Wholesale and retail	7.23	4.33	1.67	3.82	1.89
Hotels and restaurants	2.40	1.10	2.18	2.36	1.02
Finance	3.19	2.75	1.16	0.23	13.91
Real estate/housing	3.26	1.15	2.82	15.60	0.21
Leasing and commercial services		0.76		1.13	
Scientific research, technology service, and geological survey		1.41		0.36	
Water conservation, environment, and public facility management		1.53		7.68	
# Public Facility Management					
Residents and other services		0.15		0.65	
Education		11.46		1.28	
Health, safety, and welfare		5.47		0.69	
Social, sports, and entertainment		1.00		0.91	
Public administration and social organizations		12.33		1.29	

8.4 The West Region

Table 8.51 The GDPs, Total Investments in Fixed Properties, and Employment Situation of West China (2012)

Area	*Province/ Municipality/ Autonomous Region (i)*	*Regional GDP (0.1 Billion Yuan)*	*Value of Fixed Properties (0.1 Billion Yuan)*	*Employments (10,000 People)*
West China	22. Chongqing	11409.6	7409.4	1795.8
	23. Sichuan	23872.8	12496.0	2323.5
	24. Guizhou	6852.2	4164.4	800.0
	25. Yunnan	10309.5	8576.4	1492.5
	26. Tibet	701.0	708.7	130.2
	27. Shaanxi	14453.7	9915.2	1211.9
	28. Gansu	5650.2	3298.1	657.0
	29. Qinghai	1893.5	1719.3	194.9
	30. Ningxia	2341.3	2086.9	262.9
	31. Xinjiang	7505.3	5792.2	753.6
	Total	84989.1	55098.6	9263.6

Source: Data are from *China Statistics Abstracts 2013* (XLS, DOC).

Table 8.52 The Regional Economic Structures and Measures of Comparative Gains of the Provinces, Municipalities, and Autonomous Regions of West China (2012)

Area	*Province/ Municipality/ Autonomous Region (i)*	*Demand Structure $d_i(t)$ (%)*	*Employment Structure $s_i^1(t)$ (%)*	*Comparative Gain of Labor $G_i^1(t)$*	*Investment Structure $s_i^2(t)$ (%)*	*Comparative Gain of Capital $G_i^2(t)$*
West China	22. Chongqing	13.42	15.51	0.87	11.51	1.17
	23. Sichuan	28.09	25.08	1.12	22.68	1.24
	24. Guizhou	8.06	8.64	0.93	7.56	1.07
	25. Yunnan	12.13	16.11	0.75	15.57	0.78
	26. Tibet	0.82	1.41	0.58	1.29	0.64
	27. Shaanxi	17.01	13.08	1.30	18.00	0.95
	28. Gansu	6.65	7.09	0.94	5.99	1.11
	29. Qinghai	2.23	2.10	1.06	3.12	0.71
	30. Ningxia	2.75	2.84	0.97	3.79	0.73
	31. Xinjiang	8.83	8.14	1.08	10.51	0.84
	Total	100	0.1651	—	100	—

Table 8.53 The GDPs, Total Investments in Fixed Properties, and Employment Situation of Xinjiang's Three Economic Sectors and Industries (2012)

Sectors (i)	*Total Output of Sector (0.1 Billion Yuan)*	*Total Investment on Fixed Properties (0.1 Billion Yuan)*	*Employment (10,000 People)*
Total	7505.31	6158.8	288.8
Primary sector	1320.57	275.6	56.2
Secondary sector	3481.56	3197.2	79.5
Tertiary sector	2703.18	2686	153.1
Industries (*i*)			
Agriculture, forest, animal husbandry, fishery	1320.57	275.6	56.2
Mining		747.9	18.9
Manufacturing		1581.5	30.6
Production/supply of electricity, gas, and water		773.7	7.6
Construction	631.50	94.1	22.4
Communication, transportation, storage, postal service	357.90	437.2	12.6
Information transfer, computer, and software areas		43.9	2.6
# Information Transfer			
Wholesale and retail	426.65	83.7	8.2
Hotels and restaurants	108.39	38	2.8
Finance	360.40	3	8.1
Real estate/housing	194.38	1413.8	3.2
Leasing and commercial services		19.4	5.4
Scientific research, technology service, and geological survey		17.9	6.1
Water conservation, environment, and public facility management		329.7	5.4
# Public Facility Management			
Residents and other services		9.5	0.6
Education		84.2	36

Table 8.53 (continued) The GDPs, Total Investments in Fixed Properties, and Employment Situation of Xinjiang's Three Economic Sectors and Industries (2012)

Sectors (i)	*Total Output of Sector (0.1 Billion Yuan)*	*Total Investment on Fixed Properties (0.1 Billion Yuan)*	*Employment (10,000 People)*
Health, safety, and welfare		36.3	15.8
Social, sports, and entertainment		33	2.9
Public administration and social organizations		136.4	43.4

Source: The data are from *China Statistics Abstracts 2013*.

Table 8.54 The Structures and Comparative Gain Measures of Xinjiang's Three Economic Sectors and Industries ($t = 2012$)

Sectors (i)	*Demand Structure $d_i(t)$ (%)*	*Employment Structure $s_i^1(t)$ (%)*	*Comparative Gain of Labor $G_i^1(t)$*	*Investment Structure $s_i^2(t)$ (%)*	*Comparative Gain of Capital $G_i^2(t)$*
Total	100	100	—	100	—
Primary sector	17.60	19.46	0.91	4.47	3.93
Secondary sector	46.39	27.53	1.69	51.91	0.89
Tertiary sector	36.02	53.01	0.68	43.61	0.83
Industries (*i*)					
Agriculture, forest, animal husbandry, fishery	17.60	19.46	0.91	4.47	3.93
Mining		6.54		12.14	
Manufacturing		10.60		25.68	
Production/supply of electricity, gas, and water		2.63		12.56	
Construction	8.41	7.76	1.08	1.53	5.51
Communication, transportation, storage, postal service	4.77	4.36	1.09	7.10	0.67
Information transfer, computer, and software areas		0.90		0.71	

continued

Table 8.54 (continued) The Structures and Comparative Gain Measures of Xinjiang's Three Economic Sectors and Industries ($t = 2012$)

Sectors (i)	*Demand Structure $d_i(t)$ (%)*	*Employment Structure $s_i^1(t)$ (%)*	*Comparative Gain of Labor $G_i^1(t)$*	*Investment Structure $s_i^2(t)$ (%)*	*Comparative Gain of Capital $G_i^2(t)$*
# Information Transfer					
Wholesale and retail	5.68	2.84	2.00	1.36	4.18
Hotels and restaurants	1.44	0.97	1.49	0.62	2.34
Finance	4.80	2.80	1.71	0.05	98.58
Real estate/housing	2.59	1.11	2.34	0.23	11.28
Leasing and commercial services		1.87		0.32	
Scientific research, technology service, and geological survey		2.11		0.29	
Water conservation, environment, and public facility management		1.87		5.35	
# Public Facility Management					
Residents and other services		0.21		0.15	
Education		12.47		1.37	
Health, safety, and welfare		5.47		0.59	
Social, sports, and entertainment		1.00		0.54	
Public administration and social organizations		15.03		2.21	

Table 8.55 The GDPs, Total Investments in Fixed Properties, and Employment Situation of Shaanxi's Three Economic Sectors and Industries (2012)

Sectors (i)	*Total Output of Sector (0.1 Billion Yuan)*	*Total Investment on Fixed Properties (0.1 Billion Yuan)*	*Employment (10,000 People)*
Total	14453.68	12044.5	411.2
Primary sector	1370.16	496.1	3.6
Secondary sector	8073.87	4332.8	168.6
Tertiary sector	5009.65	7245.6	238.9
Industries (*i*)			
Agriculture, forest, animal husbandry, fishery	1370.16	496.1	3.6
Mining		1041.8	30
Manufacturing		2338.6	86.2
Production/supply of electricity, gas, and water		371.2	9.6
Construction	1226.46	581.2	42.8
Communication, transportation, storage, postal service	617.39	805.4	18.3
Information transfer, computer, and software areas		86.3	7.8
# Information Transfer			
Wholesale and retail	1166.90	337.5	17.3
Hotels and restaurants	312.27	171.4	7.5
Finance	551.20	29	14.6
Real estate/housing	450.12	3733.8	6.6
Leasing and commercial services		113.9	4.5
Scientific research, technology service, and geological survey		119.4	14.4
Water conservation, environment, and public facility management		989.2	8.9
# Public Facility Management			
Residents and other services		66.9	1.5

continued

Table 8.55 (continued) The GDPs, Total Investments in Fixed Properties, and Employment Situation of Shaanxi's Three Economic Sectors and Industries (2012)

Sectors (i)	*Total Output of Sector (0.1 Billion Yuan)*	*Total Investment on Fixed Properties (0.1 Billion Yuan)*	*Employment (10,000 People)*
Education		166.1	56.9
Health, safety, and welfare		133.9	21.9
Social, sports, and entertainment		91	5
Public administration and social organizations			

Source: Data are from *China Statistics Abstracts 2013*.

Table 8.56 The Structures and Comparative Gain Measures of Shaanxi's Three Economic Sectors and Industries ($t = 2012$)

Sectors (i)	*Demand Structure* $d_i(t)$ *(%)*	*Employment Structure* $s_i^1(t)$ *(%)*	*Comparative Gain of Labor* $G_i^1(t)$	*Investment Structure* $s_i^2(t)$ *(%)*	*Comparative Gain of Capital* $G_i^2(t)$
Total	100	100	—	100	—
Primary sector	9.48	0. 88	10.83	4.12	2.30
Secondary sector	55.86	41.00	1.36	35.97	1.55
Tertiary sector	34.66	58.10	0.60	60.16	0.58
Industries (*i*)					
Agriculture, forest, animal husbandry, fishery	9.48	0. 88	10.83	4.12	2.30
Mining		7.30		8.65	
Manufacturing		20.96		19.42	
Production/supply of electricity, gas, and water		2.33		3.08	
Construction	8.49	10.41	0.82	4.83	1.76
Communication, transportation, storage, postal service	4.27	4.45	0.96	6.69	0.64
Information transfer, computer, and software areas		1.90		0.72	

Table 8.56 (continued) The Structures and Comparative Gain Measures of Shaanxi's Three Economic Sectors and Industries ($t = 2012$)

Sectors (i)	*Demand Structure $d_i(t)$ (%)*	*Employment Structure $s_i^1(t)$ (%)*	*Comparative Gain of Labor $G_i^1(t)$*	*Investment Structure $s_i^2(t)$ (%)*	*Comparative Gain of Capital $G_i^2(t)$*
# Information Transfer					
Wholesale and retail	8.07	4.33	1.87	2.80	2.88
Hotels and restaurants	2.16	1.82	1.18	1.42	1.52
Finance	3.81	3.55	1.07	0.24	15.34
Real estate/housing	3.11	1.61	1.94	31.00	0.10
Leasing and commercial services		1.09		0.95	
Scientific research, technology service, and geological survey		3.50		0.99	
Water conservation, environment, and public facility management		2.16		8.21	
# Public Facility Management					
Residents and other services		0.36		0.56	
Education		13.84		13.79	
Health, safety, and welfare		5.33		1.11	
Social, sports, and entertainment		1.22		0.76	
Public administration and social organizations		12.94		3.34	

Table 8.57 The GDPs, Total Investments in Fixed Properties, and Employment Situation of Sichuan's Three Economic Sectors and Industries (2012)

Sectors (i)	*Total Output of Sector (0.1 Billion Yuan)*	*Total Investment on Fixed Properties (0.1 Billion Yuan)*	*Employment (10,000 People)*
Total	23872.80	17040.0	640.9
Primary sector	3297.21	443.0	4.0
Secondary sector	12333.28	6107.9	302.7
Tertiary sector	8242.31	10489	329
Industries (*i*)			
Agriculture, forest, animal husbandry, fishery	3297.21	443.0	4.0
Mining		465.7	24.5
Manufacturing		4327.5	144.5
Production/supply of electricity, gas, and water		1281.0	17.6
Construction	1782.75	33.7	116.1
Communication, transportation, storage, postal service	707.19	2086.6	23.6
Information transfer, computer, and software areas		66.5	6
# Information Transfer			
Wholesale and retail	1342.25	324.9	18.5
Hotels and restaurants	621.72	253.6	7.6
Finance	1303.56	46.4	23
Real estate/housing	703.51	5017.5	6.6
Leasing and commercial services		110.9	5.2
Scientific research, technology service, and geological survey		60.2	16.2
Water conservation, environment, and public facility management		1801.8	10.6
# Public Facility Management			
Residents and other services		60.3	1.1
Education		202.2	89.4

Table 8.57 (continued) The GDPs, Total Investments in Fixed Properties, and Employment Situation of Sichuan's Three Economic Sectors and Industries (2012)

Sectors (i)	*Total Output of Sector (0.1 Billion Yuan)*	*Total Investment on Fixed Properties (0.1 Billion Yuan)*	*Employment (10,000 People)*
Health, safety, and welfare		149.9	39.3
Social, sports, and entertainment		162.8	4.5
Public administration and social organizations		145.4	82.6

Source: Data are from *China Statistics Abstracts 2013* (XLS, DOC).

Table 8.58 The Structures and Comparative Gain Measures of Sichuan's Three Economic Sectors and Industries ($t = 2012$)

Sectors (i)	*Demand Structure* $d_i(t)$ *(%)*	*Employment Structure* $s_i^1(t)$ *(%)*	*Comparative Gain of Labor* $G_i^1(t)$	*Investment Structure* $s_i^2(t)$ *(%)*	*Comparative Gain of Capital* $G_i^2(t)$
Total	100	100	—	100	—
Primary sector	13.81	0.63	21.97	2.60	5.31
Secondary sector	51.66	47.23	1.09	35.84	1.44
Tertiary sector	34.53	51.34	0.67	61.56	0.56
Industries (*i*)					
Agriculture, forest, animal husbandry, fishery	13.81	0.63	21.97	2.60	5.31
Mining		3.82		2.73	
Manufacturing		22.54		25.40	
Production/supply of electricity, gas, and water		2.75		7.52	
Construction	7.47	18.12	0.41	0.20	37.78
Communication, transportation, storage, postal service	2.96	3.68	0.80	12.25	0.24
Information transfer, computer, and software areas		0.94		0.39	

continued

Table 8.58 (continued) The Structures and Comparative Gain Measures of Sichuan's Three Economic Sectors and Industries ($t = 2012$)

Sectors (i)	Demand Structure $d_i(t)$ (%)	Employment Structure $s_i^1(t)$ (%)	Comparative Gain of Labor $G_i^1(t)$	Investment Structure $s_i^2(t)$ (%)	Comparative Gain of Capital $G_i^2(t)$
# Information Transfer					
Wholesale and retail	5.62	2.89	1.93	1.91	2.95
Hotels and restaurants	2.60	1.19	2.20	1.49	1.75
Finance	5.46	3.59	1.52	0.27	20.05
Real estate/housing	2.95	1.03	2.86	29.45	0.10
Leasing and commercial services				0.65	
Scientific research, technology service, and geological survey		2.53		0.35	
Water conservation, environment, and public facility management		1.65		10.57	
# Public Facility Management					
Residents and other services		0.17		0.35	
Education		13.95		1.19	
Health, safety, and welfare		6.13		0.88	
Social, sports, and entertainment		0.70		0.96	
Public administration and social organizations		12.89		0.85	

Table 8.59 The GDPs, Total Investments in Fixed Properties, and Employment Situation of Guizhou's Three Economic Sectors and Industries (2012)

Sectors (i)	*Total Output of Sector (0.1 Billion Yuan)*	*Total Investment on Fixed Properties (0.1 Billion Yuan)*	*Employment (10,000 People)*
Total	6852.20	5717.8	269.5
Primary sector	891.91	79.0	1.6
Secondary sector	2677.54	1564.3	109.1
Tertiary sector	3282.75	4074.6	158.8
Industries (*i*)			
Agriculture, forest, animal husbandry, fishery	891.91	79.0	1.6
Mining		279.8	18.2
Manufacturing		1020.3	47.6
Production/supply of electricity, gas, and water		253.3	8.4
Construction	460.48	10.9	34.9
Communication, transportation, storage, postal service	687.45	756.2	9.1
Information transfer, computer, and software areas		6.4	2.5
# Information Transfer			
Wholesale and retail	514.49	76.3	13.2
Hotels and restaurants	266.58	47.0	3.9
Finance	365.87		7.4
Real estate/housing	176.75	1859.3	5.4
Leasing and commercial services		30.6	3
Scientific research, technology service, and geological survey		8.8	5.4
Water conservation, environment, and public facility management		1069.8	4.2
# Public Facility Management			
Residents and other services		10.5	1

continued

Table 8.59 (continued) The GDPs, Total Investments in Fixed Properties, and Employment Situation of Guizhou's Three Economic Sectors and Industries (2012)

Sectors (i)	*Total Output of Sector (0.1 Billion Yuan)*	*Total Investment on Fixed Properties (0.1 Billion Yuan)*	*Employment (10,000 People)*
Education		117.2	44.2
Health, safety, and welfare		19.6	14.6
Social, sports, and entertainment		50.7	1.6
Public administration and social organizations		22.1	43.3

Source: The data is from *China Statistics Abstracts 2013.*

Table 8.60 The Structures and Comparative Gain Measures of Guizhou's Three Economic Sectors and Industries ($t = 2012$)

Sectors (i)	*Demand Structure $d_i(t)$ (%)*	*Employment Structure $s_i^1(t)$ (%)*	*Comparative Gain of Labor $G_i^1(t)$*	*Investment Structure $s_i^2(t)$ (%)*	*Comparative Gain of Capital $G_i^2(t)$*
Total	100	100	—	100	—
Primary sector	13.02	0.59	21.92	1.38	9.43
Secondary sector	39.08	40.48	0.97	27.36	1.43
Tertiary sector	47.91	58.92	0.81	71.26	0.67
Industries (*i*)					
Agriculture, forest, animal husbandry, fishery	13.02	0.59	21.92	1.38	9.43
Mining		6.75		4.89	
Manufacturing		17.66		17.84	
Production/supply of electricity, gas, and water		3.12		4.43	
Construction	6.72	12.95	0.52	0.19	35.16
Communication, transportation, storage, postal service	10.03	3.38	2.97	13.23	0.76
Information transfer, computer, and software areas		0.93		0.11	

Table 8.60 (continued) The Structures and Comparative Gain Measures of Guizhou's Three Economic Sectors and Industries (t = 2012)

Sectors (i)	*Demand Structure $d_i(t)$ (%)*	*Employment Structure $s_i^1(t)$ (%)*	*Comparative Gain of Labor $G_i^1(t)$*	*Investment Structure $s_i^2(t)$ (%)*	*Comparative Gain of Capital $G_i^2(t)$*
# Information Transfer					
Wholesale and retail	7.51	4.90	1.53	1.33	5.63
Hotels and restaurants	3.89	1.45	2.69	0.82	4.73
Finance	5.34	2.75	1.94		
Real estate/housing	2.58	2.00	1.29	32.52	0.08
Leasing and commercial services		1.11		0.54	
Scientific research, technology service, and geological survey		2.00		0.15	
Water conservation, environment, and public facility management		1.56		18.71	
# Public Facility Management					
Residents and other services		0.57		0.18	
Education		16.40		2.05	
Health, safety, and welfare		5.42		0.34	
Social, sports, and entertainment		0.59		0.87	
Public administration and social organizations		16.07		0.39	

Table 8.61 The GDPs, Total Investments in Fixed Properties, and Employment Situation of Qinghai's Three Economic Sectors and Industries (2012)

Sectors (i)	*Total Output of Sector (0.1 Billion Yuan)*	*Total Investment on Fixed Properties (0.1 Billion Yuan)*	*Employment (10,000 People)*
Total	1893.54	1883.4	61.7
Primary sector	176.91	74.5	1.6
Secondary sector	1092.34	887.2	23.7
Tertiary sector	624.29	921.5	36.6
Industries (*i*)			
Agriculture, forest, animal husbandry, fishery	176.91	74.5	1.6
Mining		84.4	2.6
Manufacturing		505.2	11.7
Production/supply of electricity, gas, and water		235.6	2
Construction	196.45	62	7.4
Communication, transportation, storage, postal service	71.87	232	3.4
Information transfer, computer, and software areas		2	0.9
# Information Transfer			
Wholesale and retail	109.44	18.1	2.4
Hotels and restaurants	21.39	11.4	0.7
Finance	83.73	0.9	2.2
Real estate/housing	31.41	378.1	0.8
Leasing and commercial services		31.4	0.8
Scientific research, technology service, and geological survey		3.1	2.8
Water conservation, environment, and public facility management		76	1
# Public Facility Management			
Residents and other services		1.2	0.3
Education		46.2	7.6

Table 8.61 (continued) The GDPs, Total Investments in Fixed Properties, and Employment Situation of Qinghai's Three Economic Sectors and Industries (2012)

Sectors (i)	*Total Output of Sector (0.1 Billion Yuan)*	*Total Investment on Fixed Properties (0.1 Billion Yuan)*	*Employment (10,000 People)*
Health, safety, and welfare		14.9	3.7
Social, sports, and entertainment		19.3	0.8
Public administration and social organizations		86.9	9.2

Source: The data are from *China Statistics Abstracts 2013*.

Table 8.62 The Structures and Comparative Gain Measures of Qinghai's Three Economic Sectors and Industries (t = 2012)

Sectors (i)	*Demand Structure $d_i(t)$ (%)*	*Employment Structure $s_i^1(t)$ (%)*	*Comparative Gain of Labor $G_i^1(t)$*	*Investment Structure $s_i^2(t)$ (%)*	*Comparative Gain of Capital $G_i^2(t)$*
Total	100	100	—	100	—
Primary sector	9.34	2.59	3.60	3.96	2.36
Secondary sector	57.69	38.41	1.50	47.11	1.22
Tertiary sector	32.97	59.32	0.56	48.93	0.67
Industries (*i*)					
Agriculture, forest, animal husbandry, fishery	9.34	2.59	3.60	3.96	2.36
Mining		4.21		4.48	
Manufacturing		18.96		26.82	
Production/supply of electricity, gas, and water		3.24		12.51	
Construction	10.37	11.99	0.87	3.29	3.15
Communication, transportation, storage, postal service	3.80	5.51	0.69	12.32	0.31
Information transfer, computer, and software areas		1.46		0.11	

continued

Table 8.62 (continued) The Structures and Comparative Gain Measures of Qinghai's Three Economic Sectors and Industries ($t = 2012$)

Sectors (i)	Demand Structure $d_i(t)$ (%)	Employment Structure $s_i^1(t)$ (%)	Comparative Gain of Labor $G_i^1(t)$	Investment Structure $s_i^2(t)$ (%)	Comparative Gain of Capital $G_i^2(t)$
# Information Transfer					
Wholesale and retail	5.78	3.89	1.49	0.96	6.01
Hotels and restaurants	1.13	1.13	1.00	0.61	1.87
Finance	4.42	3.57	1.24	0.05	92.54
Real estate/housing	1.66	1.30	1.28	20.08	0.08
Leasing and commercial services		1.30		1.67	
Scientific research, technology service, and geological survey		4.54		0.16	
Water conservation, environment, and public facility management		1.62		4.04	
# Public Facility Management					
Residents and other services		0.49		0.06	
Education		12.32		2.45	
Health, safety, and welfare		6.00		0.79	
Social, sports, and entertainment		1.30		1.02	
Public administration and social organizations		14.91		4.61	

Table 8.63 The GDPs, Total Investments in Fixed Properties, and Employment Situation of Gansu's Three Economic Sectors and Industries (2012)

Sectors (i)	*Total Output of Sector (0.1 Billion Yuan)*	*Total Investment on Fixed Properties (0.1 Billion Yuan)*	*Employment (10,000 People)*
Total	5650.20	5145	211.3
Primary sector	780.50	171.1	5.5
Secondary sector	2600.09	2737.8	73.7
Tertiary sector	2269.61	2236.3	132.1
Industries (*i*)			
Agriculture, forest, animal husbandry, fishery	780.50	171.1	5.5
Mining		339.5	9.7
Manufacturing		970.9	33.8
Production/supply of electricity, gas, and water		634.2	7.2
Construction	529.85	793.2	23
Communication, transportation, storage, postal service	319.66	306.6	10.4
Information transfer, computer, and software areas		32.2	2
# Information Transfer			
Wholesale and retail	398.52	138.7	6
Hotels and restaurants	141.68	61.9	1.9
Finance	184.43	11	7.2
Real estate/housing	146.32	915.2	1.9
Leasing and commercial services		28.6	1.7
Scientific research, technology service, and geological survey		32.6	5.6
Water conservation, environment, and public facility management		367.5	5.2
# Public Facility Management			
Residents and other services		43.7	0.4

continued

Table 8.63 (continued) The GDPs, Total Investments in Fixed Properties, and Employment Situation of Gansu's Three Economic Sectors and Industries (2012)

Sectors (i)	*Total Output of Sector (0.1 Billion Yuan)*	*Total Investment on Fixed Properties (0.1 Billion Yuan)*	*Employment (10,000 People)*
Education		64.8	36.8
Health, safety, and welfare		38.2	11.7
Social, sports, and entertainment		52.6	2.6
Public administration and social organizations		142.7	38.7

Source: The data are from *China Statistics Abstrcats 2013* (XLS, DOC).

Table 8.64 The Structures and Comparative Gain Measures of Gansu's Three Economic Sectors and Industries ($t = 2012$)

Sectors (i)	*Demand Structure $d_i(t)$ (%)*	*Employment Structure $s_i^1(t)$ (%)*	*Comparative Gain of Labor $G_i^1(t)$*	*Investment Structure $s_i^2(t)$ (%)*	*Comparative Gain of Capital $G_i^2(t)$*
Total	100	100	—	100	—
Primary sector	13.81	2.60	5.31	3.33	4.15
Secondary sector	46.02	34.88	1.32	53.21	0.86
Tertiary sector	40.17	62.52	0.64	43.47	0.92
Industries (*i*)					
Agriculture, forest, animal husbandry, fishery	13.81	2.60	5.31	3.33	4.15
Mining		4.59		6.60	
Manufacturing		1.60		18.87	
Production/supply of electricity, gas, and water		3.41		12.33	
Construction	9.38	10.89	0.86	15.42	0.61
Communication, transportation, storage, postal service	5.66	4.92	1.15	5.96	0.95
Information transfer, computer, and software areas		0.95		0.63	

Table 8.64 (continued) The Structures and Comparative Gain Measures of Gansu's Three Economic Sectors and Industries ($t = 2012$)

Sectors (i)	*Demand Structure* $d_i(t)$ *(%)*	*Employment Structure* $s_i^1(t)$ *(%)*	*Comparative Gain of Labor* $G_i^1(t)$	*Investment Structure* $s_i^2(t)$ *(%)*	*Comparative Gain of Capital* $G_i^2(t)$
# Information Transfer					
Wholesale and retail	7.05	2.84	2.48	2.70	2.62
Hotels and restaurants	2.51	0.90	2.79	1.20	2.08
Finance	3.26	3.41	0.96	0.21	15.27
Real estate/housing	2.59	0.90	2.38	17.79	0.15
Leasing and commercial services		0.80		0.56	
Scientific research, technology service, and geological survey		2.65		0.63	
Water conservation, environment, and public facility management		2.46		7.14	
# Public Facility Management					
Residents and other services		0.19		0.85	
Education		17.42		1.26	
Health, safety, and welfare		5.54		0.74	
Social, sports, and entertainment		1.23		1.02	
Public administration and social organizations		18.32		2.77	

Table 8.65 The GDPs, Total Investments in Fixed Properties, and Employment Situation of Ningxia's Three Economic Sectors and Industries (2012)

Sectors (i)	*Total Output of Sector (0.1 Billion Yuan)*	*Total Investment on Fixed Properties (0.1 Billion Yuan)*	*Employment (10,000 People)*
Total	2341.29	2096.6	67.4
Primary sector	199.40	71.8	2.4
Secondary sector	1159.37	1037.2	25.9
Tertiary sector	982.52	988	39.1
Industries (*i*)			
Agriculture, forest, animal husbandry, fishery	199.40	71.8	2.4
Mining		151.5	6.6
Manufacturing		628.1	10.8
Production/supply of electricity, gas, and water		226.3	4.1
Construction	280.74	31.3	4.4
Communication, transportation, storage, postal service	196.49	113.3	3.6
Information transfer, computer, and software areas		11.5	0.6
# Information Transfer			
Wholesale and retail	124.66	53.6	2.8
Hotels and restaurants	42.67	13.2	0.6
Finance	167.48	0.7	3
Real estate/housing	87.51	597.2	1.1
Leasing and commercial services		8	1.8
Scientific research, technology service, and geological survey		1	1.2
Water conservation, environment, and public facility management		108.3	2
# Public Facility Management			
Residents and other services		10	
Education		29.1	8.4

Table 8.65 (continued) The GDPs, Total Investments in Fixed Properties, and Employment Situation of Ningxia's Three Economic Sectors and Industries (2012)

Sectors (i)	*Total Output of Sector (0.1 Billion Yuan)*	*Total Investment on Fixed Properties (0.1 Billion Yuan)*	*Employment (10,000 People)*
Health, safety, and welfare		11.1	4
Social, sports, and entertainment		8	0.7
Public administration and social organizations		23	9.3

Source: The data are from *China Statistics Abstracts 2013*.

Table 8.66 The Structures and Comparative Gain Measures of Ningxia's Three Economic Sectors and Industries (t = 2012)

Sectors (i)	*Demand Structure $d_i(t)$ (%)*	*Employment Structure $s_i^1(t)$ (%)*	*Comparative Gain of Labor $G_i^1(t)$*	*Investment Structure $s_i^2(t)$ (%)*	*Comparative Gain of Capital $G_i^2(t)$*
Total	100	100	—	100	—
Primary sector	8.52	3.56	2.39	3.42	2.49
Secondary sector	49.52	38.43	1.29	0.49	1.00
Tertiary sector	41.96	58.01	0.72	47.12	0.89
Industries (*i*)					
Agriculture, forest, animal husbandry, fishery	8.52	3.56	2.39	3.42	2.49
Mining		9.79		7.23	
Manufacturing		16.02		29.96	
Production/supply of electricity, gas, and water		6.08		10.79	
Construction	11.99	6.53	1.84	1.49	8.03
Communication, transportation, storage, postal service	8.39	5.34	1.57	5.40	1.55
Information transfer, computer, and software areas		0.89		0.55	

continued

Table 8.66 (continued) The Structures and Comparative Gain Measures of Ningxia's Three Economic Sectors and Industries ($t = 2012$)

Sectors (i)	*Demand Structure $d_i(t)$ (%)*	*Employment Structure $s_i^1(t)$ (%)*	*Comparative Gain of Labor $G_i^1(t)$*	*Investment Structure $s_i^2(t)$ (%)*	*Comparative Gain of Capital $G_i^2(t)$*
# Information Transfer					
Wholesale and retail	5.32	4.15	1.28	2.56	2.08
Hotels and restaurants	1.82	0.89	2.05	0.63	2.89
Finance	7.15	4.45	1.61	0.03	214.25
Real estate/housing	3.74	1.63	2.29	28.48	0.13
Leasing and commercial services		2.67		0.38	
Scientific research, technology service, and geological survey		1.78		0.05	
Water conservation, environment, and public facility management		2.97		5.17	
# Public Facility Management					
Residents and other services				0.48	
Education		12.46		1.39	
Health, safety, and welfare		5.93		0.53	
Social, sports, and entertainment		1.04		0.38	
Public administration and social organizations		13.80		1.10	

Table 8.67 The GDPs, Total Investments in Fixed Properties, and Employment Situation of Chongqing's Three Economic Sectors and Industries (2012)

Sectors (i)	*Total Output of Sector (0.1 Billion Yuan)*	*Total Investment on Fixed Properties (0.1 Billion Yuan)*	*Employment (10,000 People)*
Total	11409.60	8736.2	353.2
Primary sector	940.01	375.4	1.3
Secondary sector	5975.18	2857.8	185.3
Tertiary sector	4494.41	5503	166.5
Industries (*i*)			
Agriculture, forest, animal husbandry, fishery	940.01	375.4	1.3
Mining		167.7	10.4
Manufacturing		2268.6	82.3
Production/supply of electricity, gas, and water		372.9	7.5
Construction	994.17	48.6	85.2
Communication, transportation, storage, postal service	515.15	835.0	15.8
Information transfer, computer, and software areas	208.36	80.4	3.9
# Information Transfer			
Wholesale and retail	847.99	112.3	19.8
Hotels and restaurants	189.98	59.5	7.1
Finance	915.65	2.5	13.0
Real estate/housing	620.17	2991.4	7.8
Leasing and commercial services	211.06	92.0	6.1
Scientific research, technology service, and geological survey	94.73	15.9	5.7
Water conservation, environment, and public facility management	40.17	858.5	4.6
# Public Facility Management			
Residents and other services	105.05	68.1	0.9

continued

Table 8.67 (continued) The GDPs, Total Investments in Fixed Properties, and Employment Situation of Chongqing's Three Economic Sectors and Industries (2012)

Sectors (i)	*Total Output of Sector (0.1 Billion Yuan)*	*Total Investment on Fixed Properties (0.1 Billion Yuan)*	*Employment (10,000 People)*
Education	230.25	109.3	37.1
Health, safety, and welfare	182.33	52.7	13.5
Social, sports, and entertainment	59.06	115.3	2.6
Public administration and social organizations	274.46	110.0	28.5

Source: Data are from *China Statistics Abstacts 2013* (CLS, DOC).

Table 8.68 The Structures and Comparative Gain Measures of Chongqing's Three Economic Sectors and Industries ($t = 2012$)

Sectors (i)	*Demand Structure $d_i(t)$ (%)*	*Employment Structure $s_i^1(t)$ (%)*	*Comparative Gain of Labor $G_i^1(t)$*	*Investment Structure $s_i^2(t)$ (%)*	*Comparative Gain of Capital $G_i^2(t)$*
Total	100	100	—	100	—
Primary sector	8.24	0.37	22.51	4.30	1.92
Secondary sector	52.37	52.47	1.00	32.71	1.60
Tertiary sector	39.39	47.14	0.84	62.99	0.63
Industries (*i*)					
Agriculture, forest, animal husbandry, fishery	8.24	0.37	22.51	4.30	1.92
Mining		2.94		1.92	
Manufacturing		23.3		25.97	
Production/supply of electricity, gas, and water		2.12		4.27	
Construction	8.71	24.11	0.36	0.56	15.68
Communication, transportation, storage, postal service	4.52	4.46	1.01	9.56	0.47
Information transfer, computer, and software areas	1.83	1.12	1.64	0.92	1.98

Table 8.68 (continued) The Structures and Comparative Gain Measures of Chongqing's Three Economic Sectors and Industries ($t = 2012$)

Sectors (i)	*Demand Structure $d_i(t)$ (%)*	*Employment Structure $s_i^1(t)$ (%)*	*Comparative Gain of Labor $G_i^1(t)$*	*Investment Structure $s_i^2(t)$ (%)*	*Comparative Gain of Capital $G_i^2(t)$*
# Information Transfer					
Wholesale and retail	7.43	5.61	1.33	1.29	5.78
Hotels and restaurants	1.67	2.02	0.82	0.68	2.44
Finance	8.03	3.69	2.18	0.03	276.14
Real estate/housing	5.44	2.21	2.45	34.24	0.16
Leasing and commercial services	1.85	1.73	1.07	1.05	1.76
Scientific research, technology service, and geological survey	0.83	1.61	0.51	0.18	4.55
Water conservation, environment, and public facility management	0.35	1.30	0.27	9.83	0.04
# Public Facility Management					
Residents and other services	0.92	0.25	3.61	0.78	1.18
Education	2.02	10.50	0.19	1.25	1.61
Health, safety, and welfare	1.60	3.82	0.42	0.60	2.65
Social, sports, and entertainment	0.52	0.74	0.70	1.32	0.39
Public administration and social organizations	2.41	8.07	0.30	1.26	1.91

Table 8.69 The GDPs, Total Investments in Fixed Properties, and Employment Situation of Yunnan's Three Economic Sectors and Industries (2012)

Sectors (i)	*Total Output of Sector (0.1 Billion Yuan)*	*Total Investment on Fixed Properties (0.1 Billion Yuan)*	*Employment (10,000 People)*
Total	10309.47	7831.1	392.7
Primary sector	1654.55	204.3	9.2
Secondary sector	4419.20	2544.4	168.9
Tertiary sector	4235.72	5082.4	214.5
Industries (*i*)			
Agriculture, forest, animal husbandry, fishery	1654.55	204.3	9.2
Mining		363.0	22.3
Manufacturing		1223.4	70.5
Production/supply of electricity, gas, and water		952.0	9.7
Construction	968.48	6	66.4
Communication, transportation, storage, postal service	247.53	780.5	13.6
Information transfer, computer, and software areas		59.7	4.3
# Information Transfer			
Wholesale and retail	1039.64	234.9	23.5
Hotels and restaurants	330.67	168.2	8.5
Finance	541.18	4.9	9.8
Real estate/housing	243.03	2447.4	7.3
Leasing and commercial services		53.2	6.2
Scientific research, technology service, and geological survey		30.4	7.8
Water conservation, environment, and public facility management		787.7	5.7
# Public Facility Management			
Residents and other services		27	0.8
Education		184.9	54.9

Table 8.69 (continued) The GDPs, Total Investments in Fixed Properties, and Employment Situation of Yunnan's Three Economic Sectors and Industries (2012)

Sectors (i)	*Total Output of Sector (0.1 Billion Yuan)*	*Total Investment on Fixed Properties (0.1 Billion Yuan)*	*Employment (10,000 People)*
Health, safety, and welfare		70.5	19.4
Social, sports, and entertainment		114.3	3.4
Public administration and social organizations		118.8	49.3

Source: The data are from *China Statistics Abstracts 2013*.

Table 8.70 The Structures and Comparative Gain Measures of Yunnan's Three Economic Sectors and Industries ($t = 2012$)

Sectors (i)	*Demand Structure $d_i(t)$ (%)*	*Employment Structure $s_i^1(t)$ (%)*	*Comparative Gain of Labor $G_i^1(t)$*	*Investment Structure $s_i^2(t)$ (%)*	*Comparative Gain of Capital $G_i^2(t)$*
Total	100	100	—	100	—
Primary sector	16.05	2.34	6.85	2.61	6.15
Secondary sector	42.87	43.01	1.00	32.49	1.32
Tertiary sector	41.09	54.62	0.75	64.90	0.63
Industries (*i*)					
Agriculture, forest, animal husbandry, fishery	16.05	2.34	6.85	2.61	6.15
Mining		5.68		4.64	
Manufacturing		17.95		15.62	
Production/supply of electricity, gas, and water		2.47		12.16	
Construction	9.39	16.91	0.56	0.08	122.61
Communication, transportation, storage, postal service	2.40	3.46	0.69	9.97	0.24
Information transfer, computer, and software areas		1.10		0.76	

continued

Table 8.70 (continued) The Structures and Comparative Gain Measures of Yunnan's Three Economic Sectors and Industries ($t = 2012$)

Sectors (i)	Demand Structure $d_i(t)$ (%)	Employment Structure $s_i^1(t)$ (%)	Comparative Gain of Labor $G_i^1(t)$	Investment Structure $s_i^2(t)$ (%)	Comparative Gain of Capital $G_i^2(t)$
# Information Transfer					
Wholesale and retail	10.08	5.98	1.69	3.00	3.36
Hotels and restaurants	3.21	2.16	1.48	2.15	1.49
Finance	5.25	2.50	2.10	0.06	83.90
Real estate/housing	2.36	1.86	1.27	31.25	0.08
Leasing and commercial services		1.58		0.68	
Scientific research, technology service, and geological survey		1.99		0.39	
Water conservation, environment, and public facility management		1.45		10.06	
# Public Facility Management					
Residents and other services		0.20		0.34	
Education		13.98		2.36	
Health, safety, and welfare		4.94		0.90	
Social, sports, and entertainment		0.87		1.46	
Public administration and social organizations		12.55		1.52	

Table 8.71 The GDPs, Total Investments in Fixed Properties, and Employment Situation of Tibet's Three Economic Sectors and Industries (2012)

Sectors (i)	*Total Output of Sector (0.1 Billion Yuan)*	*Total Investment on Fixed Properties (0.1 Billion Yuan)*	*Employment (10,000 People)*
Total	701.03	670.5	25.2
Primary sector	80.38	24.2	0.6
Secondary sector	242.85	217.3	2.2
Tertiary sector	377.80	428.9	22.32
Industries (*i*)			
Agriculture, forest, animal husbandry, fishery	80.38	24.2	0.6
Mining		43	0.2
Manufacturing		44.8	0.7
Production/supply of electricity, gas, and water		90.7	0.8
Construction	187.50	38.8	0.5
Communication, transportation, storage, postal service	26.23	136.6	0.6
Information transfer, computer, and software areas		12.7	0.3
# Information Transfer			
Wholesale and retail	42.75	14.8	0.5
Hotels and restaurants	23.78	24.7	0.3
Finance	32.04	2.4	0.8
Real estate/housing	20.77	58	0.02
Leasing and commercial services		10.5	0.2
Scientific research, technology service, and geological survey		3.2	1.1
Water conservation, environment, and public facility management		46.9	0.2
# Public Facility Management			
Residents and other services		4.7	

continued

Table 8.71 (continued) The GDPs, Total Investments in Fixed Properties, and Employment Situation of Tibet's Three Economic Sectors and Industries (2012)

Sectors (i)	*Total Output of Sector (0.1 Billion Yuan)*	*Total Investment on Fixed Properties (0.1 Billion Yuan)*	*Employment (10,000 People)*
Education		18.4	4.4
Health, safety, and welfare		7.9	1.6
Social, sports, and entertainment		14.1	0.6
Public administration and social organizations		74	11.7

Source: The data are from *China Statistics Abstracts 2013*.

Table 8.72 The Structures and Comparative Gain Measures of Tibet's Three Economic Sectors and Industries ($t = 2012$)

Sectors (i)	*Demand Structure $d_i(t)$ (%)*	*Employment Structure $s_i^1(t)$ (%)*	*Comparative Gain of Labor $G_i^1(t)$*	*Investment Structure $s_i^2(t)$ (%)*	*Comparative Gain of Capital $G_i^2(t)$*
Total	100	100	—	100	—
Primary sector	11.47	2.38	4.82	3.61	3.18
Secondary sector	34.64	8.73	3.97	32.41	1.07
Tertiary sector	53.89	88.57	0.61	63.97	0.84
Industries (*i*)					
Agriculture, forest, animal husbandry, fishery	11.47	2.38	4.82	3.61	3.18
Mining		0.79		6.41	
Manufacturing		2.78		6.68	
Production/supply of electricity, gas, and water		3.17		13.53	
Construction	26.75	1.98	13.48	5.79	4.62
Communication, transportation, storage, postal service	3.74	2.38	1.57	20.37	0.18
Information transfer, computer, and software areas		1.19		1.89	

Table 8.72 (continued) The Structures and Comparative Gain Measures of Tibet's Three Economic Sectors and Industries ($t = 2012$)

Sectors (i)	*Demand Structure* $d_i(t)$ *(%)*	*Employment Structure* $s_i^1(t)$ *(%)*	*Comparative Gain of Labor* $G_i^1(t)$	*Investment Structure* $s_i^2(t)$ *(%)*	*Comparative Gain of Capital* $G_i^2(t)$
# Information Transfer					
Wholesale and retail	6.10	1.98	3.07	2.21	2.76
Hotels and restaurants	3.39	1.19	2.85	3.68	0.92
Finance	4.57	3.17	1.44	0.36	12.77
Real estate/housing	2.96	0.08	37.33	8.65	0.34
Leasing and commercial services		0.79		1.57	
Scientific research, technology service, and geological survey		4.37		0.48	
Water conservation, environment, and public facility management		0.79		6.99	
# Public Facility Management					
Residents and other services				0.70	
Education		17.46		2.74	
Health, safety, and welfare		6.35		1.18	
Social, sports, and entertainment		2.38		2.10	
Public administration and social organizations		46.43		11.04	

8.5 The Six Administrative Areas

Table 8.73 The GDPs, Total Investments in Fixed Properties, and Employment Situations of all Provinces, Municipalities, and Autonomous Regions in China (2012)

Area	*Province/Municipality/ Autonomous Region (i)*	*Regional GDP (0.1 Billion Yuan)*	*Value of Fixed Properties (0.1 Billion Yuan)*	*Employments (10,000 People)*
North	1. Beijing	17879.4	7409.6	1795.8
	2. Tianjin	12893.9	9898.4	606.8
	3. Hebei	26575.0	15244.6	1792.3
	4. Shanxi	12112.8	8223.9	1152.0
	5. Inner Mongolia	15880.6	13442.1	1034.5
	Total	85341.7	54168.6	6381.4
Northeast	6. Liaoning	24846.4	15492.1	2218.1
	7. Jilin	11939.2	9136.2	1121.3
	8. Heilongjiang	13619.6	8143.7	1487.8
	Total	50405.2	32772.0	4827.2
East	9. Shanghai	20181.7	7674.8	1781.6
	10. Jiangsu	54058.2	27258.1	5014.9
	11. Zhejiang	34665.3	15460.7	3838.8
	12. Anhui	17212.1	8855.8	1745.7
	13. Fujian	19701.8	11304.8	1927.8
	14. Jiangxi	12948.9	6513.7	1531.6
	15. Shandong	50013.2	27551.5	3564.0
	Total	208781.2	104619.4	19404.4
South	16. Henan	29599.3	22060.0	2509.0
	17. Hubei	22250.5	12554.7	2380.0
	18. Hunan	22154.2	12488.8	2019.9
	19. Guangdong	57067.9	22871.9	5089.0
	20. Guangxi	12948.9	11068.5	1229.4
	21. Hainan	2855.5	2009.9	345.9
	Total	146962.5	83053.8	13573.2
Southwest	22. Chongqing	11409.6	6341.4	1437.1
	23. Sichuan	23872.8	12496.0	2323.5
	24. Guizhou	6852.2	4164.4	800.0
	25. Yunnan	10309.5	8576.4	1492.5
	26. Tibet	701.0	708.7	130.2
	Total	53145.1	32286.9	6283.3

Table 8.73 (continued) The GDPs, Total Investments in Fixed Properties, and Employment Situations of all Provinces, Municipalities, and Autonomous Regions in China (2012)

Area	*Province/Municipality/ Autonomous Region (i)*	*Regional GDP (0.1 Billion Yuan)*	*Value of Fixed Properties (0.1 Billion Yuan)*	*Employments (10,000 People)*
Northwest	27. Shaanxi	14453.7	9915.2	1211.9
	28. Gansu	5650.2	3298.1	657.0
	29. Qinghai	1893.5	1719.3	194.9
	30. Ningxia	2341.3	2086.9	262.9
	31. Xinjiang	7505.3	5792.2	753.6
	Total	31844.0	22811.7	3080.3
National total		576479.7	329712.4	53449.8

Source: Data are from *China Statistics Abstracts 2013* (XLS, DOC).

Table 8.74 The Regional Economic Structures and Comparative Gain Measures of the Provinces, Municipalities, and Autonomous Regions of China (2012)

Area	*Province/ Municipality/ Autonomous Region (i)*	*Demand Structure $d_i(t)$ (%)*	*Employ-ment Structure $s_i^1(t)$ (%)*	*Compa-rative Gain of Labor $G_i^1(t)$*	*Investment Structure $s_i^2(t)$ (%)*	*Comparative Gain of Capital $G_i^2(t)$*
North	1. Beijing	3.10	3.36	0.92	2.25	1.38
	2. Tianjin	2.24	1.14	1.96	2.99	0.75
	3. Hebei	4.61	3.35	1.38	4.62	1.00
	4. Shanxi	2.10	2.16	0.97	2.49	0.84
	5. Inner Mongolia	2.75	1.94	1.42	4.08	0.67
	Total	14.80	11.95	1.24	16.43	0.90
Northeast	6. Liaoning	4.31	4.15	1.04	4.70	0.92
	7. Jilin	2.07	2.10	0.99	2.77	0.75
	8. Heilongjiang	2.36	2.78	0.85	2.47	0.96
	Total	8.74	9.03	0.97	9.94	0.88
East	9. Shanghai	3.50	3.33	1.05	2.33	1.50
	10. Jiangsu	9.38	9.38	1.00	8.27	1.13
	11. Zhejiang	6.01	7.18	0.84	4.69	1.29
	12. Anhui	2.99	3.27	0.91	2.69	1.11
	13. Fujian	3.42	3.61	0.95	3.43	1.00

continued

Table 8.74 (continued) The Regional Economic Structures and Comparative Gain Measures of the Provinces, Municipalities, and Autonomous Regions of China (2012)

Area	*Province/ Municipality/ Autonomous Region (i)*	*Demand Structure $d_i(t)$ (%)*	*Employment Structure $s_i^1(t)$ (%)*	*Comparative Gain of Labor $G_i^1(t)$*	*Investment Structure $s_i^2(t)$ (%)*	*Comparative Gain of Capital $G_i^2(t)$*
	14. Jiangxi	2.25	2.87	0.78	1.98	1.14
	15. Shandong	8.68	6.67	1.30	8.36	1.04
	Total	36.22	36.31	1.00	31.75	1.14
South	16. Henan	5.13	4.69	1.09	6.69	0.77
	17. Hubei	3.86	4.45	0.87	3.81	1.01
	18. Hunan	3.84	3.78	1.02	3.79	1.01
	19. Guangdong	9.90	9.52	1.04	1.26	1.43
	20. Guangxi	2.26	2.30	0.98	2.60	0.67
	21. Hainan	0.50	0.63	0.79	0.61	0.82
	Total	25.49	25.39	1.00	25.20	1.01
Southwest	22. Chongqing	1.98	2.69	0.74	1.92	1.03
	23. Sichuan	4.14	4.35	0.95	3.79	1.09
	24. Guizhou	1.19	1.50	0.79	1.26	0.94
	25. Yunnan	1.79	2.79	0.64	2.60	0.69
	26. Tibet	0.12	0.24	0.50	0.21	0.57
	Total	9.22	10.57	0.87	9.78	0.94
Northwest	27. Shaanxi	2.51	2.27	1.11	3.01	0.83
	28. Gansu	0.98	1.23	0.80	1.00	0.98
	29. Qinghai	0.33	0.36	0.92	0.52	0.63
	30. Ningxia	0.41	0.49	0.84	0.63	0.65
	31. Xinjiang	1.30	1.41	0.92	1.76	0.74
	Total	5.53	5.76	0.96	6.92	0.80

Table 8.75 The GDPs, Total Investments in Fixed Properties, and Employment Situations of the East Administrative Area of China (2012)

Area	*Province/Municipality/ Autonomous Region (i)*	*Regional GDP (0.1 Billion Yuan)*	*Value of Fixed Properties (0.1 Billion Yuan)*	*Employments (10,000)*
East	9. Shanghai	20181.7	7674.8	1781.6
	10. Jiangsu	54058.2	27258.1	5014.9
	11. Zhejiang	34665.3	15460.7	3838.8
	12. Anhui	17212.1	8855.8	1745.7
	13. Fujian	19701.8	11304.8	1927.8
	14. Jiangxi	12948.9	6513.7	1531.6
	15. Shandong	50013.2	27551.5	3564.0
	Total	208781.2	104619.4	19404.4

Source: Data are from *China Statistics Abstracts 2013* (XLS, DOC).

Table 8.76 The Regional Economic Structures and Comparative Gain Measures of the East Administrative Area of China (2012)

Area	*Province/ Municipality/ Autonomous Region (i)*	*Demand Structure $d_i(t)$ (%)*	*Employment Structure $s_i^1(t)$ (%)*	*Comparative Gain of Labor $G_i^1(t)$*	*Investment Structure $s_i^2(t)$ (%)*	*Comparative Gain of Capital $G_i^2(t)$*
East	9. Shanghai	9.67	9.18	1.05	7.34	1.32
	10. Jiangsu	25.89	25.84	1.00	26.05	0.99
	11. Zhejiang	16.60	17.78	0.93	14.78	1.12
	12. Anhui	8.24	9.00	0.92	8.46	0.97
	13. Fujian	9.44	9.93	0.95	10.81	0.87
	14. Jiangxi	6.20	7.89	0.79	6.23	1.00
	15. Shandong	23.95	18.37	1.30	26.33	0.91
	Total	100	100	—	100	—

Table 8.77 The GDPs, Total Investments in Fixed Properties, and Employment Situations in the South Administrative Area of China (2012)

Area	*Province/Municipality/ Autonomous Region (i)*	*Regional GDP (0.1 Billion Yuan)*	*Value of Fixed Properties (0.1 Billion Yuan)*	*Employments (10,000)*
South	16. Henan	29599.3	22060.0	2509.0
	17. Hubei	22250.5	12554.7	2380.0
	18. Hunan	22154.2	12488.8	2019.9
	19. Guangdong	57067.9	22871.9	5089.0
	20. Guangxi	12948.9	11068.5	1229.4
	21. Hainan	2855.5	2009.9	345.9
	Total	146962.5	83053.8	13573.2

Source: Data are from *China Statistics Abstracts 2013* (XLS, DOC).

Table 8.78 The Regional Economic Structures and Comparative Gain Measures of the South Administrative Area of China (2012)

Area	*Province/ Municipality/ Autonomous Region (i)*	*Demand Structure $d_i(t)$ (%)*	*Employment Structure $s_i^1(t)$ (%)*	*Comparative Gain of Labor $G_i^1(t)$*	*Investment Structure $s_i^2(t)$ (%)*	*Comparative Gain of Capital $G_i^2(t)$*
South	16. Henan	20.14	18.48	1.09	25.56	0.79
	17. Hubei	15.14	17.53	0.86	15.12	1.00
	18. Hunan	15.07	14.88	1.01	15.04	1.00
	19. Guangdong	38.83	37.49	1.04	27.54	1.41
	20. Guangxi	8.87	9.06	0.98	13.33	0.67
	21. Hainan	1.94	2.55	0.76	2.42	0.80
	Total	100	100	—	100	—

Table 8.79 The GDPs, Total Investments in Fixed Properties, and Employment Situations of the North Administrative Area of China (2012)

Area	*Province/Municipality/ Autonomous Region (i)*	*Regional GDP (0.1 Billion Yuan)*	*Value of Fixed Properties (0.1 Billion Yuan)*	*Employments (10,000 People)*
North	1. Beijing	17879.4	7409.6	1795.8
	2. Tianjin	12893.9	9898.4	606.8
	3. Hebei	26575.0	15244.6	1792.3
	4. Shanxi	12112.8	8223.9	1152.0
	5. Inner Mongolia	15880.6	13442.1	1034.5
	Total	85341.7	54168.6	6381.4

Source: Data are from *China Statistics Abstracts 2013* (XLS, DOC).

Table 8.80 The Regional Economic Structures and Comparative Gain Measures of the North Administrative Area of China (2012)

Area	*Province/ Municipality/ Autonomous Region (i)*	*Demand Structure $d_i(t)$ (%)*	*Employment Structure $s_i(t)$ (%)*	*Comparative Gain of Labor $G_i^1(t)$*	*Investment Structure $s_i^2(t)$ (%)*	*Comparative Gain of Capital $G_i^2(t)$*
North	1. Beijing	20.95	28.14	0.74	13.68	1.53
	2. Tianjin	15.11	9.51	1.59	18.18	0.83
	3. Hebei	30.14	28.09	1.07	28.14	1.07
	4. Shanxi	14.19	18.05	0.77	15.18	0.93
	5. Inner Mongolia	18.61	16.21	1.15	24.82	0.75
	Total	100	100	—	100	—

Table 8.81 The GDPs, Total Investments in Fixed Properties, and Employment Situations of the Northeast Administrative Area of China (2012)

Area	*Province/Municipality/ Autonomous Region (i)*	*Regional GDP (0.1 Billion Yuan)*	*Value of Fixed Properties (0.1 Billion Yuan)*	*Employments (10,000 People)*
Northeast	6. Liaoning	24846.4	15492.1	2218.1
	7. Jilin	11939.2	9136.2	1121.3
	8. Heilongjiang	13619.6	8143.7	1487.8
	Total	50405.2	32772.0	4827.2

Source: Data are from *China Statistics Abstracts 2013* (XLS, DOC).

Table 8.82 The Regional Economic Structures and Comparative Gain Measures of the Northeast Administrative Area of China (2012)

Area	*Province/ Municipality/ Autonomous Region (i)*	*Demand Structure $d_i(t)$ (%)*	*Employment Structure $s_i^1(t)$ (%)*	*Comparative Gain of Labor $G_i^1(t)$*	*Investment Structure $s_i^2(t)$ (%)*	*Comparative Gain of Capital $G_i^2(t)$*
North-east	6. Liaoning	49.29	45.95	1.07	47.27	1.04
	7. Jilin	23.69	23.23	1.02	27.88	0.85
	8. Heilongjiang	27.02	30.82	0.88	24.85	1.09
	Total	100	100	—	100	—

Table 8.83 The GDPs, Total Investments in Fixed Properties, and Employment Situations of the Northwest Administrative Area of China (2012)

Area	*Province/Municipality/ Autonomous Region (i)*	*Regional GDP (0.1 Billion Yuan)*	*Value of Fixed Properties (0.1 Billion Yuan)*	*Employments (10,000 People)*
North-west	27. Shaanxi	14453.7	9915.2	1211.9
	28. Gansu	5650.2	3298.1	657.0
	29. Qinghai	1893.5	1719.3	194.9
	30. Ningxia	2341.3	2086.9	262.9
	31. Xinjiang	7505.3	5792.2	753.6
	Total	31844.0	22811.7	3080.3

Source: Data are from *China Statistics Abstracts 2013* (XLS, DOC).

Table 8.84 The Regional Economic Structures and Comparative Gain Measures of the Northwest Administrative Area of China (2012)

Area	*Province/ Municipality/ Autonomous Region (i)*	*Demand Structure $d_i(t)$ (%)*	*Employment Structure $s_i^1(t)$ (%)*	*Comparative Gain of Labor $G_i^1(t)$*	*Investment Structure $s_i^2(t)$ (%)*	*Comparative Gain of Capital $G_i^2(t)$*
North-west	27. Shaanxi	45.39	39.34	1.15	43.47	1.04
	28. Gansu	17.74	21.33	0.83	14.46	1.23
	29. Qinghai	5.95	6.33	0.94	7.54	0.79
	30. Ningxia	7.35	8.53	0.86	9.15	0.80
	31. Xinjiang	23.57	24.47	0.96	25.39	0.93
	Total	100	100	—	100	—

Table 8.85 The GDPs, Total Investments in Fixed Properties, and Employment Situations of the Southwest Administrative Area of China (2012)

Area	*Province/Municipality/ Autonomous Region (i)*	*Regional GDP (0.1 Billion Yuan)*	*Value of Fixed Properties (0.1 Billion Yuan)*	*Employments (10,000)*
Southwest	22. Chongqing	11409.6	6341.4	1437.1
	23. Sichuan	23872.8	12496.0	2323.5
	24. Guizhou	6852.2	4164.4	800.0
	25. Yunnan	10309.5	8576.4	1492.5
	26. Tibet	701.0	708.7	130.2
	Total	53145.1	32286.9	6183.3

Source: Data are from *China Statistics Abstracts 2013* (XLS, DOC).

Table 8.86 The Regional Economic Structures and Comparative Gain Measures of the Southwest Administrative Area of China (2012)

Area	*Province/ Municipality/ Autonomous Region (i)*	*Demand Structure $d_i(t)$ (%)*	*Employ-ment Structure $s_i^1(t)$ (%)*	*Compa-rative Gain of Labor $G_i^1(t)$*	*Investment Structure $s_i^2(t)$ (%)*	*Comparative Gain of Capital $G_i^2(t)$*
South-west	22. Chongqing	21.47	23.24	0.92	19.64	1.09
	23. Sichuan	44.92	37.58	1.20	38.70	1.16
	24. Guizhou	12.89	12.94	1.00	12.90	1.00
	25. Yunnan	19.40	24.14	0.80	26.56	0.73
	26. Tibet	1.32	2.10	0.63	2.20	0.60
	Total	100	100	—	100	—

8.6 The Three Special Economic Areas

Table 8.87 The GDPs, Total Investments in Fixed Properties, and Employment Situations of the Three Special Economic Areas of China (2012)

Area	*Province/Municipality/ Autonomous Region (i)*	*Regional GDP (0.1 Billion Yuan)*	*Value of Fixed Properties (0.1 Billion Yuan)*	*Employments (10,000)*
Bohai Economic Rim	1. Liaoning	24846.4	15492.1	2218.1
	2. Hebei	26575.0	15244.6	1792.3
	3. Beijing	17879.4	7409.6	1795.8
	4. Tianjin	12893.9	9848.4	606.8
	5. Shandong	50013.2	27551.5	3564.0
	6. Shanxi	12112.8	8223.9	1152.0
	7. Inner Mongolia	15880.6	13442.1	1034.5
	Total	160102.3	97212.2	12163.5
Yangtze River Delta Economic Zone	8. Jiangsu	54058.2	27258.1	5014.9
	9. Shanghai	20181.7	7674.8	1781.6
	10. Zhejiang	34665.3	15460.7	3838.8
	Total	108905.2	50393.6	10635.3
Greater Pearl River Delta Region	11. Fujian	19701.8		1927.8
	12. Guangdong	57067.9	11304.8	5089.0
	13. Guangxi	13035.1	22871.9	1229.4
	14. Guizhou	6852.2	11068.5	800.0

continued

Table 8.87 (continued) The GDPs, Total Investments in Fixed Properties, and Employment Situations of the Three Special Economic Areas of China (2012)

Area	*Province/Municipality/ Autonomous Region (i)*	*Regional GDP (0.1 Billion Yuan)*	*Value of Fixed Properties (0.1 Billion Yuan)*	*Employments (10,000)*
	15. Hainan	2855.5	4164.4	345.9
	16. Hunan	22154.2	2009.9	2019.9
	17. Jiangxi	12948.9	12488.8	1531.6
	18. Sichuan	23872.8	6513.7	2323.5
	19. Yunnan	10309.5	12496.0	1492.5
	20. Macao		8576.4	
	21. Hong Kong			
	Total	168797.9	91493.5#	16759.6
National	Grand total	437904.4	239099.3	39558.4

Source: Data are from *China Statistics Abstracts 2013* (XLS, DOC).

Table 8.88 The Regional Economic Structures and Comparative Gain Measures of the Three Special Economic Areas of China (2012)

Area	*Province/ Municipality/ Autonomous Region (i)*	*Demand Structure $d_i(t)$ (%)*	*Employment Structure $s_i^1(t)$ (%)*	*Comparative Gain of Labor $G_i^1(t)$*	*Investment Structure $s_i^2(t)$ (%)*	*Comparative Gain of Capital $G_i^2(t)$*
Bohai Economic Rim	1. Liaoning	5.67	5.61	1.01	6.48	0.88
	2. Hebei	6.07	4.53	1.34	6.38	0.95
	3. Beijing	4.08	4.54	0.90	3.10	1.32
	4. Tianjin	2.94	1.53	1.92	4.12	0.71
	5. Shandong	11.42	9.01	1.27	11.52	0.99
	6. Shanxi	2.77	2.91	0.95	3.44	0.81
	7. Inner Mongolia	3.63	2.83	1.28	5.62	0.65
	Total	36.58*	30.76*	1.19*	40.66*	0.90**
Yangtze River Delta Economic Zone	8. Jiangsu	12.34	12.68	0.97	11.40	1.08
	9. Shanghai	4.61	4.50	1.02	3.21	1.44
	10. Zhejiang	7.92	9.70	0.82	6.47	1.22
	Total	24.87	26.88	0.93*	21.08	1.18**

Table 8.88 (continued) The Regional Economic Structures and Comparative Gain Measures of the Three Special Economic Areas of China (2012)

Area	*Province/ Municipality/ Autonomous Region (i)*	*Demand Structure* $d_i(t)$ *(%)*	*Employment Structure* $s_i^1(t)$ *(%)*	*Comparative Gain of Labor* $G_i^1(t)$	*Investment Structure* $s_i^2(t)$ *(%)*	*Comparative Gain of Capital* $G_i^2(t)$
Greater Pearl River Delta Region	11. Fujian	4.50	4.627	0.92	4.73	0.95
	12. Guangdong	13.03	12.449	1.01	9.57	1.36
	13. Guangxi	2.98	3.120	0.96	4.63	0.64
	14. Guizhou	1.56	1.654	0.77	1.74	0.90
	15. Hainan	0.65	0.803	0.75	0.84	0.77
	16. Hunan	5.06	5.055	0.99	5.22	0.97
	17. Jiangxi	2.96	4.068	0.76	2.72	1.09
	18. Sichuan	5.45	5.546	0.93	5.23	1.04
	19. Yunnan	2.35	3.673	0.62	3.59	0.65
	20. Macao					
	21. Hong Kong					
	Total	38.55	41.827	0.91*	38.26	1.01**
National	Grand total	1.000	1.000	—	1.000	—

Note: The data values labeled with * or ** are not the simple sums of the relevant data. Instead, they represent the regional means, where * stands for the means of the comparative gains of labor of the relevant regions, and ** those of comparative gains of capital.

Table 8.89 The GDPs, Total Investments in Fixed Properties, and Employment Situations of Yangtze River Delta Economic Zone of China (2012)

Area	*Province/Municipality/ Autonomous Region (i)*	*Regional GDP (0.1 Billion Yuan)*	*Value of Fixed Properties (0.1 Billion Yuan)*	*Employments (10,000 People)*
Yangtze River Delta Economic Zone	8. Jiangsu	54058.2	27258.1	5014.9
	9. Shanghai	20181.7	7674.8	1781.6
	10. Zhejiang	34665.3	15460.7	3838.8
	Total	108905.2	50393.6	10635.3

Source: Data is from *China Statistics Abstracts 2013* (XLS, DOC).

Table 8.90 The Regional Economic Structure and Measures of Comparative Gains of the Yangtze River Delta Region of China (2012)

Area	Province/ Municipality/ Autonomous Region (i)	Demand Structure $d_i(t)$ (%)	Employment Structure $s_i^1(t)$ (%)	Comparative Gain of Labor $G_i^1(t)$	Investment Structure $s_i^2(t)$ (%)	Comparative Gain of Capital $G_i^2(t)$
Yangtze River Delta Region	Jiangsu	49.64	47.15	1.05	54.09	0.98
	Shanghai	18.53	16.75	1.13	15.23	1.22
	Zhejiang	31.83	36.10	0.88	30.68	1.04
	Total	100	100	—	100	—

Table 8.91 The GDPs, Total Investments in Fixed Properties, and Employment Situations of Bohai Economic Rim of China (2012)

Area	Province/Municipality/ Autonomous Region (i)	Regional GDP (0.1 Billion Yuan)	Value of Fixed Properties (0.1 Billion Yuan)	Employments (10,000 People)
Bohai Economic Rim	1. Liaoning	24846.4	15492.1	2218.1
	2. Hebei	26575.0	15244.6	1792.3
	3. Beijing	17879.4	7409.6	1795.8
	4. Tianjin	12893.9	9848.4	606.8
	5. Shandong	50013.2	27551.5	3564.0
	6. Shanxi	12112.8	8223.9	1152.0
	7. Inner Mongolia	15880.6	13442.1	1034.5
	Total	160102.3	97212.2	12163.5

Source: Data are from *China Statistics Abstracts 2013* (XLS, DOC).

Table 8.92 The Regional Economic Structure and Measures of Comparative Gains of the Bohai Economic Rim of China (2012)

Area	*Province/ Municipality/ Autonomous Region (i)*	*Demand Structure $d_i(t)$ (%)*	*Employment Structure $s_i^1(t)$ (%)*	*Comparative Gain of Labor $G_i^1(t)$*	*Investment Structure $s_i^2(t)$ (%)*	*Comparative Gain of Capital $G_i^2(t)$*
Bohai Economic Rim	Liaoning	15.51	18.24	0.85	15.94	0.97
	Hebei	16.59	14.74	1.13	15.68	1.06
	Beijing	11.16	14.76	0.76	7.62	1.46
	Tianjin	8.05	4.99	1.61	10.13	0.79
	Shandong	31.22	29.30	1.07	28.34	1.10
	Shanxi	7.56	9.47	0.80	8.46	0.89
	Inner Mongolia	9.91	8.50	1.17	13.83	0.72
	Total	100	100	—	100	—

Table 8.93 The GDPs, Total Investments in Fixed Properties, and Employment Situations of the Greater Pearl River Delta Region of China (2012)

Area	*Province/Municipality/ Autonomous Region (i)*	*Regional GDP (0.1 Billion Yuan)*	*Value of Fixed Properties (0.1 Billion Yuan)*	*Employments (10,000)*
Greater Pearl River Delta Region	11. Fujian	19701.8	11304.8	1927.8
	12. Guangdong	57067.9	22871.9	5089.0
	13. Guangxi	13035.1	11068.5	1229.4
	14. Guizhou	6852.2	4164.4	800.0
	15. Hainan	2855.5	2009.9	345.9
	16. Hunan	22154.2	12488.8	2019.9
	17. Jiangxi	12948.9	6513.7	1531.6
	18. Sichuan	23872.8	12496.0	2323.5
	19. Yunnan	10309.5	8576.4	1492.5
	20. Macao			
	21. Hong Kong			
	Total	168797.9	91493.5#	16759.6#

Source: Data are from *China Statistics Abstracts 2013* (XLS, DOC).

Table 8.94 The Regional Economic Structure and Measures of Comparative Gains of the Greater Pearl River Delta Region of China (2012)

Area	*Province/ Municipality/ Autonomous Region (i)*	*Demand Structure $d_i(t)$ (%)*	*Employment Structure $s_i^1(t)$ (%)*	*Comparative Gain of Labor $G_i^1(t)$*	*Investment Structure $s_i^2(t)$ (%)*	*Comparative Gain of Capital $G_i^2(t)$*
Greater Pearl River Delta Region	Fujian	11.67	11.50	1.01	12.36	0.94
	Guangdong	33.81	30.36	1.11	25.00	1.35
	Guangxi	7.72	7.34	1.05	12.10	0.64
	Guizhou	4.06	4.77	0.85	4.55	0.89
	Hainan	1.69	2.06	0.82	2.20	0.77
	Hunan	13.12	12.05	1.09	13.65	0.96
	Jiangxi	7.67	9.14	0.84	7.12	1.08
	Sichuan	14.14	13.86	1.02	13.66	1.04
	Yunnan	6.11	8.91	0.69	9.37	0.65
	Macao					
	Hong Kong					
	Total	100	100	—	100	—

Afterword

At the end of this book, we like to remind the reader that because of various reasons, the conclusions contained in this book are provided for reference only.

- Although the theory and methods employed in this book are scientifically sound, due to missing data, reliability of data, and compatibility of data, some of the conclusions may not be as reliable as we like them to be; and
- This book investigates the investment and employment problems from the angle of pure economics, where problems are considered from the angle of how the finite resources can be optimized, without taking into account the other factors, such as politics, cultures, rooms of business profits, and others. Therefore, when making decisions related to investment and employment, what is concluded in this book should only be employed as a framework, beyond which other relevant factors should also be considered collectively and jointly.

References

Allen, C. 2004. *Duel in the Snows: The True Story of the Younghusband Mission to Lhasa.* London: John Murray.

Beckwith, C. I. 2009. *Empires of the Silk Road: A History of Central Eurasia from the Bronze Age to the Present.* New Jersey: Princeton University Press.

Belov, E.A. 1996. Anti-Chinese rebellion led by Babujav in Inner Mongolia, 1915–1916. *Annaly (Moscow)*, no. 2.

Bertalanffy, L. von. 1924. *Einführung in Spengler's Werk.* Literaturblatt Kolnische Zeitung, May.

Bertalanffy, L. von. 1934. *Modern Theories of Development* (Translation by J. H. Woodge). Oxford: Oxford University Press.

Betta, C. 2004. *The Other Middle Kingdom: A Brief History of Muslims in China.* Indianapolis, IN: Indianapolis University Press.

BMC Evolutionary Biology. 2011. Tracing the legacy of the early Hainan Islanders—A perspective from mitochondrial DNA. *BMC Evolutionary Biology*, February 15. Retrieved February 18, 2011.

Boland-Crewe, T. and Lea, D. 2004. *The Territories of the People's Republic of China.* East Sussex, UK: Psychology Press.

Bonavia, D. 1995. *China's Warlords.* New York: Oxford University Press.

Booz, P. Z. 1998. *Yunnan.* Odyssey Passport. McGraw-Hill Contemporary.

Borjigin, H. 2004. The history and political character of the name of "Nei Menggu" (Inner Mongolia). *Inner Asia*, vol. 6, pp. 61–80.

China Network. 2010. Reports on the Domestic Economy and Social Development of Fujian Province 2009.

Chinadaily.com 2010. Hainan officials rule out bubble burst. Chinadaily.com.cn. 2010-05-10. Retrieved September 8, 2010.

Chinadaily.com. (2011–2012). http://www.Chinadaily.com.cn/business/2011-02/14/content_12004550.htm

Chinadaily.com. 2012. Retrieved April 27, 2012.

Chinadaily.com 2013. http://www.Chinadaily.com.cn/cndy/2013-06/25/content_16654150.htm., accessed on January 13, 2014.

Chinagate 2007. Geographic location, administrative district, and the population of Tianjin. http://cn.chinagate.cn/ECONOMIC/2007-11/30/content_9328447.htm November 30, 2007, retrieved on July 17, 2012.

Chinaperspective.com 2011. Hainan Province: Economic News and Statistics for Hainan's Economy. Thechinaperspective.com. Retrieved December 11, 2011.

China Statistics Yearbook 2012, edited by National Bureau of Statistics of China, published by China Statistics Press, Beijing.

Close, U. and McCormick, D. 1922. Where the mountains walked. *National Geographic Magazine*, vol. 41, no. 5, pp. 445–464.

Cooke, S. 2008. Surviving state and society in northwest China: The Hui experience in Qinghai Province under the PRC. *Journal of Muslim Minority Affairs*, vol. 28, no. 3, pp. 401–420.

Cotterell, A. 2007. *The Imperial Capitals of China: An Inside View of the Celestial Empire.* London: Pimlico.

Dai, Y. C. 2009. *The Sichuan Frontier and Tibet: Imperial Strategy in the Early Qing*. Seattle, WA: University of Washington Press.

Danielson, E. N. 2010. *Discover Shanghai*. Singapore: Marshall Cavendish.

Danielson, E. N. 2004. *Shanghai and the Yangzi Delta*. Singapore: Marshall Cavendish/Times Editions.

Dowman, K. 1988. *The Power-Places of Central Tibet: The Pilgrim's Guide*. London: Routledge and Kegan Paul.

Edmonds, R. L. 1985. *Northern Frontiers of Qing China and Tokugawa Japan: A Comparative Study of Frontier Policy*. University of Chicago, Department of Geography; Research Paper No. 213.

Edmund, R. 1837. *Embassy to the Eastern Courts of Cochin-China, Siam, and Muscat*. New York: Harper and Brothers. p. 123.

Einstein, A. 1987. *The Collected Papers of Albert Einstein*. Princeton, NJ: Princeton University Press.

Elvin, M. 1977. Market towns and waterways: The county of Shanghai from 1480 to 1910. In: G. William Skinner (editor), *The City in Late Imperial China*. Stanford: Stanford University Press.

Engels, F. 1878. *Herrn Eugen Dührings Umwälzung der Wissenschaft* (commonly known as *Anti-Dühring*). Originally published in German in 1878. Also, see: Herr Eugen Dühring's revolution in science (anti-Dühring), Co-operative Pub. Society of Foreign Workers in the U.S.S.R 1934.

English, J. and Feng, G. F. 1972. *Tao De Ching*. New York: Vintage Books.

Erlang, A. K. 1909. The theory of probabilities and telephone conversations. *Nyt. Tidsskrift for Matematik B*, vol. 20, pp. 33–40.

Feigon, L. 1998. *Demystifying Tibet: Unlocking the Secrets of the Land of the Snows*. Chicago: Ivan R. Dee.

Findley, C. V. 2005. *The Turks in World History*. Oxford, England: Oxford University Press.

Forbes, A. and Henley, D. 2011. *Traders of the Golden Triangle*. Chiang Mai: Cognoscenti Books.

Fordham.edu. 2014. http://www.fordham.edu/halsall/source/polo-kinsay.html, accessed on January 11, 2014.

Fujian Economic News 2014. http://www.thechinaperspective.com/topics/province/fujian-province/The China Perspective|Fujian Economic News and Data, accessed on January 12, 2014.

Fukien. 2008. In *Encyclopædia Britannica*. Retrieved December 20, 2008, from Encyclopædia Britannica Online: http://www.britannica.com/EBchecked/topic/221639/Fujian

Geography.about.com. 2013. Largest Cities through History. Retrieved 2013-09-10.

Gernet, J. 1996. *A History of Chinese Civilization*. New York: Cambridge University Press.

Goldstein, M. C. 1994. Change, conflict and continuity among a community of nomadic pastoralists—A case study from Western Tibet, 1950–1990. In: R. Barnett and S. Akiner (editors), *Resistance and Reform in Tibet*. London: Hurst and Co.

Goldstein, M. C. 2007. *A History of Modern Tibet: Volume 2: The Calm before the Storm, 1951–1955*. Berkeley, CA: University of California Press.

Goodman, D. 2004. *China's Campaign to Open Up the West: National, Provincial, and Local Perspectives*. Cambridge, England: Cambridge University Press.

Gu, M. 2001. Some strategic thoughts on improving the economic status of intellectuals of China. *Treasury of Modern Chinese Thoughts*, no. 6, pp. 176–192.

Gu, M. and Zhang, X. F. 2000. Scientific definition of the concept of intellectual: The new stage and a primary research on a model of its value estimation. *Advances in Systems Science and Applications*, vol. 2, no. 1, pp. 229–233.

Guthrie, W. K. 1979. *A History of Greek Philosophy—The Presocratic Tradition from Parmenides to Democritus*. Cambridge, England: Cambridge University Press.

Harper, D. 2007. *Beijing: City Guide* (7th Edition). Oakland, California: Lonely Planet Publications.

Hendrix, H. 2001. *Getting the Love You Want: A Guide for Couples*. New York: Owl Books.

Hill, J. E. 2009. *Through the Jade Gate to Rome: A Study of the Silk Routes during the Later Han Dynasty, 1st to 2nd Centuries CE*. South Carolina: BookSurge, Charleston.

Horesh, N. 2009. *Shanghai's Bund and Beyond*. New Haven: Yale University Press.

Hutchings, G. 2003. Guizhou Province. *Modern China: A Guide to a Century of Change*. Cambridge, MA: Harvard University Press.

Interscience.wiley. 2013. Regional Inequality in China: A Case Study of Zhejiang Province-Wei-2004-Tijd schriftvooreconomischeensocialegeografie-Wiley Online Library. interscience.wiley.com. 2004-02-16. Retrieved 2013-09-10.

Jin, D., Ma, X. L., and Bender, M. 2006. *Butterfly Mother: Miao (Hmong) Creation Epics from Guizhou, China.* Cambridge, MA: Hackett Publishing.

Johnson, L. C. 1995. *Shanghai: From Market Town to Treaty Port.* Stanford, CA: Stanford University Press.

Kapp, R. A. 1974. Chungking as a center of Warlord power, 1926–1937. In: Mark Elvin and G. William Skinner (editors), *The Chinese City between Two Worlds.* Stanford, CA: Stanford University Press, pp. 143–170.

Kim, H. D. 2004. *Holy War in China: The Muslim Rebellion and State in Chinese Central Asia, 1864–1877.* Stanford, CA: Stanford Uiversity Press.

Klir, G. 1985. *Architecture of Systems Problem Solving.* New York, NY: Plenum Press.

Lao Tzu (unknown). 1990. Tao Te Ching: The classic book of integrity and the way. In: Translated, annotated, and with an Afterword by V. H. Mair, *An Entirely New Translation Based on the Recently Discovered Ma-Wang-Tui Manuscript.* New York: Bantam Books.

Lao Tzu (time unknown) 1997. *Tao Te Ching, 25th-Anniversary Edition.* Translated by Jane English and Gia-Fu Feng. New York: Vintage.

Li, L. L., Dray-Novey, A., and Kong, H. L. 2007. Beijing: From Imperial Capital to Olympic City. New York: Palgrave Macmillan.

Li, Y. 1991. *Imbalanced Chinese Economy* (1st Edition). Escanaba, MI: The Economic Daily Press, April.

Liaoning Travel Guide. 2010. *Liaoning Travel Guide: Map, History, Sightseeing, Ethnic Minority, Climate,* accessed on March 14, 2010.

Lin, Y. 1999. *General Systems Theory: A Mathematical Approach.* New York: Kluwer and Plenum Publishers.

Lin, Y. 2007. Systemic yoyo model and applications in Newton's, Kepler's laws, etc. *Kybernetes: The International Journal of Cybernetics, Systems and Management Sciences,* vol. 36, nos. 3/4, pp. 484–516.

Lin, Y. 2008. *Systemic Yoyos: Impacts of the Second Dimension. Auerbach Publications.* New York: CRC Press, an imprint of Taylor & Francis.

Lin, Y., Duan, X. J., Zhao, C. L., and Xu, L. D. 2012. *Systems Science: Methodological Approaches.* New York: CRC Press, an imprint of Taylor & Francis.

Lin, Y. and OuYang, S. C. 2010. *Irregularities and Prediction of Major Disasters.* New York: CRC Press, an imprint of Taylor & Francis.

Liu, D. Z. (editor) 1994. *Four Classes of Chinese Medicine in Modern Language.* Tianjin: Tianjin Translation Press of Science and Technology.

Liu, L. H. 1995. *Translingual Practice: Literature, National Culture, and Translated Modernity—China, 1900–1937* (illustrated, annotated ed.). Redwood City, CA: Stanford University Press.

MacKerras, C. and Yorke, A. 1991. *The Cambridge Handbook of Contemporary China.* Cambridge, England: Cambridge University Press.

MacKie, N. 2005. China's west seeks to impress investors. *BBC News,* May 4, 2005, accessed on April 28, 2010. http://news.bbc.co.uk/2/hi/business/4512015.stm

Marks, T. A. 1978. Nanchao and Tibet in South-western China and central Asia. *The Tibet Journal,* vol. 3, no. 4, pp. 13–16.

Marx, K. and Engels, F. 1979. *Selected Works of Marx and Engels* (vol. 6). Seattle, WA: The People's Publishing House, p. 492.

Maygew, B., Miller, K., and English, A. 2002. *Facts about South-West China—History. South-West China* (2 ed.). Franklin, TN: Lonely Planet.

McCarthy, S. K. 2009. *Communist Multiculturalism: Ethnic Revival in Southwest China.* Seattle, WA: University of Washington Press.

McIsaac, L. 2000. The city as nation: Creating a Wartime Capital in Chongqing. In: Joseph W. Esherick (editor), *Remaking the Chinese City, 1900–1950.* Honolulu: University of Hawaii Press.

Millward, J. A. 1998. *Beyond the Pass: Economy, Ethnicity, and Empire in Qing Central Asia, 1759–1864.* Stanford, CA: Stanford University Press.

Millward, J. A. 2007. *Eurasian Crossroads: A History of Xinjiang.* New York: Columbia University Press.

Ministry of Commerce. 2013. Doing Business in China—Survey. Ministry of Commerce—People's Republic of China, accessed on August 5, 2013.

Moore, M. 2011. China to build world's biggest airport. *The Daily Telegraph,* September 9, 2011.

Mote, F. W. 1999. *Imperial China: 900–1800.* Cambridge, MA: Harvard University Press.

National Bureau of Statistics of China. 2013. Communiqué of the National Bureau of Statistics of People's Republic of China on Major Figures of the 2010 Population Census (No. 2). 29 April 2011, accessed August 4, 2013. http://www.stats.gov.en/english/StatisticalCommuniqu/

Osnos, E. 2011. Faust, China, and nuclear power. *New Yorker*, accessed on October 12, 2011. http://www.newyorker.com/news/evan-osnos/faust-china-and-nuclearpower

Parsons, J. B. 1957. The culmination of a Chinese peasant rebellion: Chang Hsien-chung in Szechwan, 1644–1646. *The Journal of Asian Studies*, vol. 16, no. 3, pp. 387–400.

Peng, R. S. 2012. Movement of rural labors and urbanization. *Research of Rural Countryside*, vol. 11, no. 7, pp. 2010–2013.

People's Daily Online. 2012. Origins of and changes to the names of China's provinces and autonomous regions (in simplified Chinese). *The People's Daily Online*, accessed on January 26, 2012.

People.com. 2013. *People's Daily Online*—East China province leads the way in per capita GDP. English.people.com.cn, accessed on September 10, 2013. http://english.people.com.cn/200601/30/eng20060130_239357.html

Peopledaily.com 2013. Country Profile: Jiangsu province—*People's Daily Online*. English.peopledaily.com.cn, accessed on March 26, 2013.

Pomeranz, K. 2010. *Musings on a Museum: A Trip to Xibaipo. http://www.thechinabeat.org/?p=2384*, accessed on January 11, 2014.

Qian, C. 1995. *General Yue Fei*. Translator Honorable Sir T. L. Yang. Hong Kong: Joint Publishing (H.K.) Co., Ltd.

Qian, X. S. 1981. Systems science, cognitive science, and human body science. *Journal of Natur*e, vol. 1, pp. 9.

Qian, X. S. 2007. *Engineering Control Theory* (New Century Edition). Shanghai: Shanghai Jiaotong University Press.

Raa, T. T. 2009. *The Economics of Input–Output Analysis*. Cambridge, England: Cambridge University Press.

Roberts, E. 1837. *Embassy to the Eastern Courts of Cochin—China, Siam, and Muscat*. New York: Harper & Brothers.

Sage, S. F. 2006. *Ancient Sichuan and the Unification of China*. Albany, NY: State University of New York Press.

Schram, L. M. 2006. *The Monguors of the Kansu-Tibetan Frontier: Their Origin, History, and Social Organization*. Whitefish, MT: Kessinger Publishing.

Seetharaman, D. 2013. Mazda in Chongqing. Reuters.com, accessed on December 10, 2013.

Shih, C. Y. and Shi, A. Y. 2002. *Negotiating Ethnicity in China: Citizenship as a Response to the State*. Psychology Press, accessed June 28, 2010.

Smith, W. W. 2009. *China's Tibet?: Autonomy or Assimilation*. Lanham, MD: Rowman and Littlefield.

Svanberg, I. 1988. *The Altaic-Speakers of China: Numbers and Distribution*. Multiethnic Series, Center for Multiethnic Research, Uppsala University, Uppsala, Sweden.

Tao, L. X. 1996. Structural stability and normative of national economic operation system. *Journal of Quantitative and Technical Economics*, vol. 8, pp. 47–50.

Tao, L. X. 2001a. Imbalance and optimization analysis of China's regional economic structures of the east, center, and west. *Finance of Modern Time*, no. 12, pp. 45–51.

Tao, L. X. 2001b. A new look at the imbalance and optimization of China's sectorial structure. In: Shi, Z. L. and Wu, J. J. (editors), *Sectorial Development and Business Reforms in China*. Beijing: Press of Economic Management, pp. 65–73.

Tao, L. X. 2001c. On fuzzy comprehensive evaluation models of talents. *Social Science of Jiangxi*, vol. 23, no. 4, pp. 153–156.

Tao, L. X., Guo, Y. F., and Du, W. W. 2011. Criteria for omnibearing imbalance of macroeconomic system structures and their strategic optimization. *Kybernetes: The International Journal of Cybernetics, Systems and Management Sciences*, vol. 40, no. 5, pp. 848–853.

Thechinaperspective.com 2012a. Guangdong Province: Economic News and Statistics for Guangdong's Economy. Retrieved 25 April 2012.

Thechinaperspective.com 2012b. Jiangxi Province: Economic News and Statistics for Jiangxi's Economy, accessed on December 24, 2012.

Thechinaperspective.com 2013. Anhui Province: Economic News and Statistics for Anhui's Economy, accessed on April 10, 2013.

Thurgood, G. 2006. Sociolinguistics and contact-induced language change: Hainan Cham, Anong, and Phan Rang Cham. *Tenth International Conference on Austronesian Linguistics*, January 17–20, 2006, Palawan, Philippines. Linguistic Society of the Philippines and SIL International.

Tian, L., Li, J., Zhang, K., and Guest, P. 1998. Women's status, institutional barriers and reproductive health care: A case study in Yunnan, China. *Health Policy*, vol. 84, no. 2, pp. 284–297.

Tianjinplus.com 2012. Tianjin Plus. Tianjinplus.com, retrieved on December 11, 2012.

Uvista.com 2013. Suzhou -Venice in the Orient—China Travel Guide. Uvista.com, accessed on March 26, 2013.

Wigner, E. P. 1969. *Physical Science and Human Values*. Santa Barbara, CA: Greenwood Press.

Wilhalm, R. and Baynes, C. 1967. *The I Ching or Book of Changes* (3rd Edition). Princeton, NJ: Princeton University Press.

Wu, J. H. 1994. Investment in education and economic growth in Japan after WWII. *Journal of Southwest Normal University (Philosophy and Social Science Edition)*, vol. 26, no. 3, pp. 321–345.

Wu, Y. and Lin, Y. 2002. *Beyond Nonstructural Quantitative Analysis: Blown-Ups, Spinning Currents and the Modern Science*. New Jersey: World Scientific, River Edge.

Wu, X. M. 1996. *Pansystems: A Philosophy that is unfit to the God's Model.* Wuhan: Wuhan Publishing House, p. 6.

Xiaolu, Z., Jianxin, X., Tongnian, Z., and Wu, X. M. 2003. Pansystems rediscovery: 784c biosystems society economy management. *Advances in Systems Science and Application*, vol. 5, no. 1, pp. 231–280.

Zhan, J. F. and Wu, X. M. 2002. Pansystems OR principles to sociol economic systems. *Advances in Systems Science and Application*, vol. 4, no. 1, pp. 24–31.

Zhang, K. and Hu, C. S. 2006. *World Heritage in China*. Guangzhou: The Press of South China University of Technology.

Zhang, Z., Hai, R., Song, Z., Xia, L., Liang, Y., Cai, H., Liang, Y. et al., 2009. Spatial variation of *Yersinia pestis* from Yunnan Province of China. *Am. J. Trop. Med. Hyg.*, vol. 81, no. 4, pp. 714–717.

Zhao, M., Kong, Q.P., Wang, H.W., Peng, M.S., Xie, X.D., Wang, W.Z., Duan, J.G. et al., 2009. Mitochondrial genome evidence reveals successful Late Paleolithic settlement on the Tibetan Plateau. *Proc. Natl. Acad. Sci. USA*, vol. 106, pp. 21230–21235.

Index

D

E

F

H

T